Transport Geographies

Transport Geographies

Mobilities, Flows and Spaces

Edited by

Richard Knowles
Jon Shaw
Iain Docherty

BLACKWELL PUBLISHING
350 Main Street, Malden, MA 02148-5020, USA
9600 Garsington Road, Oxford OX4 2DQ, UK
550 Swanston Street, Carlton, Victoria 3053, Australia

First published 2008 by Blackwell Publishing Ltd

1 2008

Library of Congress Cataloging-in-Publication Data

Transport geographies : mobilities, flows and spaces / edited by Richard Knowles, Jon Shaw, Iain Docherty.
p. cm.
Includes bibliographical references and index.
ISBN 978-1-4051-5322-5 (hardcover : alk. paper) – ISBN 978-1-4051-5323-2 (pbk. : alk. paper) 1. Transportation. 2. Economic geography. I. Knowles, R. D. (Richard D.) II. Shaw, Jon. III. Docherty, Iain.

HE147.5.T73 2008
388.09–dc22

2007033297

A catalogue record for this title is available from the British Library.

Set in 10 on 12.5 pt Sabon
by SNP Best-set Typesetter Ltd., Hong Kong
Printed and bound in Singapore
by Markono Print Media Pte Ltd

The publisher's policy is to use permanent paper from mills that operate a sustainable forestry policy, and which has been manufactured from pulp processed using acid-free and elementary chlorine-free practices. Furthermore, the publisher ensures that the text paper and cover board used have met acceptable environmental accreditation standards.

For further information on
Blackwell Publishing, visit our website:
www.blackwellpublishing.com

Contents

List of figures vii
Notes on contributors xi
Preface xvii
List of abbreviations xix

Part 1 Fundamentals of transport geographies **1**

1 Introducing transport geographies 3
Jon Shaw, Richard Knowles and Iain Docherty

2 Transport and economic development 10
Danny MacKinnon, Gordon Pirie and Matthias Gather

3 Transport and the environment 29
Stephen Potter and Ian Bailey

4 Transport and social justice 49
Julian Hine

5 Transport governance and ownership 62
Jon Shaw, Richard Knowles and Iain Docherty

Part 2 Transport flows and spaces **81**

6 Connected cities 83
Iain Docherty, Genevieve Giuliano and Donald Houston

7 Geographies of rural transport 102
David Gray, John Farrington and Andreas Kagermeier

8 Inter-urban and regional transport 120
Clive Charlton and Tim Vowles

9 Global air transport 137
Brian Graham and Andrew R. Goetz

10 International maritime freight movements 156
Jean-Paul Rodrigue and Michael Browne

11 Individual transport patterns 179
Stephen Stradling and Jillian Anable

12 Transport, tourism and leisure 196
Derek Hall

Part 3 Future transport geographies **213**

13 Transport directions to the future 215
Glenn Lyons and Becky Loo

14 Revitalized transport geographies 227
John Preston and Kevin O'Connor

References 238
Index 278

List of Figures

Figure 2.1.	The familiar 'honeycomb' distribution of places ordered in terms of their significance, associated with Central Place Theory. After Christaller, 1933.	12
Figure 2.2.	The Taaffe, Morrill and Gould model. After Taaffe *et al.*, 1963.	13
Figure 2.3.	The manufacturing belt in the North East and Mid West of the USA. Source: Hoyle & Smith, 1998.	16
Figure 2.4.	Falling costs of air/sea transport and satellite/ telephone communication. Source: Eddington, 2006a.	18
Figure 2.5.	Japanese motor vehicle production in the I-65 and I-75 corridors of the USA. After Hoyle & Smith, 1998.	19
Figure 2.6.	Factors necessary for transport to stimulate positive economic development in a regional economy. Source: Banister & Berechman, 2001.	22
Figure 2.7.	Tanzania's transport network. The infrastructure has enabled inland produce to be exported but at the same time resulted in cheaper imports reaching inland markets. After Gould & White, 1974.	24
Figure 2.8.	Hired horse cart transporting household waste in a shanty settlement in Cape Town, South Africa. G. Pirie.	26
Figure 3.1.	Scene from Tati's *Trafic*.	37
Figure 3.2.	Actual and projected growth in UK and South East air passengers. Source: DfT, 2003a.	38
Figure 3.3	Impact of Energy Products Directive on unleaded fuel prices, 2004.	44
Figure 3.4.	London Congestion Charge area. After TfL, 2006.	45
Figure 4.1.	Access time and income level. Source: Hine & Mitchell, 2001, 2003.	54

Figure 4.2. Distance from bus networks in Bradford by level of car ownership. Source: Friends of the Earth, 2001. 57
Figure 4.3. Accessibility determined by proximity to the bus network in Greater Belfast. Source: Wu and Hine, 2002. 58
Figure 5.1. The Interstate system in the USA. After Rodrigue *et al.*, 2006. 64
Figure 5.2. (a) Railway companies merged by Act of Parliament to form one of the 'big four' in Great Britain in 1923. After Shaw *et al.*, 1998. 67
Figure 5.2. (b) The operating territories of the 'big four' railway companies, 1923–1947. These companies were combined and nationalized as British Railways in 1948. After Shaw *et al.*, 1998. 67
Figure 5.3. 'Bus wars' in central Manchester: vehicles from rival companies block the passage of a Metrolink tram. D. Hennigan. 70
Figure 5.4. The numerous TOCs operating out of Manchester Piccadilly station. R. Knowles. 74
Figure 5.5. Devolved territories in the UK. The UK government remains in London, but power over many domestic matters has been transferred by Acts of the UK Parliament to the Scottish Parliament in Edinburgh, the Welsh Assembly in Cardiff and (when not suspended due to local squabbles) the Northern Ireland Assembly in Belfast. 78
Figure 6.1. Metro-Land advertising poster, 1919. 87
Figure 6.2. Traffic in Wuhan, China, November 2004. G. Giuliano. 89
Figure 7.1. Deeside and Donside, Aberdeenshire. 104
Figure 7.2. Access to Colleges of Further Education or 'Sixth Forms' (school years 12 and 13) in Cornwall, England. 110
Figure 7.3. Traditional rural 'big bus' service approaching Aberdeen. D. Gray. 113
Figure 7.4. 'Publicar' service provided by Swiss Post. A. Kagermeier. 114
Figure 7.5. Organized hitchhiking: the CARLOS project in Switzerland. A. Kagermeier. 115
Figure 8.1. The TGV, Thalys and Eurostar networks. 124
Figure 8.2. Selected Trans-European Transport Network priority projects. 129
Figure 9.1. Principal international air traffic flows (IATA). 138
Figure 9.2. The globalization/liberalization/sustainability nexus. 141
Figure 9.3. The air transport freedoms. 142
Figure 9.4. Global alliance passenger traffic market shares, 2005, RPKm. Source: *Airline Business*, 2006b. 143
Figure 9.5 Hub-and-spoke model. 146
Figure 10.1. Ton-miles shipped by maritime transportation, 1970–2005 (billions). Data sourced from UNCTAD, 2006 and various other dates. 159

Figure 10.2. The world's 50 largest container ports, 2003. Data sourced from *Containerization International*, various dates. 167
Figure 10.3. Major port holdings, 2006. Data sourced from websites of port holding companies (www.apmholdings.com, www.dpw.co.ae, www.hph.com, www.psa.com.sg, www.ssamarine.com, www.eurogate.co.uk). 170
Figure 10.4. Three major pendulum routes serviced by Orient Overseas Container Line (OOCL), 2006. Data sourced from OOCL website (www.oocl.com). 174
Figure 10.5. Containerized cargo flows along major trade routes, 2005 (in million TEU). Data sourced from UNCTAD, 2006 and various dates. 176
Figure 11.1. Factors influencing travel behaviour. Source: Garling, 1995. 180
Figure 11.2. Broadband penetration in G7 countries since 2001. Source: OECD, 2006b. 189
Figure 11.3. ICT and e-work influences on travel demand. Source: DHC, 2006. 191
Figure 12.1. Potential tourism transport modes. Adapted and developed from Collier, 1994. 201
Figure 12.2. Tourism-related experiential continuum for selected transport modes. Adapted and developed from Lumsdon and Page, 2004. 207
Figure 12.3. Location of UK restored railways. Modified from Butcher, 2006. 210
Figure 13.1. Modal split for freight transportation in the EU, 1970–2001. After European Commission (2006c). 217
Figure 13.2. Principal countries in the Asia-Pacific region. 218
Figure 13.3. The four Intelligent Infrastructure Systems scenarios described in Table 13.3 expressed alongside 'axes of uncertainty' on a two-dimensional grid. After OST, 2006: 42. 224
Figure 14.1. Relationships between mobility and accessibility. 229
Figure 14.2. A model of an extended metropolitan region. After Healey, 2001. 231
Figure 14.3. Extended metropolitan areas of the USA. After Lang & Dhavale, 2005. 231
Figure 14.4. The three dimensions of sustainability. 235
Figure 14.5. An interdisciplinary framework for transport geography. Other disciplines can be added to this mesh and the framework morphed to suit. 236

Notes on Contributors

Jillian Anable is a Research Fellow in the Centre for Transport Policy at the Robert Gordon University. Her research focuses on transport and climate change, with particular emphasis on the application of behavioural and psychological theory to the understanding of travel mode choice. She currently leads the transport topic for the UK Energy Research Centre where her research focuses on identifying combinations of transport policies that will achieve a 60 per cent reduction in carbon emissions from the transport sector by 2050.

Ian Bailey is a Senior Lecturer in the School of Geography at the University of Plymouth and a fellow of the University's Centre for Sustainable Futures. His research focuses on environmental policy and politics, in particular the neoliberalization of national and international climate policy through the deployment of market-based instruments and voluntary agreements. He has published widely on many aspects of environmental policy, and his most recent book is *New Environmental Policy Instruments in the European Union* (2003).

Michael Browne is Professor of Logistics in the School of Architecture and the Built Environment at the University of Westminster. His research focuses on the links between public policies and logistics strategies. He is responsible for directing research activities in logistics and recent work has considered sustainable distribution, energy use for global freight transport and city logistics strategies in European cities. He is the Assistant Editor of *Transport Reviews* and the convenor of the Special Interest Group on Urban Goods Movement (part of the World Conference on Transport Research).

Clive Charlton is a Senior Lecturer in the School of Geography at the University of Plymouth. His transport research interests principally concern rail developments and policy in Europe, high-speed rail and the evolution and implementation of community rail policy in Britain. In addition, he has research interests which focus on Spain and Latin America. Clive has been involved with the Devon and Cornwall Rail Partnership at the University of Plymouth since 1990 and is the book reviews editor of the *Journal of Transport Geography.*

Iain Docherty is a Senior Lecturer in the Department of Management at the University of Glasgow and Chair of the Transport Geography Research Group of the Royal Geographical Society (with the Institute of British Geographers). His research focuses on the impacts of political systems and structures of governance on urban policy and city development strategies. Iain has written widely these issues, and is author of *Making Tracks: the Politics of Local Rail Transport* (1999) and co-editor of *A New Deal for Transport? The UK's Struggle with the Sustainable Transport Agenda* (Blackwell, 2003). He is a Non-Executive Director of Transport Scotland, and was the joint recipient of the 2006 Scottish Young Transport Researcher of the Year.

John Farrington is Professor of Transport and Environment in the School of Geosciences and Director of the Institute for Transport and Rural Research, at the University of Aberdeen. His research focuses on transport policy analysis in the contexts of society, economy and environment, and, in particular, the conceptualization and policy relevance of accessibility and its engagement with wider discourses including globalization, sustainability and social justice. Recent papers consider the conceptualization of rural accessibility, social inclusion and social justice, and the 'new narrative' of accessibility.

Matthias Gather is Professor of Transport Policy and Regional Planning at Erfurt University of Applied Sciences and Director of the *Institut Verkehr und Raum* (Transport and Spatial Planning Institute). His research interests are the spatial and environmental impacts of infrastructure development and (dis)investment, mainly in former socialist European countries; regional railways and demographic change in low-density areas; and accessibility for all. He is currently engaged in European projects on the provision of transport infrastructure and regional development between Germany, Poland and the Czech Republic. He is also co-author of the textbook *Verkehrsgeographie* (Transport Geography) (2007).

Genevieve Giuliano is Professor and Senior Associate Dean for Research in the School of Policy, Planning and Development, and Director of the METRANS Transportation Center, at the University of Southern California. She conducts research in transportation planning and policy, including transportation and land use relationships, urban development and transportation policy analysis. She has published over 120 papers, and co-authored the third edition of *The Geography of Urban Transportation* (2004) with Susan Hanson. She serves on the editorial boards of *Urban Studies* and *Transport Policy*. In 2005 she received the Transportation Research Board's W. J. Carey Award for distinguished research and service.

Andrew R. Goetz is a Professor in the Department of Geography and the Intermodal Transportation Institute at the University of Denver. His research interests include transportation, urban and economic geography; airline industry and policy; airport planning; urban, metropolitan and statewide transportation planning; intermodal transportation; transportation, land use, and urban growth; globalization and sustainable transportation. He has co-authored two books on air transportation topics, *Denver International Airport: Lessons Learned* (1997) and *Airline Deregulation and Laissez-faire Mythology* (1992). He is also the North American Associate Editor for the *Journal of Transport Geography.*

Brian Graham is Professor of Human Geography in the School of Environmental Sciences at the University of Ulster. He is a Chartered Geographer of the Royal Geographical Society (with the Institute of British Geographers) and was formerly Chair of its Transport Geography Research Group. Brian is on the editorial boards of the *Journal of Transport Geography* and *Transport Reviews* and has published widely on many aspects of air transport. His present research interests focus on the interconnections between air transport and economic development. He is the author of *Geography and Air Transport* (1995) and has acted as an advisor on aviation matters to government departments in Northern Ireland.

David Gray is Professor of Transport Policy at the Robert Gordon University, where he heads up the Centre for Transport Policy. His primary focus is rural transport and recent work has considered the nature and perception of the rural transport 'problem'; the impact of national road user charging in rural areas; and the relationship between social capital, informal lift giving and rural mobility. His other area of interest is transport policy in Scotland and he was recently seconded to the Scottish Executive to assist in the preparation of its National Transport Strategy.

Derek Hall was head of the Leisure and Tourism Management Department at the Scottish Agricultural College where he had a personal chair in Regional Development. He has been Secretary and Chair of the Transport Geography Research Group of the Royal Geographical Society (with the Institute of British Geographers). Semi-retired, he is currently Visiting Professor at HAMK University of Applied Sciences in Finland. He has around 300 publications in geography, regional development, transport and tourism, and has a particular regional interest in Central and Eastern Europe.

Julian Hine is Professor of Transport at the Transport and Road Assessment Centre in the Built Environment Research Institute at the University of Ulster. His research interests are transport policy and planning, including policy implementation, transport disadvantage and social justice, and pedestrian behaviour. He has written extensively on these issues, and his books include *Transport Disadvantage and Social Exclusion* (2003), *Integrated Futures and Transport Choices: UK Transport Policy Beyond the 1998 White Paper and Transport Acts* (2003) and *Transport, Demand Management and Social Inclusion* (2004).

Donald Houston is Lecturer in Human Geography at the University of Dundee, with interests in the social welfare implications of urban and regional change in

Britain, particularly in relation to the labour market. His recent research projects have investigated the roles of commuting, migration, housing and social policy in producing and reinforcing social disadvantage in different spatial contexts. Donald has published papers in various journals mostly in relation to the spatial mismatch hypothesis, examining transport and other spatial barriers to out-of-town employment nodes.

Andreas Kagermeier is Professor of Leisure and Tourism Geography at the University of Trier, and has served as chair of the *Arbeitskreis Verkehr* of the *Deutsche Gesellschaft für Geographie* (Transport Research Group of the German Association of Geography). His work in the field of transport geography concentrates on innovative concepts for urban and rural transport, and the impacts of land use patterns on mobility. In the leisure and tourism area, his research focuses on leisure mobility, leisure facilities and events, and cultural and bicycle tourism. He also has an interest in the regional economic impact of tourism development in the Maghreb countries.

Richard Knowles is Professor of Transport Geography at the University of Salford and a member of the Research Institute for the Built and Human Environment. He is editor of the *Journal of Transport Geography* and co-editor of Ashgate's *Transport and Mobility* monographs. His published research focuses on the effects of transport policy changes, including bus deregulation and rail privatisation, and the impacts of new transport infrastructure including Manchester's Metrolink tram system, the Channel Tunnel and the Danish/Swedish Øresundsbron. Richard's books include *Modern Transport Geography* (1992; 1998) and *Scandinavia* (1991). In 2004 he was presented with the Ullman Award in Transportation Geography.

Becky Loo is an Associate Professor in the Department of Geography at the University of Hong Kong. She specializes in research on transport and development, with a focus on Hong Kong and mainland China. In particular, she has published a series of research papers on road safety, and on people's travel behaviour and its relationship with the evolution of city structure. Her research on the Internet and the spatial movement of passengers and freight has received both local and international recognition. She was the recipient of the University of Hong Kong's Outstanding Young Researcher Award in 2005/06.

Glenn Lyons is Professor of Transport and Society and the founder and Director of the Centre for Transport and Society (CTS) at the University of the West of England. With a team of 20 transport planners and social scientists, Glenn's aim, and that of the CTS, is to improve and promote understanding of the inherent links between lifestyles and personal travel in the context of continuing social and technological change. Glenn was formerly Director of the Transport Visions Network and Chairman of the Transport Planning Society; he is currently Chairman of the UK's Universities Transport Study Group. In the field of traveller information, Glenn is an external advisor to the UK Department for Transport.

Danny MacKinnon is an economic and political geographer at the University of Aberdeen, with interests in regional economic development, devolution and state restructuring, and labour geography. He has an active interest in the relationship between transport geography and economic geography and is currently engaged in research projects on devolution and transport policy in the UK and the geography of employment relations in the privatised rail industry in Britain. He is co-author of the economic geography textbook *An introduction to economic geography: globalisation, uneven development and place* (2007).

Kevin O'Connor is Professor of Urban Planning at the University of Melbourne and a member of the Volvo Research Centre on the Government and Management of Urban Transport. His research explores the spatial organization of the economy, especially in metropolitan areas. He has investigated airlines and airline service as an influence upon the advanced service activity of Pacific Asian cities. At another level he has used journey to work data to show how local labour markets within the metropolitan area respond to industrial change. Much of his work focuses on Australian cities and is linked to local policy debates.

Gordon Pirie is Assistant Professor and Head of the Department of Geography and Environmental Studies at the University of the Western Cape. He works mainly on air, road and rail transport in Southern Africa and is particularly interested in the social aspects of transport. Gordon has written extensively on transport and travel under apartheid and has completed a major research project on civil aviation in the inter-war British Empire. He is presently researching road and airport provision in late colonial sub-Saharan Africa.

Stephen Potter is the Open University's Professor of Transport Strategy. His research concentrates upon transport policy and sustainability issues, and includes work on the interplay between technical and behavioural change approaches to reduce environmental impacts such as eco-reforms to transport taxation, the development of travel plans and the diffusion of low carbon cars and other eco-technologies. His recent publications include *Unfare Solutions . . . Local Earmarked Charges to Fund Public Transport* (SPON Press, 2004).

John Preston is a Professor in the Transportation Research Group of the School of Civil Engineering and the Environment, University of Southampton. He was previously Director of the Transport Studies Unit at the University of Oxford and a Tutorial Fellow in Geography at St Anne's College. He has published over 160 articles, book chapters, conference and working papers on transport demand and cost modelling, regulatory studies, and land-use and environment interactions. He also co-edited *Integrated Transport – Implications for Regulation and Competition* (2000) and *Integrated Futures and Transport Choices: UK Transport Policy Beyond the 1998 White Paper and Transport Acts* (2003).

Jean-Paul Rodrigue is an Associate Professor in the Department of Economics and Geography at Hofstra University. He was Chair of the Transport Geography

Specialty Group of the American Association of Geographers between 2004 and 2006. Among his transport research interests, the topics of the transport geography of logistics, inland freight distribution, gateways and freight corridors have received particular attention. Jean-Paul has recently written two books, *The Global Economic Space: Advanced Economies and Globalisation* (2000), which won the PricewaterhouseCoopers best business book award, and *The Geography of Transport Systems* (2006), which he co-authored with Claude Comtois and Brian Slack.

Jon Shaw is Reader in Human Geography and Director of the Centre for Sustainable Transport at the University of Plymouth. He is also the UK and Ireland Associate Editor of the *Journal of Transport Geography*. His research focuses upon issues associated with mobility, accessibility and transport policy and governance. He is the author of *Competition, Regulation and the Privatisation of British Rail* (2000), and co-edited *All Change: British Railway Privatisation* (2000) and *A New Deal for Transport? The UK's Struggle with the Sustainable Transport Agenda* (Blackwell, 2003). He jointly won the award of Scottish Young Transport Researcher in 2006 (a notable achievement for an Englishman!).

Stephen Stradling is Professor of Transport Psychology at Napier University's Transport Research Institute in Edinburgh. He was previously a Senior Lecturer in the Department of Psychology at Manchester University. His research in traffic psychology has covered violations, errors and lapses, applications of the theory of planned behaviour in aberrant drivers, and speeding drivers and speed enforcement. His research in transport psychology covers car dependence and modal shift, focusing on the psychological attractiveness or otherwise of cars and other transport modes.

Tim Vowles is Visiting Assistant Professor in the Department of History at Colorado State University. His research interests include air transportation, particularly the impacts of low-cost carriers, intermodal transportation planning and world cities networks. He has published in journals such as *Professional Geographer*, *Journal of Transport Geography*, *Transportation Law Journal* and *Transport Reviews*. Tim is also on the editorial board of *Asia Pacific Viewpoint* and on the Panel of Reviewers of the *Journal of Air Transportation*.

Preface

Transport matters have increased greatly in significance in recent times. Climate change, congestion and the state of transport infrastructure are only three amongst a catalogue of critically important issues that have come to the forefront of debates – at local, regional, national and global scales – about how and why we travel. At the same time, the academic literature on transport and mobility has burgeoned, as researchers from a range of backgrounds seek to understand better the underlying economic, environmental and social processes that contribute towards continually changing transport patterns. Central to this work is the recognition and reaffirmation of transport's role in defining the function and character of spaces and places, and the flows and interactions within and between them. In other words, transport is fundamental to geography, and geography is fundamental to transport.

Our aim in this book is to provide a general introduction to key ideas, concepts and themes of the various geographies of transport, and how they relate both to the discipline of geography more broadly and to complementary areas of study; amongst these are economics, engineering, environmental studies, political science, psychology, spatial planning, sociology and transport studies. Our approach is to present these ideas in a manner that is both empirically informed and theoretically robust, and in so doing we seek to contribute towards the development of a more effective understanding of the impacts of transport so that the needs and demands of future travel can be better addressed.

The generally qualitative approach of the book draws upon a range of case study material – much of which has a strong policy focus – from across Europe and beyond and has been designed to complement and add richness to the substantial body of quantitative and model based analyses traditionally associated with transport geography. The contributing authors' wealth of experience brings with it a vast array of evidence and insight, and our job as editors has been to weave these into a coherent and compelling narrative.

The idea for this book emerged from discussions with colleagues in the Transport Geography Research Group (TGRG) of the Royal Geographical Society (with the Institute of British Geographers). Over the last 35 years, members of the TGRG have played key roles in developing the sub-discipline of transport geography, nurturing international links with counterpart research groups, especially in the USA, Germany, France and Italy. They have also engaged with researchers from other disciplines who are interested in the spatial aspects of travel and transport. The Group published its own series of research conference proceedings from 1978 to 1994, and since 1990 its members have also produced a wide range of research-based books with external publishers, including two editions of *Modern Transport Geography*, the forerunner of this volume. In parallel, the *Journal of Transport Geography* (*JTG*) was launched in 1993 with the support of the Transportation Geography Specialty Group of the Association of American Geographers. The *JTG* has developed into a highly rated international outlet for top quality research and is abstracted in Thomson Scientific's *Social Science Citation Index*.

A project of this nature is complex and time consuming and we owe debts of gratitude to many people. We should firstly like to thank our fellow contributors for writing incisive and authoritative chapters. We are also grateful to Blackwell, and in particular Ben Thatcher, for help and support throughout the writing and editing process. Steve Bennett, Richard Gibb, Danny MacKinnon, John Farrington and Richard Yarwood provided valuable advice on some or all of the text, and the Meehan and England/Oster households provided valuable office and resting space during Jon's temporary relocation to the USA. Finally, special thanks are most certainly due to Jamie Quinn, of the School of Geography's Cartographical Resources Unit at the University of Plymouth, for producing all of the artwork contained within the book.

Richard Knowles
University of Salford

Jon Shaw
University of Plymouth

Iain Docherty
University of Glasgow

List of Abbreviations

A	Airbus
AE	Aspiring environmentalist
APD	Air passenger duty
AVE	*Alta Velocidad Español* (Spanish high-speed train)
B	Boeing
BBC	British Broadcasting Corporation
BRE	Building Research Establishment
CA	Car aspirer
CC	Car complacent
CEC	Commission of the European Communities
CfIT	Commission for Integrated Transport
CO	Carbon monoxide
CO_2	Carbon dioxide
COMEAP	Committee on the Medical Effects of Air Pollutants
CREA	Cape Railway Enthusiasts' Association
CS	Car sceptic
DEFRA	Department for the Environment, Food and Rural Affairs
DETR	Department of the Environment, Transport and the Regions
DfEE	Department for Education and Employment
DfES	Department for Education and Skills
DfT	Department for Transport
DHC	Derek Halden Consultancy
DHD	Die-hard driver
DPW	Dubai Ports World
DRT	Demand responsive transport
ECMT	European Conference of Ministers for Transport
EFA	Ecological footprint analysis
EIA	Environmental impact assessment

EIB	European Investment Bank
E-PRTR	European Pollutant Release and Transfer Register
ERTMS	European Rail Traffic Management System
ETE-ENE	European Train Enthusiasts – Eastern New England
EU	European Union
EU 15	The 15 member states of the EU, 1995–2004
EU 25	The 25 member states of the EU, 2004–2006 (there are 27 as of 2007)
GCC	Global commodity chains
GDP	Gross domestic product
HPH	Hutchison Port Holdings
HSR	High speed rail
I	Interstate highway
IATA	International Air Transport Association
IGE	*Interessengemeinschaft Eisenbahn* (Rail enthusiasts' community)
IMF	International Monetary Fund
IMT	Intermediate means of transport
IRFC	Indian Railways Fan Club
IURT	Inter urban and regional transport
JTG	*Journal of Transport Geography*
LCA	Life cycle analysis
LCC	Low-cost carrier
LGV	*Lignes à Grande Vitesse* (High-speed lines)
LNG	Liquid natural gas
M	Motorway
MM	Malcontented motorist
MNC	Multinational corporation
NAFTA	North American Free Trade Area
NBS	*Neubaustrecke* (New-build railway lines)
NEG	New economic geography
NGO	Non-government organisation
NO_X	Nitrogen oxides
OECD	Organisation for Economic Cooperation and Development
ONS	Office for National Statistics
OOCL	Orient Overseas Container Line
OPEC	Organisation of Petroleum Exporting Countries
OST	Office of Science and Technology
P&O	Peninsular & Orient
PAH	Polycyclic aromatic hydrocarbons
PSA	Port of Singapore Authority
RCEP	Royal Commission on Environmental Pollution
RPK	Revenue passenger kilometre
RR	Reluctant rider
SA	Sustainability Appraisal
SACTRA	Standing Advisory Committee on Trunk Road Assessment
SACU	Southern Africa Customs Union

SEA	Strategic environmental assessment
SEU	Social Exclusion Unit
SO_2	Sulphur dioxide
TEN	Trans European Network
TEN-T	Trans European Transport Network
TEU	Twenty-foot equivalent unit
TfL	Transport for London
TGV	*Train à Grande Vitesse* (High-speed train)
UK	United Kingdom
UNESCAP	United Nations Economic and Social Commission for Asia and the Pacific
USA	United States of America
VAT	Value added tax
VLCC	Very large crude carrier
WHO	World Health Organization
WTO	World Tourism Organization
WTO	World Trade Organization

Part 1 | Fundamentals of Transport Geographies

1 | Introducing Transport Geographies

Jon Shaw, Richard Knowles
and Iain Docherty

The importance of travel and transport to the functioning of our economies and societies is hardly in doubt, but the very ordinariness of transport systems often means that they are taken for granted. In the developed world at least, transport networks and systems generally work well and often it is only when something goes badly wrong that headlines are made. In the weeks during which we have been finalizing the manuscript for this book, for example, newspapers have made much of various North West European transport systems being compromised by bad weather and threatened by industrial action that caused British Airways to cancel hundreds of flights before a last-minute deal was struck and the strikes averted (BBC, 2007).

The tendency to take transport for granted has also been evident in the world of academic geography (Goetz *et al.*, 2003). Certainly it was recognised as an important element of geographical inquiry in the nineteenth and early twentieth centuries and was again popular in the wake of the quantitative revolution in the 1960s and early 1970s (Hay, 2000). But various factors, including the advent of cheap oil and a reluctance on the part of transport geographers to engage in significant theoretical debates, led to too many human geographers significantly downplaying transport matters in their analyses of social and economic patterns and systems (Hall *et al.*, 2006; Keeling, 2007; although see Harvey, 1982). In reality, the significance of travel and transport in the objects of geographical inquiry never diminished: greater levels of mobility are an attribute of an increasingly globalized world space economy (Hoyle and Knowles, 1998). Johnston *et al.* (1995: 13) have noted that 'when the history of the late 20th century is written there seems little doubt that mobility . . . will be one of its touchstones'.

Increasing numbers of human geographers are now returning to this view, most notably in joining with scholars from other disciplines reacting against 'static' social science to posit a 'new mobilities paradigm' (Sheller & Urry, 2006; see also, for example, Crang, 2002; Cresswell, 2006; Kesselring, 2004; Larsen *et al.*, 2006;

Thomsen *et al.*, 2005; and Hall *et al.*, 2006 for a collection of papers bringing together transport and economic geography). The central argument here is that social science has in the past 'trivialized or ignored' the movement of people (and other things) to the point that transport became a 'black box' of neutral processes and technologies that permit but do not really explain social and economic phenomena (Sheller & Urry, 2006: 208). Indeed, Sheller and Urry continue to suggest that 'accounting for mobility in its fullest sense challenges social science to change both the objects of its inquiries and the methodologies for research'. The rising significance of transport flows and spaces within academia offers perhaps the most promising opportunity in recent years to reposition transport geography at the heart of the mainstream human geography endeavour.

And, of course, the nature of transport flows and spaces is constantly changing. Whereas once people travelled by foot or public transport almost exclusively out of necessity, nowadays vastly more travel is by private car or aeroplane for leisure purposes. Many journeys are arguably unnecessary or could be made by other modes. The emergence of widespread Information and Communications Technologies (ICTs) is also having a considerable effect on contemporary journey patterns. For many activities, ICTs are the ultimate expression of time-space convergence: will we travel as much if we can effectively bring other places into our homes by substituting e-mail, videophones and 'e-tailing' for commuting, visiting friends and going to the supermarket?[1] Evidence to date suggests the answer to this question is both yes and no – ICTs lead us to make more and different/increasingly complex journeys, and/or to work and interact from home at least part of the time (Banister and Stead, 2004; Helminen and Ristimaki, 2007; Kwan, 2006). As such, location remains important, not only because access to the Internet remains geographically and structurally uneven but also because virtual encounters can be rather soulless compared with the tangible reality of being together (Knowles, 2006; Urry, 2002).

Transport geography

So what is transport geography and why is transport of interest to human geographers? Transport geography is in essence the study of the spatial aspects of transport (see Black, 2003; Goetz *et al.*, 2003; Hanson and Giuliano, 2004; Hensher *et al.*, 2004; Hoyle and Knowles, 1998; Keeling, 2007; Nuhn and Hesse, 2006; Rodrigue *et al.*, 2006; Tolley and Turton, 1995; White and Senior, 1983). Transport is inherently spatial – it develops because people and goods have to get places. People are rarely located in the same places as the things they want or need, and transport systems are, at their most basic, an expression of a need to link supply and demand; they are the manifestation of people's desire to access goods, services and each other.

There are two aspects of the nexus between transport and geography that have traditionally attracted study. One is the geography of transport systems themselves. These occupy a large amount of space, their form, layout and extent being determined by a range of factors such as topography (mountains, rivers, etc.), economic conditions, technological capability, sociopolitical situations and the spatial distribution of the places they link together. The USA, for example, has an extremely large amount of road space (Pucher and Lefevre, 1996), reflecting (among other things) its wealth and an emphasis on individual freedom that promotes the private car as the dominant means of transport. The East of the USA has more and denser infrastructure than the West, largely because it has more people and the terrain is easier to build on. In contrast, the former Soviet bloc has far less well developed road networks but significantly more extensive public transport systems, the result of a socialist culture that promoted communal travel. This is now changing after the fall of the Iron Curtain, subsequent economic growth and political and social transformation. Medium-term outcomes of this change are likely to be more roads and contracting public transport networks, and this new transport geography will in turn impact on the social, economic and environmental geographies of Eastern Europe and beyond (Kovacs and Spens, 2006; Taylor, 2006).

Indeed, the impact of transport is the second traditional area of study for transport geographers. A core interest of many geographers is explaining the location of phenomena over time and across space, and transport is one of the most powerful factors affecting and explaining the distribution of social and economic activity (Hoyle and Knowles, 1998; White, 1977). There has, for example, been a longstanding interest in the relationship between transport and economic development, although the precise nature of this relationship remains elusive (Banister and Berechmann, 2001; Eddington, 2006a; SACTRA, 1999; see also Black, 2001 on this and other issues). How much will the construction of a new road lead to the economic regeneration of – that is, the location of new economic activity in – a given area? How far are differences in the quality of transport infrastructure responsible for uneven development between cities, regions or countries? Part of the difficulty in answering such questions lies in the complexity of social, political and economic circumstances surrounding individual cases. We know that transport improvements can be an enabling factor when deployed as part of a range of complementary initiatives, but in certain circumstances the opening of, say, a new road link can have negative consequences for the economy and society of a peripheral place or less developed economy if it enables much easier access to the wider variety of goods and services available in the 'core' (see Hilling, 1996).

At a much broader scale, concern about the transport sector's contribution to global warming has become a popular topic among social and natural scientists (Baggott *et al.*, 2005; Black, 1998; Maddison *et al.*, 1995; HM Treasury, 2006). Until as late as the 1980s it was not widely understood that the environmental impact of transport stretched much beyond the local nuisances of noise, poor air quality and the like, but a series of scientific papers presented at the 1989 meeting of the European Conference of Ministers of Transport (ECMT) pointed to the growing threat from greenhouse gas emissions, particularly from private car traffic (Docherty, 2003; Goodwin, 1999). Whilst most governments have accepted the link between increasing CO_2 emissions and global warming, the scepticism of the Bush administration in the USA – the country with the world's highest levels of car, lorry and aeroplane usage – is not encouraging for long-term environmental sustainability.

Moving beyond these two traditional areas of study, developments such as the new mobilities paradigm offer a wealth of opportunity for transport geographers to engage with other core concerns of contemporary human geographers. Among these are the nature and production of space and place, and a flurry of scholarly activity is now exploring these ideas with specific reference to travel spaces (e.g. Brown & O'Hara, 2003; Dodge & Kitchin, 2004; Jain, 2004; Laurier, 2004; Letherby & Reynolds, 2005; Lyons & Urry, 2005; Lyons *et al.*, 2007; Massey, 2005; Sheller & Urry, 2006). Many of these authors have taken the transport networks and systems traditionally studied by transport geographers and 'unpacked' them by focusing on their component flows and spaces to reveal nuanced or counterintuitive qualities and attributes previously overlooked in transport geography research.

Work on the functions of travel space, for example, challenges the approach of judging potential transport improvements in terms of the time saved by a new investment. Whereas conventional thinking assumes that travel time is wasted time, the arrival of ICTs has brought into clear focus the rather obvious point that travel time can be useful time: the travel space can also be a highly productive working space (laptops, mobile phones, blackberries) and leisure space (portable DVD players and iPods) in addition to fulfilling more general functions as a place to read, socialize or rest. Jain and Lyons (2008) go so far as to suggest that we might regard travel time as a 'gift'. There are implications with giving policy makers the idea that they can cut back on large investment in new transport infrastructure as this – along with making better use of that which we already possess (Eddington, 2006) – remains crucial to the efficient functioning of economies and societies. If, however, work on travel spaces leads to greater efforts on the part of governments and transport providers to improve journey quality, reliability and safety, it will have been of clear practical as well as intellectual benefit.[2]

Mobility and accessibility

And the new mobilities' increasing rejection of 'social science research [that] has been "a-mobile"' (Urry, 2003: 156) offers renewed opportunity to transport geographers not just because it embeds transport and travel at the heart of a vibrant intellectual endeavour. Also of significance is its emphasis on the role of innovative – often qualitative – research methods. Although qualitative methods have often been employed by transport geographers (e.g. Bird, 1982; Bird *et al.*, 1983; Docherty, 1999; Hoyle, 1994; Pooley *et al.*, 2005; Shaw, 2000) and others in significant works which might be regarded as transport geography (e.g. Cronon, 1991; Herod, 1998; Meinig, 1986), the sub-discipline is commonly associated with positivist assumptions, methods of data collection and modelling (Goetz *et al.*, 2003; Hay, 2000). Indeed, *Modern Transport Geography* (Hoyle and Knowles, 1998: 5) stressed 'the pursuit of objectivity and truth' in transport analysis. Quantitative methods obviously retain a legitimate and necessary place within academic inquiry, but harnessing the power of qualitative techniques to enrich understanding of the specific and the subjective can bring, either on their own or as part of a 'mixed methods' research strategy, added depth and rigour to research undertakings (Baxter and Eyles, 1997; Philip, 1998).

A good example of this in transport geography is the consideration of personal mobility (see Pooley *et al.*, 2005) and accessibility (see Farrington, 2007). Quantitative data sets gathered at a high level of aggregation point to vastly increasing personal mobility as a result of increasing car ownership (see, for example, Department for Transport, 2006a; Hoyle and Knowles, 1998). For a considerable period this was taken as a good thing by governments (and many academics), leading to policies such as 'predict and provide'. This approach to transport policy is highly car-centric, seeking as it does to estimate future traffic demand and build road capacity accordingly. One consequence is that the needs of the minority who have no access to a car can all too easily be overlooked in comparison with those of the majority who do. As Goodwin (1999: 658) neatly summarizes, since 'private car use would increase... it was necessary to increase road capacity. And public service use would decline, therefore it would be logical to reduce service levels'.

The realization that 'predict and provide' would fail in its own terms – it simply would not be possible to build enough road capacity to cope with demand – and with this failure bring negative economic (more traffic jams), environmental (much more pollution) and social (dwindling public transport) consequences, led to the emergence of a 'new realism' in transport thinking which prioritises demand management and public transport provision (Goodwin *et al.*, 1991). Among British transport geographers (and others) one outcome was to renew interest in the concept of accessibility as a means of addressing mobility deprivation, especially in terms of its uneven distribution both spatially and structurally. Accessibility should not be confused with mobility: it refers to the extent to which something is 'get-at-able' (Moseley, 1979: 56; Chapter 4) and being mobile is only one of a number of ways of reaching services, facilities and social networks (others being the telephone and the Internet, for example). Equally, people are not accessibility-deprived solely as a function of lack of mobility: accessibility may also be limited on account of poverty, gender, race or other factors which might compromise access to services. Good examples are an inability to afford university tuition fees and prejudice which precludes access to certain social networks and / or situations (although see Weber, 2006).

Large-scale, mixed methods research programmes using travel diaries, in-depth interviews and focus groups in addition to 'traditional' questionnaires have been undertaken in the UK to determine both people's accessibility needs and the extent to which these are met (e.g. Farrington *et al.*, 1998, 2004). Subsequent work has sought to provide, among other things, a philosophical foundation for the concept of accessibility 'rights' (Farrington and Farrington, 2005); more sophisticated modelling techniques to take account of individual rather than aggregate geographical circumstances (Preston and Rajé, 2007); and a teasing out of the role of social capital and networks in providing mobility and accessibility for those without a car in rural areas (Gray *et al.*, 2006). In other words, a shift in focus from the general to the specific, attempting to understand subjective perceptions rather than seeking universal truths, has played a significant part in linking the practice of transport geography not just to broader geographical and social science endeavours, but also to 'real world' policy needs and aspirations.

Mobilities, flows and spaces

The following chapters pick up these and many other ideas and themes germane to transport geographies. The contributions are both wide-ranging and innovative, although inevitably we have been constrained by space and should state from the outset that the text is neither comprehensive nor definitive. Much of the discussion is focused on the developed world, for example, although some consideration is also given to topics of particular importance to developing countries and regions – such as the relationship between transport and economic advancement. Similarly, detailed examination of freight transport is restricted to an analysis of the rapid advances in bulk and containerized shipping which are key enablers in regionalized and global economies.

We have divided the book into three parts. The first of these sets out some 'fundamentals' in transport geography. In addition to this brief introduction, we include contributions based around the three 'pillars' of sustainability – the economy, the environment and society – as this encapsulates and exemplifies many of the key linkages between transport and other significant areas of geographical research. The centrality of governance to the geographies of transport, not least as states and other authorities grapple with the complexities of transport and its impacts, is reflected by the chapter dealing with the state's role in regulating and sometimes owning transport systems and services.

The remainder of the book draws upon these fundamentals. The second part considers in more detail the principal transport flows and their geographical consequences apparent at different spatial scales. Issues of mobility and accessibility are recurring themes in these chapters, as are economic, environmental and social concerns evident at local, regional and international levels; sometimes these are comparable and inter-related, sometimes they are wildly different in character. We are aware that the selection of spatial scales to be covered is potentially problematic and contestable – we have, for example, chosen to focus on the 'urban' and the 'rural' rather than the 'local' – but those chosen strike a balance between the often-competing demands of the transport-related and geographical elements of the subject matter. Rounding off the section are analyses of transport choices and activities taking place across space from the local to the global: investigations into why we travel the way we do and transport in tourism exemplify two more ways in which transport and geographical concerns are being linked with other research agendas.

In part three, attention turns to the future. The authors briefly reflect on current transport trends and consider possible future trajectories of both global transport trends and the sub-discipline of transport geography. Seemingly inexorable increases in the mobility of the human race, allied with the continuing refinement of our understanding about travel and transport behaviour, the emergence of ICTs and the impacts and geographies of these developments, provide a wealth of opportunities for transport geographers to influence the intellectual and policy agendas improving transport experiences around the world.

Notes

1 And similar questions were asked with the advent of the postal service, the telegraph and the telephone (see Mokhtarian, 2000).
2 Interestingly, *Scotland's National Transport Strategy* (Edinburgh: Scottish Executive, 2006a) identifies improving journey quality as an explicit policy goal.

2 | Transport and Economic Development

Danny MacKinnon, Gordon Pirie
and Matthias Gather

There is a clear and widely accepted association between the quality of transport infrastructure and the level of economic development within a particular country or region. In general, transport infrastructure and services are superior and more diverse in wealthy districts, countries and regions than in less developed ones. Modes of transport are better connected, their geographical reach is greater, and fewer places are inaccessible. Whilst such contrasts can be identified clearly, explaining how they have occurred is a far more demanding task, requiring a sophisticated appreciation of 'the rich complexity of the transport-development interface' (Leinbach, 1995: 338). In particular, the question of the direction in which the linkage operates is crucial: 'does transport investment promote economic growth or does growth encourage more demand for transport, and thus further investment?' (Banister & Berechman, 2001: 214). The conventional view is that the relationship is two-way, with transport acting as an important facilitator of economic development and providing an important outlet for capital investment as economies grow (Hoyle & Smith, 1998; Simon, 1996; Vance, 1986). Recent research has demonstrated the complexity of the relationship, indicating that the impacts of investment tend to be socially and spatially uneven, favouring some social groups and places over others (Banister & Berechman, 2001).

This chapter aims to bring together key strands of transport geography and economic geography, two fields which have been become divorced from one another in recent decades, in spite of the manifold connections between them on the ground (Hall *et al.*, 2006). It is also informed by relevant insights from development studies. In contrast to the economic geography and transport geography literatures respectively, it takes neither mobility nor the existing spatial distribution of economic activity for granted, assessing the dynamic linkages between the two. We view the relationship between transport and economic development as a two-way symbiosis; in a circular manner, each influences the other. Our primary concern is with the secondary or additional effects of transport investment in altering economic conditions rather than its direct impact in terms of reduced journey times and increased accessibility (Banister & Berechman, 2001). The next section revisits some major theoretical frameworks that have been used to interpret and understand the nexus between transport and economic development. We then discuss the relationship between transport and spatial development in developed countries, considering the role of transport in the shift to post-Fordist production systems and assessing the economic effects of investments in transport infrastructure. A discussion of transport and economic development in developing countries precedes our brief conclusion, which summarises the main points of the chapter.

Theoretical frameworks

Spatial analysis

The principle of the spatially determining effects of transport systems on patterns of spatial economic organization has been a key theme of location theory since von Thünen's seminal work in the mid-nineteenth century. Traditional location theory is characterized by a deductive method of analysis, beginning with the assumption of a flat, featureless plain (an isotropic plain) on which economic activity is located. The focus is then on ascertaining the effects of distance on location, with transportation costs viewed as a key expression of distance (Knox & Agnew, 1994). According to von Thünen's theory of land use, the value of a location is determined by its access to the marketplace, reflecting its geographical position, particularly in relation to major transportation routes.

In his groundbreaking work on central places, Christaller (1933: 53) also emphasized the spatially differentiating effect of transport infrastructure: 'Better transport connections result in a reduction of economic distance, a reduction not only in costs, but also in wasted time and the psychological inhibitions which impede frequent purchase of essential goods on uncomfortable, dangerous and sometimes impassable roads with bad traffic conditions'. Based on the assumption of economic rationality, central place theory offers an account of the size and distribution of settlements within an urban system. The need for shop owners to select central locations produces a hexagonal network of central places, organised into a distinct hierarchy of lower- and higher-order centres (Figure 2.1).

Whilst location theory generates neat models which real spatial patterns can be measured against, the assumptions upon which they are based are questionable (Massey, 1984). From a transport perspective, the notion that travel costs are equal in every direction is clearly at odds with the simple reality that transport networks and services sculpt landscapes of differential accessibility and land value (Knowles, 2006a). Motorway junctions are highly favoured by commercial property developers, for instance, for whom access trumps noise in manufacturing, wholesale and storage premises. Subway stations generate considerable passing traffic and nearby sites are prized by retailers seeking high volumes of consumer footfall. Lead transport plans and investments have been facilitators of urban regeneration in places such as London's docklands (Eddington, 2006b).

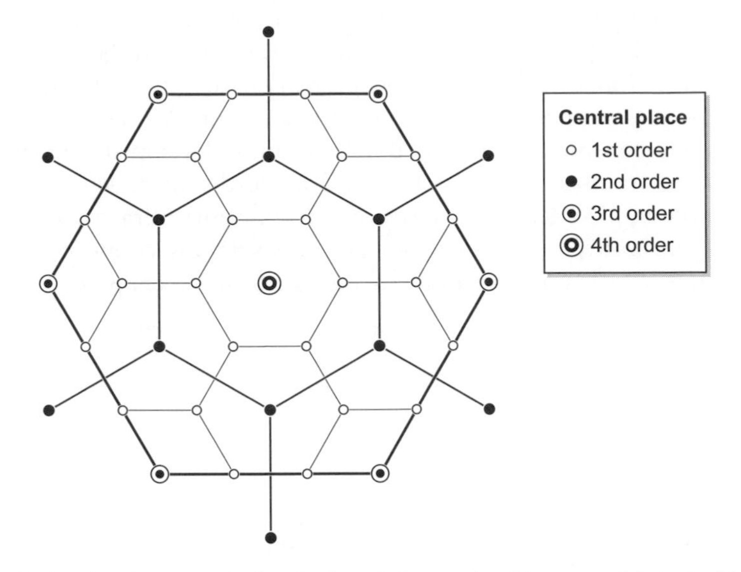

Figure 2.1. The familiar 'honeycomb' distribution of places ordered in terms of their significance, associated with Central Place Theory. After Christaller, 1933.

The 'New Economic Geography' (NEG) associated with the economist Paul Krugman seeks to explain the existence of agglomeration advantages and regional disparities, often taking a core–periphery form, within the system of economic equilibrium. One key assumption of the NEG is the 'iceberg' formulation of transport costs (Krugman, 1991), which states that a part of a good on its way from producer to consumer 'melts away' during transportation. This iceberg model is a mere analytical device which clearly cannot be observed in reality, acting as a convenient way of operationalizing distance whilst maintaining the overall properties of economic equilibrium. A key conclusion of the NEG is that reduced transport costs favour a concentration of manufacturing in a small number of centres rather than a more even dispersal across the economic landscape. This supports the earlier findings of the spatial polarization theorists (for example, Perroux), as economies of scale and scope, together with market-size effect, ensure that major agglomerations and growth poles gain certain competitive advantages over other locations. At the same time, countervailing forces such as immobile factors of production like land and labour and high rents and wages in central locations set limits on agglomeration and can, under certain conditions, encourage dispersal. Contrary to conventional assumptions regarding the benefits of transport improvements for peripheral regions, the NEG has shown that better and cheaper transport will generally promote the further concentration of economic activity in favoured locations (Eckey & Kosfeld, 2004).

Modernization theory

Modernization theory assumed that development involved developing countries following the western experience of large-scale industrialization, expressed by

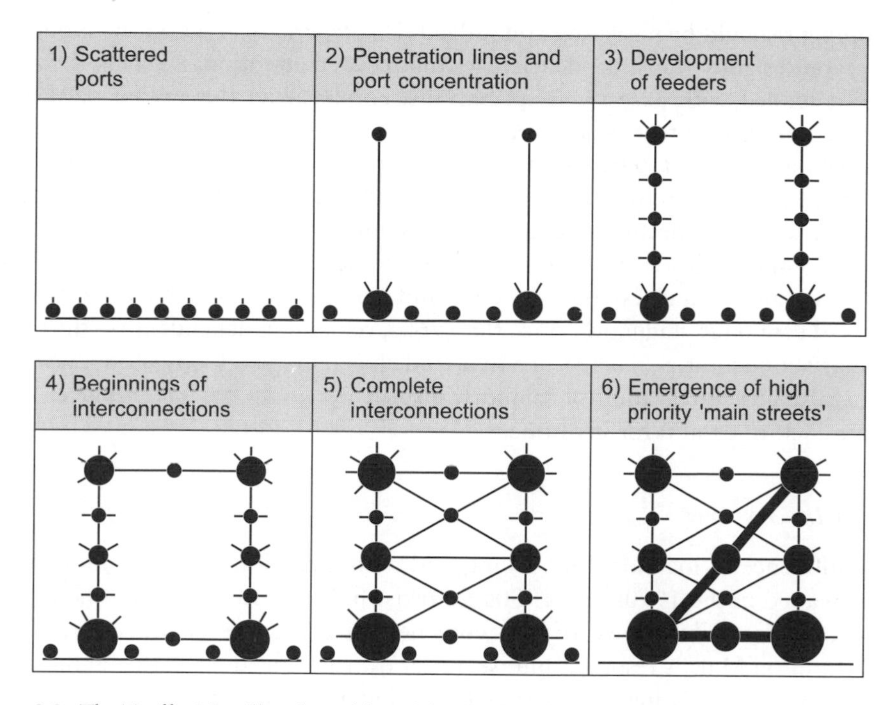

Figure 2.2. The Taaffe, Morrill and Gould model. After Taaffe *et al.*, 1963.

Rostow (1960) in his famous 'stages of economic growth' model. Using a transport metaphor appropriate to the twentieth century, he argued that transport was a precondition for the 'take off' of economic growth. Rostow's work confirmed the key role that transport investment in canals and railways had played in Western economies during the Industrial Revolution of the nineteenth century. The model lent itself to arguments for 'catch-up' transport investment in developing countries. At a moment of apparently unlimited technical promise and cheap energy in the late 1950s and 1960s, economic development consultants and advisers seized on the leverage that transport appeared to offer.

Perhaps the best-known model of the geography of transport in less developed countries is that of Taaffe, Morrill and Gould (Figure 2.2). Devised on the basis of fieldwork in Ghana and Nigeria in the 1960s, the model shows how, at successive stages of economic growth, transport investment makes particular places relatively more accessible. A transport network hierarchy and an urban hierarchy emerge simultaneously, each one reinforcing the other. In the early stages, a number of small, unconnected trading ports form the basis for the introduction of inland transport routes (Hoyle & Smith, 1998). The establishment of arterial railways and roads and associated feeder lines leads to gradual integration. Finally, an intricate route network serves a complex urban system: many minor routes connect small settlements and act as feeders to a smaller set of intercity trunk routes that link a few major centres.

A mark of the influence of the Taaffe, Morrill and Gould model is its longevity and its being venerated almost as an ideal against which actual transport network development should be measured and judged. Discrepancies between the model and reality would signify transport deficiencies that needed attention. As such, the model has been loaded with excessive expectations: emphasizing the spatial patterns of transport development, it was never intended to capture the complex social processes behind transport network investment and transport service delivery. Recent research in sub-Saharan Africa has highlighted the empirical limitations of the model (Pedersen, 2003), reflecting the in-built assumptions of modernization theory. Instead of rapid industrialization and economic development along western lines, most developing countries (outside East and South-East Asia) have experienced prolonged economic stagnation since the 1960s, remaining dependent on the export of primary commodities. This means that trade has remained focused on a few large ports whilst little integration of inland transport networks has taken place, partly due to a lack of capital for investment.

Critical theory

As the influence of locational modelling and modernization theory in human geography waned in the 1970s and 1980s, Marxist political economy became increasingly prominent. Whilst most of this work neglected the role of transport, Harvey (1982) built on Marx's resonant phrase about transport leading to the 'annihilation of space by time', relating this process to an underlying contradiction between the geographical fixity and motion of capital. Fixity of capital in one place for a sustained period – creating a built environment of factories, offices, houses, transport infrastructures and communication networks – is crucial in enabling production to take place. As economic conditions change, however, these infrastructures can themselves become a barrier to further expansion, growing increasingly obsolete in the face of more attractive investment opportunities elsewhere. In these circumstances, capital is likely to abandon existing centres of production and establish a new 'spatial fix' involving investment in different regions.

The deindustrialization of many established centres of production in the 'rustbelts' of North America and Western Europe since the late 1970s and the growth of new industry in 'sunbelt' regions and the newly-industrializing countries of East Asia can be understood in this light. Thus, while transport networks enable capital to 'annihilate space by time', linking distant sites of production, extraction and consumption, this can only be achieved through the production of fixed and immobile infrastructures that subsequently become vulnerable to devaluation as economic conditions change and other locations present more profitable opportunities for investment (Harvey, 1982).

A key development in human geography and the social sciences from the early 1990s was the cultural 'turn', which emphasizes the importance of beliefs, identities and values in shaping social action and behaviour. From a transport perspective, the cultural turn suggests a focus on transport users, examining how their attitudes, identities and values shape transport behaviour, something that has been neglected by the dominant perspectives derived from economics and engineering (Chapter 11). One of the key legacies of the cultural 'turn' for transport studies is the increased interest in travel across the social sciences, giving rise to the 'new mobilities paradigm', based on the notion that heightened mobility has become a defining characteristic of contemporary life. Whilst this paradigm' (Sheller & Urry, 2006) certainly defines a new interdisciplinary research agenda with which transport geographers can engage, an exaggerated emphasis on the novelty of mobility risks a blindness to continuities between the past and present, and its one-sided celebration of movement and fluidity risks oversimplifying the complex relationships between mobility and fixity, or 'flows' and 'places' (Harvey, 1982; see also Castells, 1989).

In this section, we have highlighted the major theoretical frameworks adopted by human geographers since the 1960s, particularly in terms of their consideration of the relationships between transport and economic development. The engagement between transport geography and critical theories in particular has been limited, reflecting the former's broadly positivist orientation, evident in its prolonged infatuation with location theory and formal models (Hall *et al.*, 2006) (Chapter 14). Whilst these should remain an important part of the subdiscipline's conceptual toolkit, political economy and cultural theories offer potentially valuable insights into important aspects of transport–economy relations. In what follows, political economy provides the integrating framework for our consideration of transport and spatial development in both developed and developing countries. Elements of traditional location theory are utilised to inform our analysis of the spatial effects of transport investment later in the chapter.

Transport and spatial development in developed countries

Production systems and transport networks

Industrialization during the nineteenth and early twentieth centuries gave rise to a distinct pattern of regional sectoral specialization, involving certain regions becoming specialized in particular industrial sectors. Characteristically, all the main stages of production from resource extraction to final manufacture were carried out within the same region. As indicated by the NEG, new transport networks, based on canal systems and particularly railways, were important in facilitating increased concentration and specialization, liberating factories from dependence on local resources and enabling them to serve larger markets. For instance, North-East England and West Central Scotland accounted for 94 per cent of shipbuilding employment in Britain in 1911 (Slaven, 1986). The completion of a continental transport network based on railroads facilitated the growing concentration of industry in the USA's manufacturing belt in the North East and Mid West, resulting in the increased specialization of individual cities (Figure 2.3).

The development of many industrial centres was linked to their position in relation to major transport networks, reflecting the wider tendency for important trading settlements to be located at highly accessible points such as the confluence of rivers, break-of-bulk points along coastlines, the end or mid-point of a rail line, or at the foot of mountain passes. In this way, the very existence of some major towns is explained by their so-called 'transport function'. The city of Chicago, for

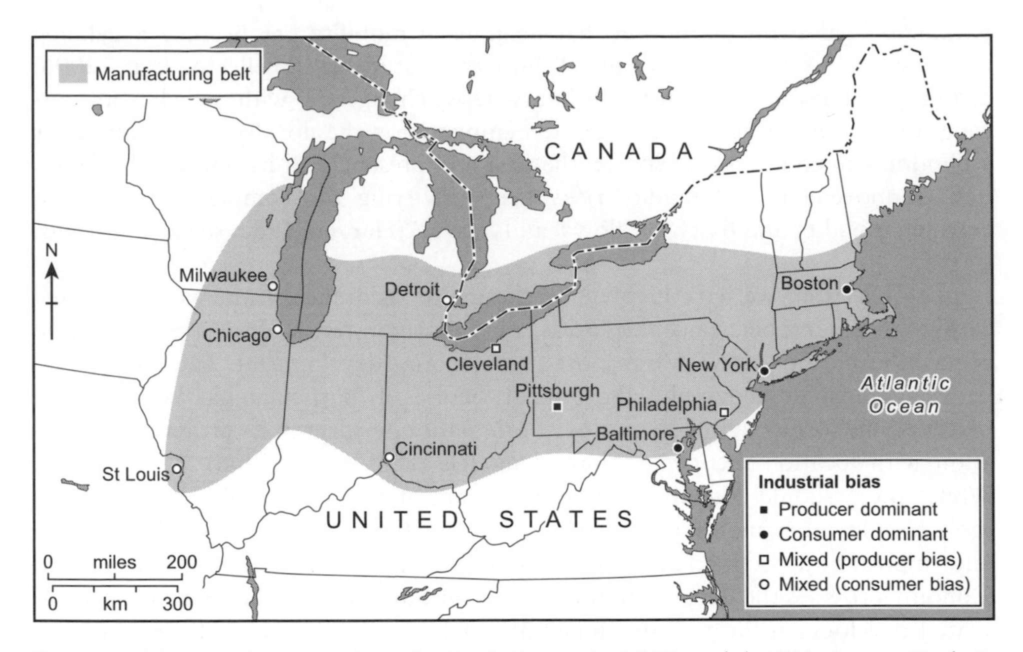

Figure 2.3. The manufacturing belt in the North East and Mid West of the USA. Source: Hoyle & Smith, 1998.

instance, owed its explosive growth from the 1840s to its role as the major transportation hub where the western- and eastern-orientated railway lines and Great Lakes shipping routes converged (Cronon, 1991). This enabled it to become a key agricultural market and processing centre, linking the resources of the vast American interior to the markets of the East coast and Europe. Grain, lumber and meat were channelled, processed and exported through Chicago and the city also operated as the key centre for the distribution of manufactured goods throughout the interior (Cronon, 1991).

The pattern of regional sectoral specialization began to break down from the 1920s, being replaced by a new Fordist system involving the mass production of consumer durables. The growth of Fordism was closely associated with the emergence of new transport technologies based on the private car and investment in road networks. This facilitated mass consumption, by providing a market for industries such as automobiles and electronics and encouraging an increased spatial separation between home and work through the growth of suburbs, particularly in North America (Walker, 1981) (Chapter 6). Suburban lifestyles became closely associated with mass consumption, with every household requiring its car, washing machine and lawnmower (Goss, 2005).

By the late 1960s, a new phase of 'neo-Fordism' was apparent as mass production technologies became increasingly routine and standardized. This created a new 'spatial division of labour' as different parts of the production process were carried out in different regions, reflecting underlying geographical variations in the cost and qualities of labour (Massey, 1984). Companies were concentrating headquarters and research and development functions in core regions where there are large pools of highly educated and skilled workers, whilst routine assembly and production was located increasingly in peripheral regions and places where costs (especially wage rates) are lowest. This dispersal of routine production has also occurred on an international scale through the 'new international division of labour' as multinational corporations (MNCs) based in western countries have shifted assembly and processing operations to developing countries (Froebel *et al.*, 1980).

International divisions of labour have become increasingly complex and intricate since the 1980s, involving an increased number of actors in different industries. In the semiconductor industry, for example, research and development functions might be based in Silicon Valley in California, skilled production carried out in the Central Belt of Scotland (the so-called 'Silicon Glen'), assembly and testing in the likes of Hong Kong and Singapore and routine assembly in low-cost locations in the Philippines, Malaysia and Indonesia (Knox *et al.*, 2003). Such arrangements are predicated on the existence of advanced transport networks which allow materials to be easily and rapidly moved between factories, although this is rarely considered in accounts of globalisation (Hall *et al.*, 2006). They require a large number of intermediate inputs and materials (as well as raw materials and finished goods) to be transported over long distances within global production networks, often controlled by large MNCs.

The increased globalization of production systems over time has been facilitated by successive revolutions in transport and communications technologies (Leyshon, 1995). The concept of time-space convergence emphasizes how, 'as a result of transport innovation, places approach each other in time-space' through reductions in the travel time between them (Janelle, 1969: 351). It takes just over one hour to travel between London and Edinburgh by jet aircraft today, for example, compared to four days by stagecoach in 1776 and eight hours by train in the late nineteenth century. The associated concept of time-space compression (Harvey, 1989) emphasises how the development of new technologies has dramatically reduced transport and communication costs, resulting in the 'annihilation of space by time' (Figure 2.4). As Knowles (2006a) reminds us, the process of time-space convergence is socially and spatially uneven, occurring primarily between key nodes within the world economy and benefiting wealthy groups such as global business executives and middle-class tourists rather than low-income people. In many respects, the 'shrinking' of space between key centres such as the world cities of London, New York and Tokyo coincides with a 'widening of space' between economically marginal locations such as sub-Saharan Africa, much of Latin America and the former Soviet Union (Leyshon, 1995).

Transport and the 'new regionalism'

Since the 1980s, the parallel processes of globalization and localization have encouraged the rise of a 'new regionalism' in economic geography. This emphasizes the increased importance of regions as economic units within a globalized economy, compared to the postwar model of integrated national economies (Storper, 1997). In particular, the success of dynamic growth regions such as the City of London

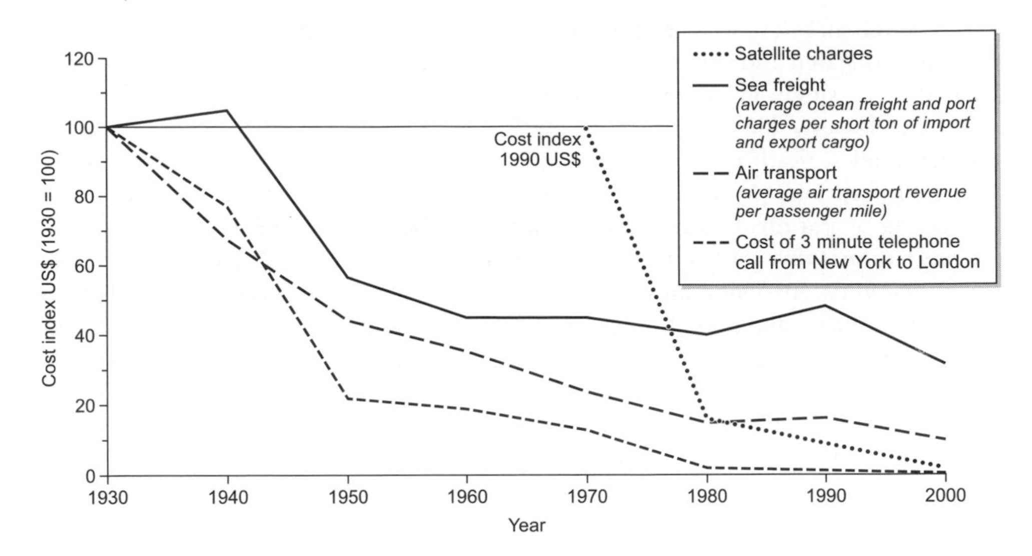

Figure 2.4. Falling costs of air/sea transport and satellite/telephone communication. Source: Eddington, 2006a.

(financial and business services), Silicon Valley (advanced electronics), Southern Germany (vehicles and electronics) and North Eastern Italy (machine tools, textiles) is rooted in the specialized production systems that have flourished there. The new regionalism examines the effects of internal factors and conditions within regions – for example, skills, rates of knowledge transfer and innovation, entrepreneurship and institutions – in helping to promote or hinder economic growth (Storper, 1997). Transport systems play an important role in facilitating economic growth within such regions, not least in terms of enabling rapid movement of materials between suppliers and manufacturers or service providers, according to the dictates of 'just in time' systems. In addition, the devolution of political power from the national level means that regional authorities have gained direct control over transport investment, allowing this to be linked more directly to regional economic needs (Chapter 5).

In some cases, high-technology clusters have grown along particular transport arteries, with examples including 'Route 128' near Boston and the 'M4 corridor' in Southern England. Good road links allow the rapid supply of both components to manufacturers and service providers and finished products to customers as required, reducing inventory costs. The locational pattern of inward investment is strongly influenced by transport networks with Japanese investment in the automobile industry in the American Mid-West concentrated along the I-75 and I-65 corridors (Figure 2.5), known as 'kanban' or 'just in time' highways (Hoyle & Smith, 1998). The distinctive pattern of clustering along the major highways can be explained by the need for close contact and collaboration between manufacturers and suppliers, granting them the flexibility to serve an increasingly diverse and fragmented

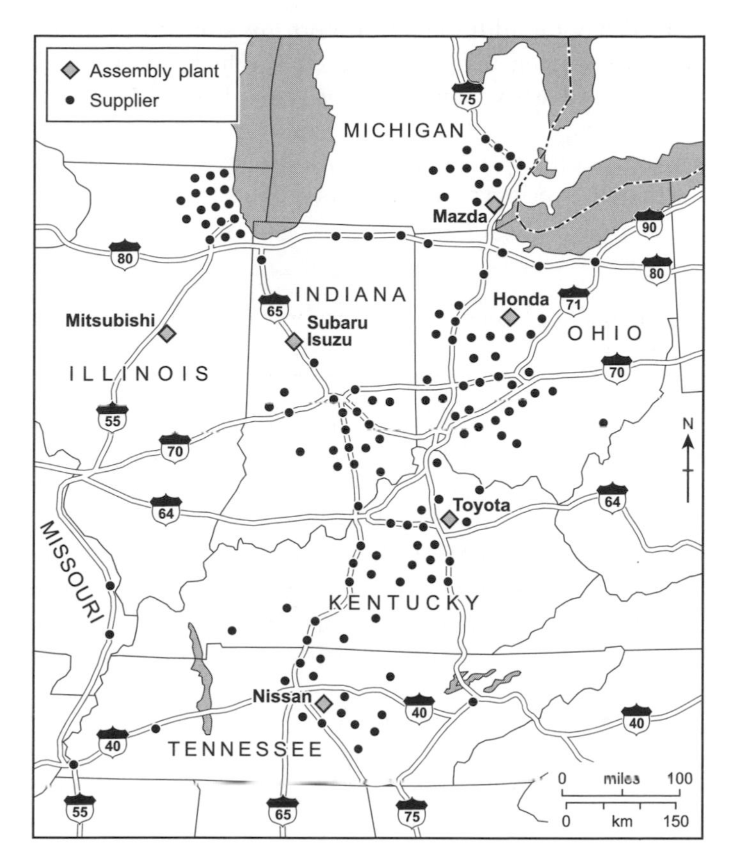

Figure 2.5. Japanese motor vehicle production in the I-65 and I-75 corridors of the USA. After Hoyle & Smith, 1998.

market by producing a range of niche products, necessitating the rapid supply of particular types and volumes of materials as required.

The spatial effects of transport investment

It is generally accepted that improved transport systems are beneficial from a national economic perspective: better roads mean faster transport, better exchange of goods and services, the utilization of comparative cost advantages and thus the enhancement of a highly specialized economy. On the whole – without regarding the external costs of transport – a national economy will benefit from a good transport system. Far more ambiguous, however, are the economic effects of the further provision of transport infrastructure in developed societies, which tend to already have high-capacity transport networks. In general, research suggests that the scope for substantial impacts on the economy is relatively limited in such cases, compared to earlier stages of development (Banister & Berechman, 2001; Eddington, 2006b). This reflects the diminishing benefits of transport investment in developed countries

(Box 2.1). The tendency for additional transport investment to simply induce additional traffic by encouraging people to use their vehicles more is also now well understood (Standing Advisory Committee on Trunk Road Assessment (SACTRA), 1994).

Focusing on the secondary effects of transport investment, statistical analyses have indicated that a 1 per cent increase in public investment can generate an

Box 2.1. The diminishing impact of transport infrastructure improvement on regional development.

In highly developed countries new transport infrastructure tends to have a diminishing impact on regional development as the economy matures. Reasons for this tendency are:

- Regional accessibility is already high. In general, industrialized nations already have a well-developed transport network, meaning that the level of accessibility is high. Therefore, further improvements of the transport infrastructure will result in only minor reductions in travel time and will not open up new areas or markets.
- Transport costs become less important. Due to economic changes such as the shift towards services, the relative importance of transport-intensive sectors is decreasing. In contrast to traditional activities such as manufacturing or mining, the growing service sector or the so-called 'new economy' does not rely as heavily on effective transport systems. Thus, transport costs become less important as a location factor, although the quality and efficiency of transport networks may become more important in line with shift to just-in-time production systems, for instance.
- Proximity is better than speed. Geographical proximity to major economic centres and clusters as a precondition of economic growth cannot be fully substituted by new transport facilities – thus, peripheral regions tend to remain remote and do not substantially gain from improved accessibility. Indeed, in some cases, further transport improvements may result in externally-located firms penetrating local markets more effectively and in local residents spending more of their income externally.
- Disparities may be deepened. An improvement in the connection of peripheral regions with central regions always works in both directions. According to the New Economic Geography, due to agglomeration effects – the advantages derived from the spatial concentration of large number of firms, suppliers, workers and consumers – central regions benefit most from such an improvement whereas peripheral regions are likely to be drained with regard to purchasing power or skilled labour. In particular, transport improvements may facilitate increased migration from peripheral to core regions.

increase in GDP of around 0.2 per cent, although such conclusions are subject to a host of important qualifications (Eddington, 2006b). For instance, they do not disentangle transport from public investment more broadly or factor in the wider economic, social and environmental impacts of transport. Most importantly, the ambiguity about cause and effect remains unresolved by such research: do transport improvements generate economic growth or vice versa? As such, the difficulties of establishing any significant correlation between transport investment and regional growth have become increasingly apparent (SACTRA, 1999). In the 1980s in West Germany, a study found the long-standing assumption that the construction of national roads fostered spatial integration and economic development – upon which a key strand of transport policy had been based – to be untenable (Lutter, 1980). Positive regional economic development was discovered to be discernible only where peripheral, rural labour markets achieved improved internal accessibility and became larger and more independent from core regions due to tangential routing. A further development of radial long-distance road connections, linking large cities and clusters, tends only to intensify the draining effect in rural areas, enabling consumers, for instance, to spend more of their income outside the region.

In a review of surveys on transport infrastructure and regional economic development in Europe, Linneker (1997) distinguishes between the spheres of consumption and production. Improved accessibility relating to consumption definitely leads to an improvement in welfare for the population, with increased competition resulting in lower prices. For the sphere of production, however, after making allowances for regional disparities, the question remains open, allowing very different answers to be advanced. Here, recent academic discussion has perhaps become too dependent on analyses of large-scale infrastructural projects in growth regions, particularly the impact of the M25 in Greater London and the Channel Tunnel (Vickerman, 1991). Reflecting the essentially enabling role of transport, Linneker (1997: 60) concludes that 'Whether further development towards higher or lower levels of economic development potential is realised . . . is determined by a large number of other factors outside the transport sector.'

This point is developed by Banister and Berechman (2001) who identify a series of necessary conditions that must be in place for transport investment to stimulate regional economic development in developed countries. The three key conditions are (i) positive economic externalities, that is, a well-functioning local economy, particularly in terms of the links between firms and suppliers and the operation of the labour market; (ii) investment factors, referring to the availability of funds, the quality of the overall network and the timing of the investment; and (iii) a favourable political environment, in terms of other supporting policies and a generally enabling policy framework. All three factors must in place for transport investment to have a positive impact on the regional economy (Figure 2.6). If only one or two of these factors are present at the time of investment, certain effects such as an improvement of accessibility may occur, but there will be no regional growth.

Rather than building new infrastructure to stimulate economic growth, one of the major transport issues requiring attention in developed countries is the reliability of transport networks and services (Eddington, 2006a). This represents the other side of the transport–economic development relationship in terms of the impact of

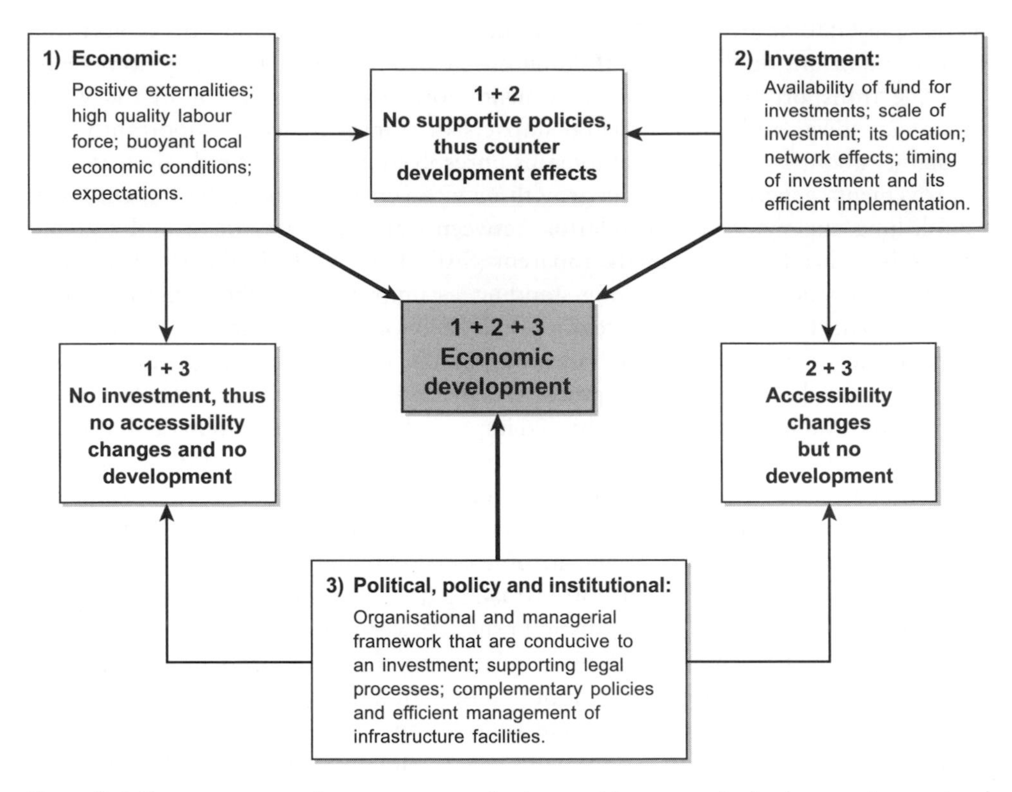

Figure 2.6. Factors necessary for transport to stimulate positive economic development in a regional economy. Source: Banister & Berechman, 2001.

rapid growth on infrastructure, creating problems when networks are unable to cope with increased demand, causing bottlenecks and congestion around key nodes and centres. These problems can constrain economic growth if left unchecked, impeding the movement of goods, information and labour and making an area less attractive to investors. Increased tendencies towards the geographical agglomeration or concentration of production in distinct clusters, coupled with the move to just-in-time supply systems, have compounded this problem.

As a result, enhancing the capacity and efficiency of transport networks through demand management measures has become a key preoccupation for policy makers. Foremost among these is congestion charging, where the authorities charge users to travel on the roads within a particular area, allowing the funds to be spent on related measures such as public transport improvements. In the UK, for instance, a report commissioned by the Treasury and Department for Transport recommended that policy should concentrate on enhancing reliability and efficiency. It identified three strategic priorities for action: congested and growing urban areas, key inter-city corridors and major international gateways such as the leading ports and airports (Eddington, 2006b), rejecting the notion that the construction of large-scale new infrastructure is required. Congestion charging works best as part of a suite of measures utilized to make better use of existing infrastructure and induce intelligent solutions for a sustainable transportation in line with market requirements (Deloitte Research, 2003). The political difficulties of introducing such an ostensibly unpopular measure in a country when unrestricted private car travel has come to be regarded as a basic right remain substantial, however, despite the success of the congestion charging scheme in central London (Chapter 3).

Transport and economic development in developing countries

The simple association between levels of economic development and transport provision is reflected in the fact that the world's poorer countries perform worst on most measures of transport availability, use and investment. This can be expressed by the vast gap in per capita passenger car numbers between developed and developing countries (Table 2.1). The process of development also highlights some stark contrasts. In China, for instance, the world's fast-growing economy, explosive rates of development have been accompanied by a great wave of motorway, railway and airport construction (Harvey, 2005), whilst much of sub-Saharan Africa remains characterized by an inadequate transport infrastructure. At the same time, the recent development of India demonstrates that rapid growth can occur in the absence of increased transport investment, although a lack of adequate capacity may subsequently impose restrictions on growth (Eddington, 2006b).

Conventional thinking has viewed the relationship in terms of transport facilitating and supporting development. Modern transport infrastructure has been widely regarded as a key driver and symbol of development with transport accounting for as much as 40 per cent of all public expenditure in developing countries in recent

Table 2.1. Passenger cars per thousand people, selected countries 2003. Source: World Bank, 2006, World Development Indicators, 2006.

Developing countries	*Cars per 1,000 people*	*Developed countries*	*Cars per 1,000 people*
Bangladesh	0.5	New Zealand	633
Ethiopia	1	Canada	561
Sierra Leone	2	Germany	545
Uganda	2	Italy	545
Bolivia	3	Switzerland	511
Gambia	6	Austria	501
India	6	France	495
Pakistan	7	United States	770
Kenya	8	Belgium	470
Afghanistan	9	Spain	455
Philippines	9	Switzerland	455
China	10	Slovenia	446
Senegal	11	United Kingdom	439
Syrian Arab Republic	12	Finland	433
Sri Lanka	13	Japan	433

Figure 2.7. Tanzania's transport network. The infrastructure has enabled inland produce to be exported but at the same time resulted in cheaper imports reaching inland markets. After Gould & White, 1974.

decades (Leinbach, 2000). Conventional transport investment has often failed to deliver the expected benefits, however, reflecting a failure to take local social needs into account within a model of development derived from the experience of industrialization in developed countries during the nineteenth century (Njenga & Davis, 2003). Proponents of replica western transport modernization may have envisaged entirely positive economic developmental effects but, in practice, benefits have been sectional and skewed. This generally reflects the legacy of colonialism, meaning that development has continued to be accompanied by underdevelopment (Frank, 1967). In some cases, transport networks established to facilitate the export of raw materials also encouraged cheaper imports that undermined local agricultural and manufacturing production (Figure 2.7). In one historic twist, sections of southern Africa's colonial railway network were laid out deliberately to bypass African tribal areas in favour of servicing the farms of European settlers. More broadly, public transport investment in sub-Saharan Africa was associated with proletarianization, racism and exclusion (Pirie, 1982).

The spatial patterns of transport development established under colonialism have tended to persist after independence with the sheer expense of new transport infrastructure meaning that investment generally remains concentrated in just a few key metropolitan areas and in airport, railway or road-based inter-city corridors (e.g. Gleave, 1991). Dualistic transport characterizes and helps to perpetuate a pronounced urban–rural (and rich–poor) divide in developing countries (Porter, 2002). The reliance on conventional road investment, for instance, has tended to benefit the wealthiest strata of the population who can afford access to private cars (Leinbach, 2000). National capital cities generally benefit most from transport infrastructure and service upgrades: suburban road paving, bus system improvements (for example, Curitiba, Brazil) and taxi termini upgrades are typical first steps. In both the colonial and post-colonial eras, backward linkages into transport equipment manufacture and engineering consultancy have tended to benefit metropolitan countries in which the relevant technology and expertise is concentrated. There is also evidence, particularly from Asian countries such as Malaysia and Sri Lanka, that infrastructure improvements have encouraged out-migration from rural areas (Leinbach, 2000).

The economic benefits of transport investment have been thought to be greater in developing economies where the establishment of basic connectivity is regarded as an important facilitator of growth. In reality, however, such benefits are far from automatic given that economic and social conditions in developing countries are very different from those prevalent in the industrialized countries in the nineteenth century, not least in terms of legacy of external control and the existence of entrenched poverty. As Leinbach (2000) observes, the economic effects of transport investment are crucially dependent on the capacity of the local or regional economy to respond effectively, a property that has been absent in many developing countries, particularly in sub-Saharan Africa. In one vital respect this echoes the identification of necessary conditions for transport to induce growth in developed economies (Banister & Berechman, 2001). Investment and political factors would seem to be equally important in the developing world, particularly in countries suffering from prolonged economic stagnation.

The problems of urban transport in the developing world are huge, not least in the burgeoning megacities such as Bangkok, Mexico City and Cairo, where the basic infrastructure is simply overwhelmed by population growth (Gwilliam, 2003). The streets of many such cities are among the most congested, most polluted and least regulated on the planet. Road congestion is endemic, reflecting the inadequacy of the infrastructure – road space typically takes up 10–12 per cent of land in Asian megacities, compared to 20–30 per cent in cities in the USA (Gwilliam, 2003: 202) – and the rapid growth of private vehicle traffic from a low base. Public transport remains inaccessible and unaffordable to the majority of the poor, particularly those condemned to eking out a meagre existence in the sprawling shanty towns around the urban fringe (Figure 2.8). In the developing world, then, the 'fundamental paradox of urban transport', namely the coexistence of excess demand for road space and the underfunding of other modes (Gwilliam, 2003: 212), is compounded by population growth, inadequate infrastructure and mass poverty. In rural areas,

Figure 2.8. Hired horse cart transporting household waste in a shanty settlement in Cape Town, South Africa. G. Pirie.

basic road infrastructure coverage is often poor, with problems of maintenance and upkeep greatly accentuated by the seasonality of the climate in tropical countries. Rural transport services in Africa are typically uncompetitive, high cost and undiversified whereas Asian transport is competitive, low cost, frequent and diversified in terms of routes, aided by a high density of demand (Njenga & Davis, 2003).

Increased awareness of the failures of past initiatives has encouraged a new approach to transport provision, reflecting also the refocusing of development policy around the goals of poverty reduction and greater local participation (Mawdsley & Rigg, 2003). Accordingly, governments no longer fall headlong for prestigious, capital-intensive transport projects selected by external organizations and corporations. Whereas the objective of transport development was to enhance gross national product for much of the postwar era, there is now a greater appreciation of the qualitative limitations of growth at any cost as the attendant ambiguities, contradictions and hidden costs have become apparent, accentuating the geographical unevenness of development.

The increased audibility of local voices in development planning has added weight to the growing impetus for transport interventions that meet new outcomes and process criteria. Resurrecting the old question of 'development for whom?' (Potter *et al.*, 2004), development agencies and government are now viewing transport from the point of view of users and providers, adopting more consultative methods that take their needs into account (Njenga & Davis, 2003). The aim is to improve the accessibility and mobility of people in ways that are affordable, equitable and sustainable (Leinbach, 2000). At relatively low cost, for example, decentralized road programmes, involving rural road rehabilitation, maintenance and upgrading, can lead to significant improvements in agricultural productivity, food security, tourism and public health (see Airey, 1985; Minten & Kyle, 2002; Yunusa & Shaibu-Imodagbe, 2002). Building all-season local roads would improve basic

Table 2.2. Percentage of people in 15 sub-Saharan countries living within 2 km of an all-season road. Source: World Bank Sub-Saharan Africa Transport Policy Programme.

Benin	32	Ethiopia	27	Mali	51
Burkina Faso	19	Ghana	44	Niger	52
Burundi	19	Kenya	44	Nigeria	47
Cameroon	20	Madagascar	67	Tanzania	38
Chad	5	Malawi	38	Zambia	51

access to public transport in much of sub-Saharan Africa (Table 2.2), although funding constraints and political difficulties have tended to frustrate this ambition (Porter, 2002). At the village level, 'intermediate means of transport' (IMTs), such as bicycles, animals and carts, have been promoted. These allow more rapid planting, harvesting and fertilizing and the easier collection of firewood and water, committing fewer women and children to daily drudgery. A rebalancing of the gender-specific benefits of transport is particularly overdue (Porter, 2002). IMTs should not be regarded as a panacea, however, lying beyond the means of many rural households in Africa, and requiring a certain 'critical mass' in terms of demand and supporting infrastructure (Njenga & Davis, 2003).

Affordable and achievable transport improvements such as street realignment and widening, street traffic intersection control, and segregation of pedestrians and vehicles are also crucial to enhancing mobility and economic productivity in cities. The problems of the residents of urban slums in terms of accessing employment opportunities and basic services remain particularly intractable, requiring a range of measures to make transport more affordable and accessible as well as 'non-transport interventions' which locate public facilities and services in more accessible areas (Olvera *et al.*, 2003). A general lesson from the world's most wretched urban and rural areas is that relatively minor transport improvements can make a considerable difference to mobility and access to public health clinics, schools and employment. Whether the profusion of such small-scale local schemes is sufficient to generate real development that lifts people out of poverty is perhaps questionable, however, taking continuing funding constraints and political difficulties into account. Small-scale projects could be said to be largely ameliorative, treating the symptoms of poverty rather than alleviating the causes. Yet such a view seems unduly harsh and reflective of conventional 'top-down' thinking, given the benefits of participation and the incorporation of local needs. This debate returns us to the perennial question of 'development for whom' and an approach that incorporates both the conventional and non-conventional sectors through a suite of appropriately targeted schemes would seem to offer the best way forward, promising incremental progress.

Conclusion

In this chapter, we have sought to examine the relationship between transport and economic development, bringing together the separate literatures of transport geography and economic geography. Whilst the reduced transport costs facilitated by cheap energy have encouraged economic geographers to take high levels of mobility for granted since the 1970s, the current era of 'peak oil' threatens to undermine this assumption, requiring a reappraisal of the role of transport within theories of globalization (Hall *et al.*, 2006). Conventional thinking has regarded transport as an important facilitator of development, but the complex and multi-faceted nature of the relationship between transport and economic development has become increasingly apparent in recent years. One basic finding is that the capacity of transport investment to generate benefits is crucially dependent on the strength of the wider national or regional economy, something that can be understood in terms of economic, investment and political/institutional conditions (Banister & Berechman, 2001). Moreover, rather than spatial integration through transport networks fostering regional convergence, it is often associated with regional divergence, deepening pre-existing differences between core and peripheral areas. For transport investment to contribute to the reduction of regional disparities it would need to be linked to a strong regional policy targeted on the worst-performing regions and matched by curbs on growth in the best-performing ones. Such 'spatial Keynesianism' (Martin, 1989) is deeply unfashionable amongst both policy makers and academics, having been abandoned in the late 1970s and 1980s as governments sought to promote national economic growth and competitiveness in an increasingly turbulent economic environment.

Over the last decade or so, these new insights have begun to inform transport policy, supporting a greater emphasis on efficiency, sustainability and effective targeting on particular groups and areas. In the developed world, whilst new transport infrastructure continues to be linked to economic growth by business groups and commentators in particular, enhancing the capacity and efficiency of the existing infrastructure has become a major concern (Eddington, 2006b). At the same time, the sustainability agenda has emphasized the need to mitigate the negative environmental impacts of transport (Chapter 3). As such, the introduction of demand management measures such as congestion charging is becoming a key issue. In the developing world, transport infrastructure improvements remain prominent in the objectives of supra-national bodies such as the World Bank and the European Union. But the established belief that transport modernization is inherently beneficial is giving way to a more nuanced understanding of its socially and spatially differentiated effects. Poverty reduction rather than economic growth per se has become the key focus of development policy, heralding a new appreciation of the needs of transport users and providers. The benefits of local road upgrading and maintenance schemes have become more apparent, alongside the growth of IMTs. Such measures offer a means of enhancing the mobility and accessibility of the poorest sections of the population, although substantial problems remain in both rural areas and rapidly growing cities.

In developed and developing countries alike, then, the assumption that the construction of new infrastructure will automatically generate economic benefits is being replaced by a growing appreciation of the need to maximize the capacity of existing networks and a realization that incremental improvements can generate substantial gains if appropriately planned and targeted.

3 | Transport and the Environment

Stephen Potter and Ian Bailey

Transport, whether by road, rail, sea, inland waterway, air, bicycle, or on foot, offers many economic and social benefits but also has a wide range of direct and indirect environmental impacts. Large amounts of finite resources in the form of fuels and materials are needed to construct vehicles and transport infrastructure. Transport activities account for over 30 per cent of all energy use by final consumers and are widely predicted to be the largest contributor to the growth of carbon dioxide emissions in the twenty first century (European Environment Agency, 2005; Haq, 1997). Road transport is also a major source of other noxious emissions, notably carbon monoxide, nitrogen oxide and particulates, as well as having a number of detrimental environmental effects related to noise and the severance of communities and ecosystems (Hensher & Button, 2003).

This chapter reviews the main types of environmental problems created by transport activities, their relationship with the discipline of transport geography, and some of the approaches developed to address transport's environmental consequences. It first defines the meaning of environmental issues in relation to transport and explains why environmental issues are of growing interest to transport geographers. The main social and natural environmental impacts of transport are then explored, together with various approaches used to lessen these impacts. Finally, some of the key political and geographical challenges involved in promoting more sustainable transport systems are discussed.

Environment

Achieving a straightforward definition of the term 'environment' and thus an unambiguous understanding of its relationship with transport (and transport geography) is rather difficult (Kessel, 2006; Lubchenco, 1998) since environmental and social concerns about transport are often closely related. Much of the research into the environmental effects of transport also covers its economic and social effects, while studies on the social dimensions of transport routinely discuss its environmental impacts on particular social groups or geographical areas. Although environmental and social concerns overlap to a large degree, they often arise from different causes, produce different kinds of effects and thus require different policy responses. Another important advantage of separating out the social and natural components of the environment is to ensure that the 'natural' environment is properly represented in decision making processes, rather than being 'constructed away' by economic interests because they are difficult to value in an immediate material way (Pedynowski, 2003). None the less, although this chapter considers the social and natural environmental effects of transport separately, it should be remembered that this is a somewhat artificial division.

Either way, the development of sustainable transport systems requires an understanding of a range of environmental, economic and social systems and the input of many academic disciplines, politicians, planners, developers and transport users (Purvis, 2004). In terms of geography's contribution to this agenda, the movement of goods and people lies at the heart of geography's preoccupation with place, spatial distribution and scale. Thinking geographically can encourage a breadth of vision and a capacity for synthesis that are essential to gain a holistic understanding of the environmental challenges posed by transport. Some of the environmental problems caused by transport are unquestionably global in scale (e.g. changes to the global climate system caused by the combustion of carbon based fuels); some are regional (e.g. freight traffic damage to ecologically sensitive areas in the Alps) (Lauber, 2002); and some are local (e.g. traffic-related smog in cities such as Athens, Mexico City and Los Angeles). Understanding and making connections between these different spatial scales are fundamental to the development of integrated transport strategies. Equally, geography's tradition of research on how governance systems influence economic and social organization is particularly suited to the study of the complex exchanges involved in the transport of goods and people (Liverman, 2004) (Chapter 5). These interests and skills provide geographers with a solid analytical base for critiquing strategies for sustainable development, providing empirically and conceptually informed insights on the consequences of different strategies, and highlighting the dangers of over reliance on technical or managerial measures that treat the symptoms of problems without dealing with their underlying causes or spatial implications (Purvis, 2004).

Transport's environmental impacts

The division between the natural and social environmental impacts of transport is roughly mirrored by the division of these impacts into first and second order effects. First order effects broadly cover direct physical impacts such as vehicle emissions (local pollution, health, global warming), noise from vehicles, land take by roads, airports and railways, as well as the extraction of materials to manufacture vehicles and the waste produced when vehicles are scrapped. Second order impacts relate to how societies and economies create and adapt to increasingly transport-intensive lifestyles. These include spatial changes in activity and settlement patterns caused by rising car ownership, dispersed working arrangements and shopping and leisure trips, all of which over time reconfigure land use patterns and increase transport dependence. Likewise changes in transport technology and costs have led to second order effects in terms of where goods are produced and marketed, such as the air freighting of 'out of season' vegetables from southern Africa to European markets. These second order effects can have long-term, entrenched and geographically differentiated social impacts. For example, health effects are not restricted to traffic accidents or even respiratory illnesses caused by traffic pollution. They also contribute to an increasingly sedentary and obese society. Not all natural impacts are first order, nor are all social impacts second order (and the boundary between the two is imprecise), but there is a pattern and the relative impacts tend to be clustered this way.

A commonly used method to evaluate the first order environmental aspects of a product or activity is Life Cycle Analysis (LCA). LCA seeks to create a systematic inventory of the resources consumed (energy, materials, water), emissions to air, land and water and waste produced during four key life cycle stages: raw material production and processing, manufacturing and distribution, use of a product or service system and disposal and/or recycling back to earlier stages.

The main materials used in the manufacture of vehicles, for instance, are steel, non-ferrous metals (e.g. aluminium and copper), plastics, glass, and, recently, composite materials. Extracting and processing these materials involves large amounts of energy and water, the production of emissions to air and water, and the creation of solid waste. During manufacture and assembly, further resources are required and more emissions and wastes are generated. For transport infrastructure, aggregates and bitumen are extracted for road building and considerable amounts of concrete are used for constructing bridges and associated structures. Similar materials are used for airports, while railway infrastructure also requires steel rails.

All forms of transport consume large amounts of energy during their use phase. The main fuel sources – petrol, diesel and kerosene – are all sources of carbon dioxide (CO_2) and air pollutants such as carbon monoxide, unburnt hydrocarbons and nitrogen oxides that are harmful to human health and/or the environment. Many parts and components also require replacement and disposal during a vehicle's use, and at the end of its life the vehicle is normally dismantled with some materials being recycled to reduce energy consumption and emissions compared with using virgin materials. The remaining, non-recycled materials (usually glass, rubber and fabrics) are either disposed of or burnt, producing further emissions.

A number of LCAs on cars and other vehicles conclude that over 70 per cent of their environmental impacts are concentrated during their 'use' phase (Funazaki *et al.*, 2003; Hughes, 1993; Organisation for Economic Cooperation and Development (OECD), 1993; Teufel *et al.*, 1999), though these studies only examined the environmental impacts of vehicles themselves and so excluded the infrastructure needed to operate transport systems. Table 3.1 shows the results from Funazaki *et al.*'s study.

As well as analysing the life cycle stages of transport activities, it is important to categorize environmental impacts according to where they occur. For example,

Table 3.1. Life Cycle Analysis for a Japanese car. Source: Funazaki *et al.*, 2003.

	Production	*In Use*	*End of life*	*Total*
CO_2 (tonnes)	5.6	23.3	0.1	29.0
NO_x (kg)	9.1	31.0	0.3	40.4
SO_2 (kg)	7.8	14.0	0.0	21.9
CFC12 (g)	–	150	418	568
Energy (GJ)	78.7	353.7	1.2	433.6
Global warming (Tonnes CO_2 equivalent)	5.6	24.2	3.0	33.0
Global warming (% distribution)	16.9	73.9	9.2	100.0

different types of airborne emissions from petrol cars will undergo varying degrees of dispersal, creating micro-, meso- and macro-level geographical variations in impact. Similarly, emissions from electric vehicles may be minimal at the point of use but highly concentrated around power stations. Altitude is also important because the CO_2 released by aircraft at altitude has a global warming effect 2.7 times that of ground-level emissions (HM Treasury and Department for Transport (DfT), 2003). These spatial characteristics are critical to understanding the impacts of transport activities on environmental and social equity, but are extremely difficult to incorporate into LCA equations, strengthening the case for interdisciplinary analysis of transport's environmental impacts. Four main geographical levels of impact can be identified (Table 3.2).

First order impacts

Health

Mounting concern about the detrimental health effects of road traffic emissions in the UK has led to the establishment of a government 'think tank', the Committee on the Medical Effects of Air Pollutants (COMEAP). Similarly, in 2006, the European Union (EU) introduced the European Pollutant Release and Transfer Register (E-PRTR) to monitor air emissions from a range of sources, including transport, to implement the United Nations PRTR Protocol signed in 2003 by the EU and 36 countries. The UK government's 1998 transport policy White Paper (Department of the Environment, Transport and the Regions (DETR), 1998: 16), quoting COMEAP (1995), notes that 'up to 24,000 vulnerable people are estimated to die prematurely each year (in the UK), and a similar number are admitted to hospital, because of exposure to air pollution, much of which is due to road traffic'. This death toll is six times higher than that for road accidents.[1] Overall, COMEAP attribute 12,500 premature deaths to ground-level ozone (from all sources), 8,100 from particulate emissions and 3,500 to sulphur dioxide emissions, while hospital admissions for these conditions account for 2.7 per cent of UK hospital admissions for respiratory diseases in urban areas. A European Commission study (Bjerrgarrd *et al.*, 1996) notes that deaths, hospitalization, sick leave and other health effects

Table 3.2. Locational distribution of transport's environmental impacts.

Location of impact	*Sources*
Local, e.g. smell, air quality, health effects, accidents, noise, severance	Exhaust emissions of particulates, volatile organic compounds, carbon monoxide, sulphur dioxide, ozone; vehicle noise; physical infrastructure.
Regional, e.g. land use effects and waste disposal	Depletion of finite resources, land take of infrastructure; urban sprawl; disposal of scrap tyres, engine oil, chemicals etc.
Continental, e.g. acid rain; changes in crops produced, location of production	Vehicle exhaust emissions – nitrogen oxides, sulphur dioxide. Land take for new cash crops, erosion of wilderness
Global, e.g. climate change, ozone depletion	Exhaust emissions of carbon dioxide, CFCs in air conditioning.

attributable to traffic pollution amounts to at least 0.4 per cent of the EU's gross domestic product (GDP).

Concern about the health impacts of transport activities has resulted in many countries developing policies to reduce traffic emissions. In Britain, the National Air Quality Strategy specifies 'safe' targets for the eight main health-threatening air pollutants in the UK (benzene, polycyclic aromatic hydrocarbons (PAHs), 1,3-butadiene, carbon monoxide (CO), particulates (especially small PM10 type particles), nitrogen oxides (NO_x), ozone and sulphur dioxide (SO_2). Transport, particularly road traffic, is a major (if not the majority) source of all these pollutants except SO_2.

Noise pollution

Surveys by the Building Research Establishment (BRE) have identified road traffic and aircraft noise as the two most widespread forms of noise disturbance in the UK (BRE, 1993; DfT, 2005). In the 1999/2000 BRE survey, 30 per cent of people reported being affected by road traffic noise, compared with 17 per cent by aircraft and 4 per cent by trains, with the noisiest areas tending to be near airports, main roads and railway lines. Sleeping is the main activity reported to be disturbed by road traffic, and medical evidence highlights that night-time noise high enough to cause sleep interference and reduced deeper ('REM') sleep can alter moods, impair learning, and reduce workplace performance (Jones, 1990). Further medical research suggests that even low-level noise, such as is commonly produced by road traffic, is a significant contributor to raised blood pressure, minor psychiatric illnesses and other stress related illnesses (World Health Organisation (WHO), 1995). Clearly, such impacts are not felt evenly across the population but, rather, create geographical inequalities in physical and mental health which need to be understood and managed.

Climate change

The contribution of transport activities to global climate change has emerged as a major environmental concern in recent years. Under the 1997 Kyoto Protocol, the UK, along with most other industrialized nations, agreed to binding reductions in their CO_2 and other greenhouse gas emissions, in the UK's case to 12.5 per cent below 1990 levels by 2008–12. Between 1993 and 2003, the contribution of domestic transport to UK CO_2 emissions rose from 25 per cent to 27 per cent (DfT, 2004a). The UK's National Atmospheric Emissions Inventory (Baggott *et al.*, 2005) estimates that CO_2 from transport has increased by 7 per cent since 1990 and current projections (DfT, 2004a) are for a further rise of 25 per cent by 2030 unless additional policy measures are taken.

Nearly three-quarters of the UK's transport CO_2 emissions come from road transport (Table 3.3), a figure similar to the European average of 80 per cent (Commission of the European Communities (CEC), 1992). Although passenger cars remain the biggest source of CO_2, road freight emissions have risen by 23 per cent over the last 10 years compared with roughly static emissions from passenger cars. This may seem to contradict the earlier claim that transport emissions are rising as a proportion of total emissions until one considers that CO_2 emissions from most other sources, particularly industry, have reduced since 1990 (Bailey and Rupp, 2005). Rail produces under 1 per cent of transport's CO_2 emissions and, despite a 26 per cent rise in passenger kilometres and a 15 per cent rise in freight tonne kilometres over the last decade, its emissions have also declined.

CO_2 emissions from aviation, meanwhile, have grown by 60 per cent in the last decade and now account for a fifth of all transport's CO_2 emissions. The 2004 transport White Paper (DfT, 2004b) noted that because emissions at altitude have a greater global warming effect, these now represent 11 per cent of the UK's total climate change impact. Recent research also suggests that seasonal variations can change the greenhouse warming potential of flights and that night flights may be

Table 3.3. UK CO_2 emissions by source. Source: DfT, 2005.
Million tonnes of carbon

Source	*2003*	*per cent*
Passenger Cars	19.8	45.2
Light duty vehicles (vans etc.)	4.4	10.0
Buses	1.0	2.3
Heavy Goods Vehicles	7.2	16.4
Mopeds and motorcycles	0.1	0.2
ALL ROAD	(32.5)	(74.2)
Railways	0.3	0.7
Domestic aviation	0.6	1.4
International aviation	8.1	18.5
Domestic shipping	0.9	2.1
International shipping	1.4	3.2
TOTAL	**43.8**	**100.0**

much more damaging than daytime ones because vapour trails act as a warming blanket, preventing heat from escaping (Stuber *et al.*, 2006). At currently predicted growth rates the aviation sector will contribute around 33 per cent of the UK's total climate change impact by 2050 if other sectors meet CO_2 reduction targets. Although air travel is clearly a key environmental issue for the twenty-first century (Bishop and Grayling, 2003; Button 2001), international air travel (and international shipping) are currently excluded from the Kyoto Protocol because of difficulties in assigning international emissions to individual countries.

Second order impacts

Transport's first order environmental impacts will be immediately familiar to most readers, but its second order impacts are no less significant and are likely to be more structurally embedded in economic and social systems.

Lifestyle health impacts

In addition to the direct health impacts of pollution and noise, research from the 1990s began to detect subtle but cumulative changes in behaviour and lifestyles caused by an increased dependency on motorized travel (Davis, 1996). Today, issues such as obesity, poor fitness and heart problems among both children and adults are all recognised to be important health issues. Quoting the 2003 Health Survey for England, the Department for Education and Skills' (DfES) (2003: 4) report *Travelling to School: an Action Plan* notes that:

> The amount of daily exercise taken by children has decreased in recent years, which has contributed to the growing proportion of children who are overweight and obese. Childhood obesity – now affecting 8.5 per cent of 6 year olds and 15 per cent of 15 year olds – often leads to obesity in adulthood. Adults who maintain their correct weight and are physically active have a reduced risk of chronic conditions such as Type 2 diabetes and heart disease.

Although western society's car-dependent culture is far from being the sole cause of such problems, it plays an important part in increasing the number of unhealthy 'couch potatoes'. These health impacts are again subject to major geographical variations related to, among other things, income, age, education, and place of residence (for example, urban versus rural living, proximity to major transport links and/or other services). These relationships are neither straightforward nor static, however, and so make for a complex background against which to target decisions and actions to improve the health of vulnerable groups (Griffiths & Fitzpatrick, 2001).

Activity patterns

The direct take of land for transport infrastructure has frequently stirred public protests over road or runway construction projects. But transport developments also produce major second order changes in transport behaviour which have profound spatial effects by influencing land use patterns and producing environmental impacts (for example, metropolitan decentralization and low-density urban sprawl generated by mass car ownership). Changes in transport costs and availability resulting from these processes can lead to relatively swift changes in personal behaviour within existing land use patterns, and can also produce cumulative changes in the spatial organization of areas. When the London Congestion Charge (see below) was introduced there was much concern that diverting traffic around the charge zone would increase congestion in areas at its periphery, thereby displacing rather than solving the problem. Retailers are now increasingly concerned about the diversion of trips away from shops in the zone (Schmocker *et al.*, 2006), creating a new type of problem by addressing an old one. These may be relatively minor effects in central London but if national or regional road user charging were to be introduced, the rearrangement of activity and congestion patterns could be substantial. Congestion charges may then need to be altered to accommodate these shifts, sparking off another round of activity pattern shifts. Again, geographical analysis has a major part to play in understanding and finding appropriate solutions to these problems by offering spatially integrated analysis which also understands the unique character and circumstances of particular places.

Land use effects

Over time, transport developments will result not just in changes to activity patterns but will also influence where developers and businesses locate new houses, shops, offices and industry. Of course, transport has always affected the spatial organization of societies and economies. Canals permitted the transport of raw materials to centres of manufacturing and made possible the development of the nineteenth-century industrial cities, though travel within towns and cities was still predominantly on foot. The creation of suburbs was initially stimulated by developments in rail, buses and trams, but the arrival of mass car ownership greatly accelerated this trend (Hunter *et al.*, 1998). This is typified by the low-density urban and semi-urban developments in California, Australia and South Africa which can also be seen in urban fringe housing and retail developments in Britain and continental Europe. A striking example of this phenomenon in Australia is the so-called 'sea and tree' change movement, where large numbers of metropolitan dwellers are relocating wholly or partly (via weekend homes) to rural or semi-rural locations (Gurran *et al.*, 2005). In 2005, the Australian Bureau of Statistics released data indicating that growth in coastal areas in 2004 was 60 per cent higher than the national average growth rate for Australia. The population of Mandurah on the Western Australian coast, for example, is predicted to grow by 55 per cent between 2006 and 2016, creating new environmental and transport stresses within the town and its surrounding region, and a corresponding decline in population, services and employment in many inland communities (Tonts, 2000).

Unlike changes in activity patterns, changes in land use build up over time and result in travel behaviour that becomes entrenched and difficult to alter. If someone, after acquiring access to a car, now shops at a more remote supermarket, a reverse

Figure 3.1. Scene from Tati's *Trafic*.

in behaviour (e.g. due to rising fuel costs) is possible by shifting back to shopping locally. But the growing suburbanization of cities and associated outmigration towards the rural fringe inherently requires high car use. Returning to a lower level of car dependence is very difficult.

Such developments are often seen as synonymous with increasing affluence, mobility and consumer choice, but are they truly a sign of progress? Back in 1971, the French film maker and social critic, Jacques Tati, epitomized the impacts of such second-level effects in his film *Trafic*, while also questioning whether the apparent freedoms created by the private car represented liberation or a new form of dependency and social control (Figure 3.1). He commented:

> In society's rush for progress and constant pressure we have created our own demise. In our rush to go somewhere we inevitably end up in a traffic jam stuck with no place to go and no control to do anything about it. Cars continue around a circle on a never-ending journey getting nowhere. Modern society is not progressing, merely riding around in a circle. Paradoxically, technology has complicated life instead of simplifying it.

Globalization and air travel

Indirect land use effects do not just relate to personal travel. For example, economic globalization and an increased emphasis on 'free' trade[2] has led to spiralling demand in the industrialized nations for 'out-of-season' produce from developing nations. CO_2 emissions from international air travel are now rising rapidly in parallel with the growth in air travel. The rise of the low-cost 'budget' airlines has become legendary (Chapters 9 and 11), but in the last 20 years long-haul air travel has also become much more popular, with over two-thirds of air travel being for leisure. As Figure 3.2 shows, the number of air passenger trips from UK airports has more than doubled since 1990 and is projected by the DfT to nearly double again by 2020. A significant proportion of this increase relates to domestic flights within the UK

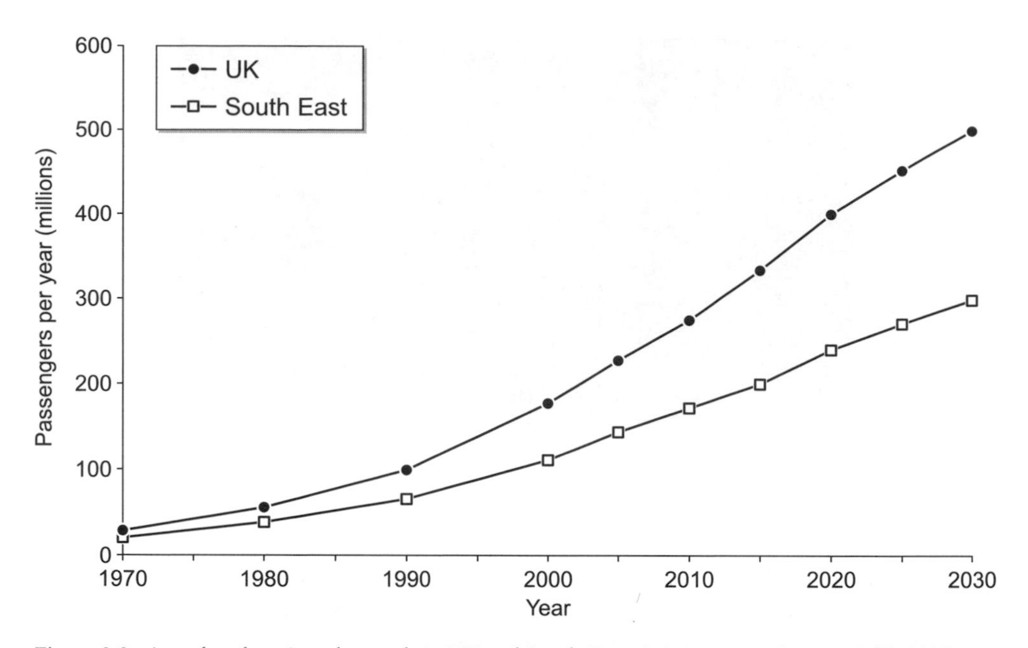

Figure 3.2. Actual and projected growth in UK and South East air passengers. Source: DfT, 2003a.

as more and more travellers opt to avoid Britain's congested motorways and unreliable railways. These trends are being mirrored in other regions of the world, for instance, with the emergence in 2003 of Air Deccan as India's first low-cost airline. Such has been Air Deccan's success that more than a dozen other low-cost airlines now operate in this market. On a global basis, air passenger kilometres are expected to grow by between anywhere from 90 to 280 per cent between 1995 and 2015, with growth of 450–820 per cent by 2050. Air freight is also expected to grow with demand for goods which can be met by global supply systems rather than local suppliers. A more detailed discussion of global aviation trends is contained in Chapter 9.

Assessing and managing transport's environmental impacts

This section examines the various types of policy used to address transport's major first and second order environmental impacts. As pointed out in Chapter 5, the governance of transport – how and whether to address economic, social or environmental needs – is intensely geographical as it involves policy making and planning from the global to the local scale. How transport is governed varies from country to country but, generally speaking, national governments provide the broad objectives and framework of transport policy, with detailed implementation taking place at regional and local levels via spatial planning strategies (Banister, 2005). We focus mainly on policies developed within Europe as a general illustration of how advanced economies are attempting to tackle these issues. The EU also provides the framework for much of UK policy dealing with the environmental aspects of transport, though substantial variation exists in the policies used in developed countries. Developing nations are generally less advanced in integrating environmental considerations into transport policy, but many are now beginning to adopt similar measures, often in modified form, to suit their environmental and economic circumstances.

Broadly speaking, policies to address the environmental impacts of transport can be divided into three categories: those dealing with transport technologies and transport emissions; those which attempt to integrate environmental considerations into transport planning; and those that seek to modify traveller behaviour by creating incentives to travel less or use less-polluting transport modes. Within these categories, five main types of measure can be identified: traditional 'command-and-control' legislation, integrated assessment techniques, voluntary agreements, market-based instruments and infrastructural investment programmes.

Legislation

Although the EU had no direct authority to act on environmental issues until the Single European Act in 1986, the Community began to develop directives covering airborne emissions from transport as early as 1970 as part of a general commitment to protect public health. The main foci of EU legislation have been the prohibition or stipulation of certain technologies (e.g. the phasing out of leaded petrol and phasing in of catalytic converters for new cars) and the creation of emissions standards for vehicles (McCormick, 2001). The EU has since developed an array of legislation covering exhaust emissions and noise, the main elements of which are summarized in Table 3.4.

Despite the undoubted ambition of this legislative programme, a number of difficulties exist with legislation as a way of controlling the environmental impacts of transport. First, legislation only targets certain pollutants and gives vehicle manufacturers few incentives to go beyond prescribed standards (Lenschow, 2002). This can be a particular problem where there is dispute over what constitutes a 'safe' level of pollution. Secondly, the amount of time required to negotiate and implement new laws can significantly hinder their ability to keep pace with technological and market developments. Laws can also only mandate technologies that have already achieved market viability (although pollution standards can stimulate innovation). Thus, the EU was able to ban leaded fuel and require manufacturers to fit new petrol vehicles with catalytic converters, but has been unable to provide comparable legal support for biofuels, hybrid vehicles and hydrogen fuel cells. Transport's dispersed nature also makes monitoring and enforcing new laws very costly and time consuming.

A final but crucial weakness of legislation is its limited capacity to influence travel behaviour. Although technological advances have cut emissions per kilometre travelled, these have failed to keep pace with demand for motorized travel. It has long been accepted that demand for travel should be influenced by policy; however, the use of legislation to achieve this is hampered by the political difficulties of introducing unpopular measures such as restrictions on car use and air travel (Hunter *et al.*,

Table 3.4. Key EU environmental legislation on transport. Source: CEC, 2006a.

Directive	*Description*
70/220	Limits for emissions of carbon monoxide and unburned hydrocarbons from petrol engines (excluding tractors and public service vehicles)
70/157	Permissible sound levels and exhaust systems of motor vehicles. 97/24 introduced similar noise standards for two- and three-wheeled vehicles.
80/779	Limits on ground-level concentrations of sulphur dioxide and suspended particulates.
80/51	Limits on noise emissions from subsonic aircraft. Extended by 92/14 (on the limitation of the operation of aeroplanes) and 2002/30 (on operating restrictions at Community airports).
85/210	Directive on lead in petrol. Builds on an earlier directive (78/611) setting maximum limits for lead content in petrol and requiring all member states to make unleaded petrol available.
88/77	Directive on emission of gaseous pollutants from diesel engines for use in vehicles. Similar to directive 70/220 covering heavy vehicles
96/96 supplemented by 2000/30	Approximated laws on roadworthiness tests for motor vehicles and introduced speed limiters and new standards for exhaust emissions from commercial vehicles.
98/69 and 98/70 amended by 2003/17	Limits for sulphur content of petrol and diesel at 50 p.p.m. (parts per million) and phases in diesel and petrol with a sulphur content of 10 p.p.m. from 2005. Requires all petrol sold in the member states to be unleaded from 2002.
2002/49	Brings together EU noise legislation to provide a common basis for tackling noise problems across the EU.

1998). This has encouraged politicians to adopt more indirect approaches to transport policy, including assessment procedures for transport developments.

Integrated environmental assessment

The planning system has long been used to curb the environmental effects of road transport, for example, the designation of 'green belts' around conurbations, restrictions on out-of-town shopping centres, the prioritization of 'brownfield' redevelopments for housing, and the use of 'Section 106' planning agreements in the UK to force developers to include measures to minimize the traffic impact of new developments. Two further techniques to improve the integration of environmental issues into planning decisions are Environmental Impact Assessment (EIA) and Strategic Environmental Assessment (SEA). Both processes, like LCA, involve analysis and recording of the likely environmental effects of development proposals, but also utilize public consultation and comment to ensure decision making remains accountable.

The EU adopted its first EIA directive (85/337/EC) in 1985 to coordinate how the member states assessed development proposals that have significant environmental impacts. The directive included two categories of project, the first where EIA was mandatory and the second where it was recommended that the member states conduct EIA where the project's environmental characteristics warranted assessment. In the transport sector, the mandatory list included new motorways, express roads, long-distance railway lines, large airport developments, trading ports and some inland waterways, while the discretionary list included other road developments, harbours, and smaller airfields (CEC, 1985). The Commission, however, received a large number of complaints during the 1990s about: failures by planning authorities to conduct EIAs on discretionary projects; unsatisfactory application of EIA procedures; insufficient weight being given to EIA findings in final decisions; and developments beginning before EIA was completed (Barnes and Barnes, 1999). The EIA directive was amended in 1997 to curb these 'sharp practices' and EIA continues to play an important role in assessing the environmental impacts of transport (CEC, 1997).

A major limitation of EIA is that it only deals with impacts on a case-by-case basis, and has no formal mechanism for assessing the cumulative effects of transport policies and plans consisting of a number of projects. Similarly, the EIA methodology only requires planners to consider first order environmental impacts and not the more ingrained second order impacts. Planners have therefore tended to overlook second order impacts, first, because they are difficult to quantify, and secondly, because if they were considered fully, they might undermine the chances of projects gaining planning approval. Finally, EIA obliges developers to identify and mitigate the impacts of proposed developments, but it does not require them to consider alternative strategies (e.g. the construction of a light rapid transit system rather than a dual carriageway) (Therivel and Partidario, 1996).

Strategic Environmental Assessment (SEA), by contrast, promotes a more preventative and high-level approach to assessment by examining the environmental impacts of policies, plans and programmes. The EU introduced its SEA Directive in 2001 (2001/42/EC) following lengthy negotiations, during which some states resisted proposals to subject national policies to mandatory SEA, forcing the Commission to restrict the scope of the directive to plans and programmes (CEC, 2001a; Noble, 2000). Despite this limitation, SEA provides a much more robust procedural framework for evaluating the cumulative impacts of transport plans and alternative strategies than EIA. The UK DfT conducted a series of pilot studies during 2004 to consider how SEA would affect local transport plans in the West Midlands, the East Riding of Yorkshire and Somerset, and has issued guidance to local authorities on the appraisal process (DfT, 2004c). SEA has been further extended in the UK by the introduction of mandatory Sustainability Appraisal (SA) to local plans under the Planning and Compulsory Purchase Act 2004. SA adopts a similar methodology to SEA, but integrates the social, environmental and economic dimensions of sustainable development into the preparation of Regional Spatial Strategies and Development Plans, in essence reintegrating social and natural environmental issues and, to an extent, first and second order impacts. There are no plans currently to introduce a European directive on sustainability appraisal.

Voluntary agreements

Voluntary agreements have become a popular approach in recent years for promoting cooperative working between governments and industry on environmental issues. The first such initiative in the transport sector in the EU was the Auto-Oil programme, which was introduced in 1996 emulating a similar programme in the USA (Wurzel, 2002). The programme was intended to combine the resources and expertise of the automobile and oil industries with those of the European Commission to produce 'better informed' directives on emissions and fuel quality. Auto-Oil was replaced in 1997 by Auto-Oil II in order to assess future emissions and air quality trends and to establish a consistent framework for evaluating policy options. Its scope included the introduction of measures by 2005 to help the EU meet its air quality objectives for 2010 and detailed assessment of the costs and benefits of reducing emissions (CEC, 2000). The programme has been criticized, however, for allowing industry too much influence over decision making, though the Commission's self-appraisal of Auto-Oil has been predictably optimistic, arguing that road transport emissions would have been 50–100 per cent higher by 2010 without Auto-Oil I.

Another set of voluntary agreements was signed in 1998–1999 with the European, Japanese and Korean automobile manufacturers' associations to reduce emissions of CO_2 from new passenger cars sold in the EU. These were to be achieved through technological developments and market changes linked to these developments, though the agreements have again been criticised for the increasingly optional nature of emissions reduction targets into the future (Table 3.5) and their lack of transparency on CO_2 reduction strategies (World Resources Institute, 2005).

Despite these criticisms, these agreements, if fully implemented, have the potential to encourage significant long-term cooperation on the reduction of transport-based CO_2 emissions (Bongaerts, 1999). In practice, however, the 2003 targets were not met and the automotive groups have since begun to argue that greater 'eco-driving' measures are needed to supplement the programme. When it emerged in 2006 that

Table 3.5. CO_2 emissions targets agreed between the Commission and European, Japanese and Korean car manufacturers. Adapted from CEC, 1999.

Every effort to achieve CO_2 emission target of 165–170 g/km by 2003
Collectively achieve CO_2 target of 140 g/km on new cars by 2008
Evaluate in 2003 potential for moving towards 120 g/km CO_2 by 2012

75 per cent of European manufacturers were likely to miss their voluntary target to reduce CO_2 emissions by 2008–2009, the Commission responded by tabling a proposal for a mandatory limit of 130g/km on carbon emissions by 2012, a lower target than the 120g/km set by the voluntary agreement but none the less a binding cut (EurActiv, 2007). These developments indicate the difficulties of using voluntary agreements to address this policy area and that stronger measures are likely to be needed to accelerate the development of cleaner automotive technologies.

Market-based instruments

Concerns about the inflexibility of legislation and the deficiencies of voluntary agreements have led to a marked shift in the past decade towards market based pricing instruments as a means to control transport demand. Economists have long argued that the core problem, in transport and other areas, is the tendency for unregulated markets to underprice the environmental externalities caused by human activities (Pearce, 1993). This implies that more effective management of transport demand can be achieved through the use of environmental taxes to encourage changes in traveller behaviour (Quinet & Vickerman, 2004). Potter and Parkhurst (2005) summarize recent eco-reforms of vehicle taxation, noting the importance of the positioning of tax measures in the transport system. This can be at three main points: the initial purchase of a vehicle, vehicle ownership (annual registration tax and company car taxation) and vehicle use (fuel, road space and parking).

Purchase measures will affect vehicle choice, but have little impact on vehicle use. Most EU states have a specific car purchase tax, the two main exceptions being the UK and Germany. In some countries this is graded according to the power of a car, while in the Netherlands purchase tax is partly counterbalanced by fixed allowances which reduce purchase taxes significantly for smaller and more fuel-efficient cars and raise the price of larger and less fuel-efficient vehicles. As well as a registration tax, Italy has higher rates of value added tax (VAT) on cars with an engine capacity of greater than 2,000cc (2,500cc for diesels). Circulation taxes can also influence vehicle choice. Britain has operated a Vehicle Excise Duty ('Car Tax') based on CO_2 emissions since 2001, though the range of payments is relatively narrow. Italy, by contrast, has a much wider-ranging power-based circulation tax that levies ten times as much on the most powerful category of car compared with the most fuel-efficient and least powerful cars. This partly explains why the Italian car fleet is around 20 per cent more fuel efficient than its British counterpart.

Taxes on fuel, road space and parking have most impact on second order transport effects related to travel behaviour and activity patterns. Between 1992 and 1999, the UK operated the 'fuel duty escalator' under which fuel duty was increased above the rate of inflation. Fuel demand elasticity studies (e.g. Glaister & Graham, 2000; Goodwin, 2002) suggest that these tax increases resulted in 10 per cent less demand for fuel in 2000 than if the duty rates had only increased at the same rate as inflation. Following blockages of oil refineries by lorry drivers and farmers in 2000, however, petrol and diesel duty was cut, the fuel duty escalator was abandoned and its impact has faded. The use of fiscal incentives has, nevertheless, resulted in some significant changes in driver behaviour. In the UK, for example,

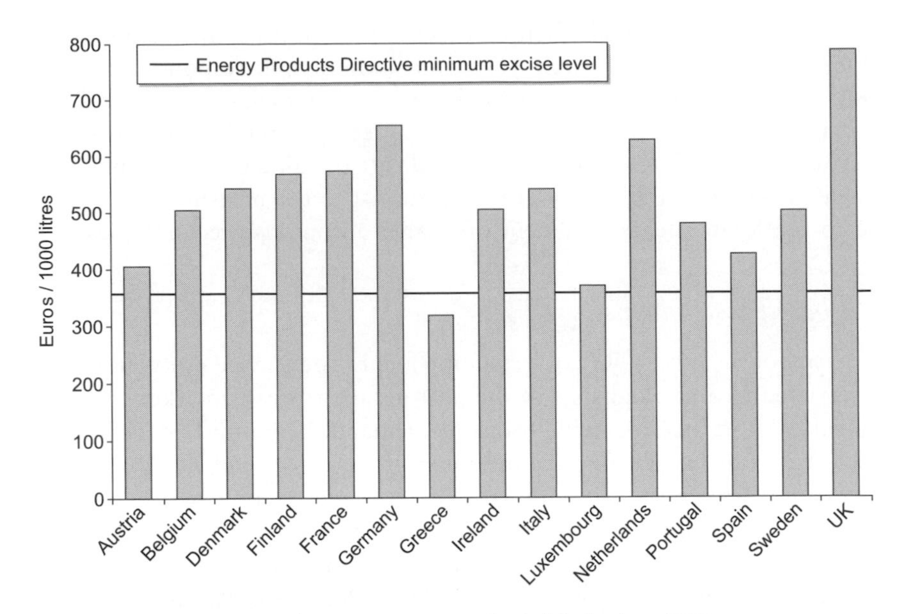

Figure 3.3. Impact of Energy Products Directive on unleaded fuel prices, 2004.

unleaded fuel shifted from 5 per cent to over 50 per cent of total British fuel consumption in just three years when differential petrol duty was introduced on leaded and unleaded fuel, though arguably this simply accelerated a shift in the market that was already occurring.

In the early 1990s, the European Commission proposed an EU-wide carbon/energy tax covering all forms of fuel and energy production, but the proposal stalled following opposition from some member states to the tax's potential economic impacts and the *de facto* surrender of tax-setting powers to the EU (Zito, 2000). The debate was revived in the late 1990s, leading to the adoption of the Energy Products Directive in 2004, a measure which introduced minimum (as opposed to harmonized) fuel duties. The directive's impact on fuel prices is modest in most member states and demonstrates the difficulties of gaining meaningful cooperation by governments to address the environmental impacts of transport, particularly where this potentially compromises economic interests or national sovereignty. Of the old EU 15, only Greece was obliged to increase taxes on unleaded petrol to comply with the directive (Figure 3.3), though Austria, Belgium, Greece, Luxembourg, Portugal, Spain and Sweden were all forced to increase national fuel duties on diesel.

Direct charges for road use have traditionally taken the form of road and bridge tolls, and are commonly used for motorways in some European countries. Toll roads have been an integral part of the French autoroute system for decades, while Germany introduced mandatory tolls for heavy trucks using the Autobahn system in 2005. The UK introduced its first and to date only toll motorway, the M6 Toll to the north east of Birmingham, in 2003.[3] Such schemes have tended to draw strong resistance from road users but, along with other market-based instruments, are now

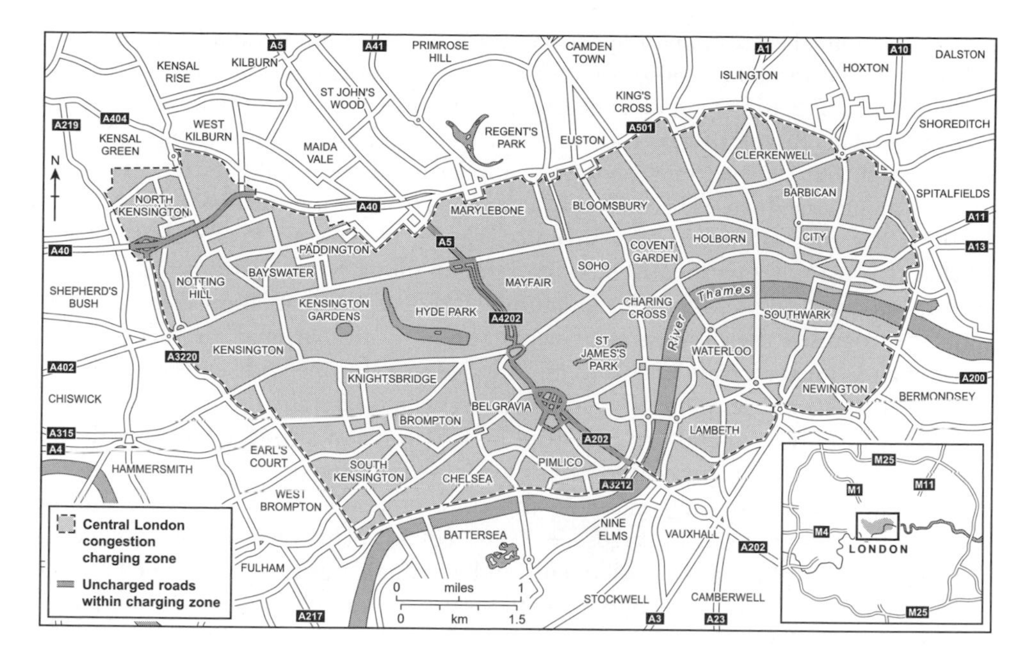

Figure 3.4. London Congestion Charge area. After TfL, 2006.

firmly established in the repertoire of approaches used to address the behavioural aspects of road transport.

City road user charging is another innovation to be adopted widely across Europe following early schemes in Singapore (1975) and Norway (1980s) (Ieromonachou *et al.*, 2006). London's Congestion Charge is now the best-known example of this type of road user charge (Figure 3.4). Introduced in 2003, it levies a charge of £8 per day to drive in central London during the scheme's hours of operation. The charge has been accompanied by measures designed to make public transport more attractive, including the provision of 300 extra buses. Compared with the last few weeks before the introduction of charging, the number of cars entering the zone has dropped by 20 per cent and congestion by up to 30 per cent (Richardson *et al.*, 2004; Transport for London (TfL), 2006). Durham has also introduced a successful scheme, but other UK cities have been hesitant to adopt this measure. From January to July 2006, Stockholm tested a sophisticated traffic management system where drivers were charged different amounts depending on the time of day they travelled within affected areas of the city. As a consequence, traffic volume fell by between 9 and 26 per cent at the major tollgates. Following the end of the trial, Sweden's new Alliance government said it would reintroduce congestion charging and use the money to build a relief road around the capital. Athens is the latest European city to consider introducing congestion charging in response to the failure of other schemes to make significant headway against its chronic congestion problems.

Despite its increasing profile, progress to control the environmental impacts of air transport has been more sporadic. The aviation industry has been largely exempted from noise, pollution and congestion charges, principally because of concerns about the economic impacts of taxes on fares and demand for air travel (HM Treasury and DfT, 2003), though pressure for greater regulation of aviation activities has intensified amid concerns about the environmental impacts of air transport. The Royal Commission on Environmental Pollution (RCEP) (2002a) commented that the fuel tax exemption applied to domestic flights represented a penalty on other industries because only emissions from domestic flights are included in the UK's total domestic CO_2 inventory under the Kyoto Protocol (the same applies to most other nations). Thus, the aviation industry is effectively 'free-riding' the efforts of other sectors of the economy to reduce CO_2 emissions. At the same time it recognised the difficulties of negotiating an international air fuel tax, but recommended that aircraft emissions be included in a Europe-wide emissions charge that airports would levy on all aircraft, passenger or freight flights to and from European airports. In 2005, the European Commission produced proposals for further measures to regulate the environmental impacts of transport, including the inclusion of aircraft CO_2 emissions in the EU emissions trading scheme and the ending of tax exemptions on kerosene (CEC, 2005a). This received overwhelming support from the European Parliament, and although initially condemned by the airline industry as ignoring 'economic realities' (European Parliament, 2006), recent reports suggest that some airlines are warming to emissions trading as a way of avoiding direct taxation on flights.

Infrastructural investment

A final measure to reduce the environmental impacts of transport is the use of infrastructural investments to promote the smooth running of transport systems and more environmentally friendly transport modes (e.g. cancelling road-building programmes and investing in rail upgrades). The EU's Trans European Network (TEN-T) was launched in 1996 to encourage cooperation between the member states in the design of integrated transport networks (Barnes and Barnes, 1999) (Chapter 8). 14 projects were sanctioned in the first round of TEN-T, with a further 30 in 2003, at an estimated overall cost of €225 billion. The majority of these projects involve inter-modal strategies, usually with upgraded rail links, and only three are exclusively road based. The TEN-T demonstrates some of the possibilities of policy integration to achieve multiple policy objectives, in this case the improved functioning of the European single market, lower-cost and safer travel, regional development, and the partial reduction of environmental impacts caused by road transport. At the same time the danger is that the TEN-T will contribute to a further escalation in transport demand and environmental impacts by making travel easier, while other developments such as rail liberalization linked to privatization have the potential to unravel many of the environmental benefits of TEN-T. The House of Lords Select Committee on the European Communities (1997) estimated that the first phase of TEN-T would only absorb one or two years' growth in road traffic. Nevertheless, when combined with assessment techniques like EIA, SEA and local planning controls, infrastructural investments provide another useful technique for controlling first and some second order environmental problems of road transport.

Conclusions

The relationship between transport and the environment is undeniably complex and multifaceted. In this chapter we have argued that understanding this relationship requires a detailed appreciation not only of overlaps between transport's environmental and social effects but also of its first and second order impacts on environmental quality, behaviour and land use patterns. First and second order impacts both have important geographical aspects, but with the notable exception of climate change second order impacts generally involve wider-scale and longer-term structural transport and social problems. For example, land take and severance from road building are local, but the second order effects of urban sprawl and dispersal of activity patterns are regional. Similarly, emissions arising from the long-haul transport of foodstuffs are a first order impact with wide, even global, environmental implications, but their second order geographical impacts in terms of shifting crop production patterns, the conversion of wilderness to agricultural land, and the global dependencies created by these activities are equally, if not more, extensive and difficult to reverse.

As these environmental impacts have become more apparent, so the number and range of policies developed to address them have also expanded. Early policies tended to be heavily dependent on legal mechanisms specifying emissions standards or lower-impact technologies, and the use of local planning controls to deal with the spatial aspects of transport policy. These have been complemented in recent times by techniques like EIA, SEA, and voluntary agreements, all of which contribute towards a more systematic analysis of environmental considerations across all scales of transport planning and more cooperative working between policy makers and industry on transport issues.

Both techniques can be criticized for focusing too heavily on the first order environmental impacts of transport at the expense of second order impacts, however. It is becoming more widely accepted that, of the various solutions discussed in this chapter, market-based instruments have perhaps the greatest potential to exert a systemic influence on both first and second order impacts by integrating the full social and environmental costs of transport activities into transport pricing to produce changes in traveller expectations and behaviour. It is too soon to judge whether market-based instruments will succeed in stemming modern society's rapacious appetite for car and air travel. Initiatives like the London Congestion Charge and other purchase and road user taxes provide some grounds for optimism, but a common theme in assessing all the various transport policies discussed is their failure to achieve the level of integration required to make a substantial difference to the environmental problems caused by transport. This should not be surprising given the complex nature of transport's environmental effects. Rather than expecting any single policy approach to solve all problems, integrated policy packages are needed that combine different measures, each targeting the specific aspects of the wider problem they are best equipped to deal with. Whether governments have the foresight or leeway to develop such integrated transport policies remains to be seen.

Another critical theme emerging from this review is that of geographical inequality, both in the distribution of environmental problems caused by transport and the policies used to alleviate them. One reason for this diversity is that politicians rarely enjoy the luxury of being able to take an entirely dispassionate view of transport problems. Even the most well intentioned and reasoned of policies are almost guaranteed to provoke the wrath of at least one influential interest group – be it car manufacturers, oil companies, airlines, road user associations, or environmental NGOs – or to produce some unintended consequences, whose impacts vary from country to country and region to region. Solving transport's environmental problems therefore requires not just technical acumen in issues like road user pricing but also a clear understanding of the spatial and social implications of different strategies and an ability to connect general principles to the settlements and communities they affect.

Notes

1 Though over one million fatal road accidents occur each year globally, with the majority and greatest increase taking place in less developed countries. Again the causes of this are not straightforward, but include: poor vehicle and road maintenance, poor driver education, and relaxed attitudes towards road-traffic laws.
2 More accurately referred to as international trade, as the terms of trade between nations are, in fact, highly regulated through the World Trade Organisation (WTO) and regional blocs like the EU, the North American Free Trade Area (NAFTA), and the Southern Africa Customs Union (SACU).
3 Plans to extend the M6 toll road between north Birmingham and Cheshire were scrapped in 2006 in favour of a scheme to widen the existing non-toll motorway. Announcing the decision, the government claimed that a toll road would be too expensive and require too much land.

4 | Transport and Social Justice

Julian Hine

The concept of social justice is contested (Boucher and Kelly, 1998) and is 'grounded (according to its protagonists) not on a concrete way of life but on rationality and need' (Minogue, 1998: 254). It is a hotly debated topic in political philosophy: this discipline generates views of social justice which do not explicitly consider transport but within which we can set a transport perspective. For this purpose, Minogue's (1998: 254) formulation of social justice is useful: 'Social justice is the belief that it is the duty of government to redistribute the wealth of a society so that each person enjoys the right at least to a basic minimum and so that, poverty having been abolished, certain equalities prevail.' This position addresses income as the means of achieving distributive social justice. Whilst at first glance appearing rather simplistic (which it certainly is not – there is a great deal of argument underlying the statement), we can insert a transport perspective by arguing that if equalities are a goal of social justice, then achieving them is not just a structural problem of income, but also a geographical problem involving distance, movement and access.

Transport thus has a key role to play in promoting social justice (Foley, 2004). Numerous studies have demonstrated that the link between transport disadvantage and poor access to goods and services can contribute to social exclusion, making it difficult for people to participate fully in society (Department of the Environment, Transport and the Regions (DETR), 2000; Hine & Mitchell, 2001, 2003; Lucas *et al.*, 2001; Social Exclusion Unit (SEU), 1998, 2003). Other factors contributing to social exclusion include differentials in education and training opportunity and attainment, socio-economic circumstances, the local environment, and access to information, and physical accessibility to a wide range of opportunities including employment, shopping and recreation.

This chapter begins with a detailed definition of accessibility and how this relates to debates about transport disadvantage and social justice. Accessibility is recognised as an essential requirement for the achievement of social justice, through its contribution to reducing social exclusion (Farrington, 2007; Farrington &

Farrington, 2005; Halden, 2002; Hine & Mitchell, 2003). Following the discussion of accessibility, social exclusion and social justice are considered in turn. The chapter then goes on to look at trends and policy assumptions and how these are being used to inform and develop transport's role in promoting social justice. This discussion looks at the role of public transport and how recent trends in public transport provision have had an impact upon inner city and rural areas. As will become clear both in this chapter and elsewhere in the book (Chapters 6 and 7), greater understanding of the geographical/spatial nature of social justice is critical to the delivery of transport solutions (Hine & Grieco, 2003).

Accessibility, social exclusion and social justice

Accessibility

Accessibility can be simply defined as how 'get-at-able' a particular place or location is (Moseley, 1979). It is often confused with mobility, which is concerned only with the ability of individuals to move around (Hillman *et al.*, 1976). There is no doubt that mobility is an important aspect of accessibility, but it is only one such aspect. Internet access, for example, can make it straightforward for people to use banking, shopping and even medical services without leaving their own home. Structural factors such as poverty, lack of education and ethnicity are also important to consider. Deprivation of access to justice for black people in the 'southern' states of the USA prior to the equality legislation of the 1960s is a good example. All of this means that it is possible to think of accessibility in terms of *places* – i.e., how 'get-at-able' they are – or in terms of *people*, making reference to their ability to access goods and services.

Halden (2002) notes from a user viewpoint that accessibility definitions include three key elements: (1) the category of people under consideration; (2) the activity supply point; (3) the availability of transportation. Farrington and Farrington (2005) indicate that as with any normative concept, consideration of such definitions in policy terms often focuses on recommended levels of accessibility, such as walking distance to a bus stop. In the UK, much progress has been made in specifying such accessibility targets and also in mapping distances to the nearest key facilities for different groups in the population. This has been driven in part from a desire to better integrate land use decisions and transport investment, but also to ensure greater social inclusion for the transport disadvantaged. Such a development has been welcome because transport policy makers were slow to recognise the linkage between access, the process and nature of social exclusion, and social justice.

For a long time this was reflected in a lack of suitable indicators with which the links between mobility and access could be measured and monitored. Developing such indicators has been difficult because the relationship between the different dimensions of exclusion is unclear and it is also difficult to measure spatially. The problem here is that the unit of spatial measurement used will influence the geographical distribution that is observed (Church, 2000). To illustrate, although indices of multiple deprivation widely employed in public policy are useful, they are compromised because their accessibility data are related neither to public transport service levels nor car or vehicle ownership data. They are instead straight and direct measures of geographical distance from very basic services (Grieco *et al.*, 2000). Developments in geographic information system (GIS) and transport models represent a step forward in this area where origin and destination, mode choice and the main aspects of journey time can be mapped in low income neighbourhoods (Church *et al.*, 2000; Halden, 2002). Most recently, the employment of quantitative accessibility analysis has prompted a renewed interest in how access to key facilities is made possible (SEU, 2003) (Chapter 7).

Social exclusion

Social exclusion refers to the loss of an 'ability (by people or households) to both literally and metaphorically connect with many of the jobs, services and facilities that they need to participate fully in society' (Church and Frost, 1999: 3). In transport terms the argument can be made that a lack of access to effective transport has an impact upon the extent to which individuals can get to health facilities, local job markets and leisure activities. There is broad agreement in the literature that social exclusion represents a conceptual shift away from traditional forms of explaining disadvantage, and should not have equivalence with older terms and definitions previously applied to individuals, groups and processes, that are considered to exist and operate outside a certain social norm. Poverty, deprivation and the underclass are examples of these (Bhalla & Lapeyre, 1997; Hine & Mitchell, 2001; Lee & Murie, 1999). It has also been argued that social exclusion is not 'not social inclusion' and *vice versa* (Cameron, 2006) and clearly this brings an added complexity to the definition of social exclusion.

Although there is no common definition of the dimensions and factors which make up social exclusion, the approaches to its study taken by various authors broadly overlap (see Burchardt *et al.*, 1999; Gaffron *et al.*, 2001; Lee & Murie 1999). As with accessibility, there is general recognition that the ability of a group or individual to participate in society can be affected by a number of factors, including individuals' own characteristics and life events, characteristics of the area in which they reside and dominant social, civil and political institutions. In relation to transport, seven categories of exclusion have been identified (Church *et al.*, 2000, 2001):

- Physical exclusion – where physical barriers inhibit the accessibility of services which could be experienced by mothers with children, elderly or frail, the disabled, those encumbered by heavy loads or those who do not speak the dominant language of the society;
- Geographical exclusion – where poor transport provision and resulting inaccessibility can create exclusion, not just in rural areas but also in areas on the urban fringe;
- Exclusion from facilities – the distance of facilities (e.g. shopping, health, leisure, education) from people's homes, especially from those with no car, make access difficult;

- Economic exclusion – the high monetary or temporal costs of travel can constrain access to facilities or jobs and thus income;
- Time-based exclusion – which refers to situation where other demands on time such as caring restrict the time available for travel;
- Fear-based exclusion – where worry, fear and even terror influence how public spaces and public transport are used, particularly by women, children and the elderly; and
- Space exclusion – where security and space management strategies can discourage socially excluded individuals from using public transport spaces.

In addition, three types of processes that influence the relationship between social exclusion and transport were proposed: first, the nature of time-space organization in households; secondly, the nature of the transport system; and thirdly, the nature of time-space organization of the facilities and opportunities individuals are seeking to access. The particular form of each of these processes differs according to gender, age, cultural background, level of ability and economic circumstances. And although transport deprivation, as we have seen, is only one of a number of factors that can exclude, it is important to remember that a lack of transport can compound matters. Accordingly, there remains a need to identify at a local level a selection of indicators that reflect the processes linked to social exclusion, and, in particular, the role of transport in those exclusionary processes.

Social justice

We have already seen that social justice is a contested term. Indeed, Smith (1994: 23) has suggested that 'as a socially constructed abstraction that motivates human conduct' there is an issue about whether a universally accepted term or notion of social justice is possible. The Commission on Social Justice (1994: 18) has, none the less, argued that it can be defined in terms of a hierarchy of four ideas:

> First, the belief that the foundation of a free society is the equal worth of all citizens, expressed most basically in political and civil liberties, equal rights before the law and so on. Second, the argument that everyone is entitled, as a right of citizenship, to be able to meet their basic needs for income, shelter and other necessities. Basic needs can be met by providing resources or services, or helping people to acquire them: either way, the ability to meet basic needs is the foundation of a substantive commitment to the equal worth of all citizens. Third, self-respect and equal citizenship demand more than meeting basic needs: they demand opportunities and life chances. Finally, to achieve the first three conditions of social justice, we must recognise that not all inequalities are unjust (a qualified doctor should be paid more than a medical student), but unjust inequalities should be reduced and where possible eliminated.

Some authors (e.g. Hay, 1995; Farrington and Farrington, 2005) go further and specify the need to make a link between access and space in any discussion of equity; addressing social exclusion through measures to increase accessibility – not least with mobility-based approaches – is clearly of significance in this context. Yet local and national governments that (implicitly) accept social justice as influencing normative policy goals for transport often do so with little thought of the consequences for policy choices and their implementation. In practice, while the goal of, for example, promoting modal shift is seen as creating travel choices, it may run counter to the desires and travel patterns of non-car owners if it concentrates resources on a few heavily used corridors to the detriment of a wider spatial network.

Patterns of transport disadvantage

A number of groups in society are disadvantaged by existing forms of transport provision. Those most likely to be affected are those on low incomes, ethnic minorities, women, the elderly and disabled, and children (Gant, 2002; Hine and Mitchell, 2003; DETR, 2000). Evidence also indicates clearly that there is a high correlation between transport deprivation and factors such as low incomes, low levels of car ownership and public sector housing. People from households on low incomes in the UK, for example, make fewer journeys overall but about twice as many journeys on foot and three times as many journeys by bus as those households in the two highest income deciles (Grayling, 2001; Hine and Mitchell, 2003). By comparison, higher income groups make more journeys by car and tend to travel further. Walking remains the dominant mode of transport for people from households on low incomes, but in particular for non-car-owning households in the lowest income quintile. People living in households without cars used public transport for 25 per cent of their journeys, and this was as much as seven times greater than those households with cars. Taxi and minicab usage are also higher amongst non-car-owning households (DETR, 1998).

Although over the period 1989/91 to 2004 the number of carless households in the lowest income quintile declined by 20 per cent (Department for Transport, 2005), a substantial proportion in this income group (54 per cent) still has no car (Table 4.1). Over the same period the availability of bus services[1] has changed little, but low income families are increasingly buying cars as a response to both rising public transport fares and poorer levels of public transport accessibility to different

Table 4.1. UK household car ownership by income band, 1989/91 and 2004. Source: Department for Transport, 2005.

	1989/91			*2004*		
	None	*One*	*Two or more*	*None*	*One*	*Two or more*
Lowest real income quintile	73	24	3	54	38	7
Second quintile	48	44	8	37	47	15
Third quintile	25	55	20	20	52	28
Fourth quintile	12	55	33	11	46	43
Highest real income quintile	7	46	47	8	40	52

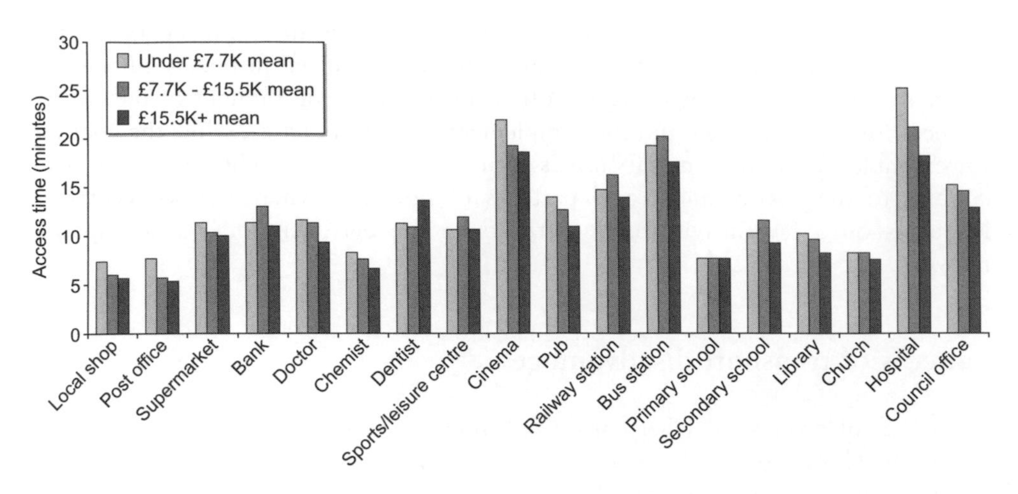

Figure 4.1. Access time and income level. Source: Hine & Mitchell, 2001, 2003.

labour and housing markets in peripheral locations (Donald and Pickup 1991) (see also Chapters 6 and 7). Income level can impact on access times to goods and services – for example, it often takes longer for people in lower income groups to access hospitals in peripheral locations (Figure 4.1).

Race and ethnicity can also be important indicators of transport access and social disadvantage (Lucas, 2004; Rajé *et al.*, 2004) and there are marked differences in car access across different racial and ethnic groups. In the USA, around 5 per cent of white households do not own cars, compared to over 20 per cent of black households, whilst in the UK, people of black and mixed ethnic origins are less likely to drive to work (Lucas, 2004; Wu and Hine, 2002). Women also experience exclusion in a number of ways which relate to patterns of travel, patterns of employment, income, caring responsibilities, access to cars and poor public transport services (Grieco *et al.*, 1989; Hamilton *et al.*, 2000). There are also differences amongst women in terms of the experiences of specific groups – older women, disabled women, women from ethnic minorities, women living in rural areas and single mothers tend to be more transport deprived, although patterns differ around the world (Hanlon, 1996; Grieco and Turner, 1997; Polk, 1996; Rosenbloom, 1996). The design of the infrastructure can mitigate against the use of a local transport system, especially for women with young children. Personal safety when using or trying to access transport infrastructure is also a major consideration for all women (DETR, 1999; Hamilton and Jenkins, 1992).

Finally, the disabled also feature in discussions surrounding the link between transport and social exclusion (DETR, 2000; Hine and Mitchell, 2001). They suffer because for a variety of reasons they find it difficult to access public services. These reasons include low incomes, the physical layout of infrastructure and design of vehicles, and the location of stops (Harris, 1971; Martin *et al.*, 1988; Mitchell, 1988; Oxley and Benwell, 1985).

Consequences of transport disadvantage

Barriers to employment

Together with the structure of the local economy, a lack of knowledge of the local job market, an unwillingness to travel outside the locality and the cost and availability of child care (Burkett, 2000; Department for Education and Employment (DfEE), 1999; Gorter *et al.*, 1993; Lawless, 1995), a lack of access to transport is an important barrier to employment opportunities (Audit Commission, 1999; Hine and Mitchell, 2001; Rural Development Commission, 1999). In disadvantaged housing estates the combination of childcare difficulties and the availability, length and cost of travel to work can sharply define the economic reality (e.g. Pawasarat & Stetzer, 1998). For individuals confronted by low wages and an increasing proportion of part-time work, travel to work costs may make taking a low income job an uneconomic option (McGregor *et al.*, 1998); the House of Commons Select Committee on Education and Employment (1998) has recommended the introduction of travel subsidies as a solution. The increasing relocation of certain types of jobs to out of town locations has also had an impact on car-less households where public transport does not easily accommodate the so-called 'reverse commute', from inner-city neighbourhoods to dispersed peripheral sites. Timetables that do not accommodate 'non-standard' forms of employment (e.g. shift work) make access problematic and temporal as well as spatial barriers to job markets are created (Hine & Scott, 2001).

Similar barriers to employment are evident in rural areas. In a study of young people in North Yorkshire, Rugg and Jones (1999) found that most respondents needed their own transport to keep a job, while public transport was seen to be unreliable and timetables did not match up to work schedules. Similar findings have been reported from other research in the UK (Monk *et al.*, 1999; Moseley, 1979; Stafford *et al.*, 1999) and in Australia (Currie, 2006).

Exclusion from services

Lack of readily available transport, whether private or public, has a clear impact on the accessibility of particular goods and services (SEU, 2001, 2003). In the case of public transport, a problem – especially for communities with low levels of car ownership – can be that services operate along transport corridors rather than in central locations. In addition, there are documented difficulties for women in low income groups in accessing healthcare facilities due to low levels of car ownership and reliance on public transport (Young, 1999). Lack of transport and the cost of public transport have been cited as factors creating a significant barrier to further education (Callender, 1999; DfEE, 1998). Also reported are the problems experienced by those living in disadvantaged areas, in terms of their ability to access financial services (Leyshon and Thrift, 1995). A recent review by the Centre for Research into Socially Inclusive Services (Sinclair, 2001) found that there was a need to understand more fully issues associated with financial inclusion and access to local goods and services.

Fear and perceptions of fear

Perceptions of fear and safety can have significant effects on levels of personal mobility. Generally speaking, people feel safer when walking around their neighbourhood during the day when compared to walking around after dark, but women feel less safe than men and are more worried about being victims of street crime. Older people, women and those from ethnic communities are also more likely to fear crime whilst using public transport (Crime Concern, 1999). A fear of interchange facilities and stations in the dark and at off-peak periods, and the need for a security presence in these locations, is also evident (Hine and Scott, 2001). Younger people also experience anxiety when using public transport. Young women in particular feel very unsafe after dark when using public transport (Crime Concern, 1999). The consequence of these fears is that trips are either not made or that alternative arrangements are made where it is possible to avoid these situations (e.g. Atkins, 1989; Pain, 1997). An investigation into perceptions of the public transport journey in Scotland found that switches from bus to car – in some instances taxis – were often a result of these fears (Hine and Scott, 2001; see also Hine and Mitchell, 2001; Pain, 1997).

Policies and practices

Local services and facilities as well as new sites for employment, health care, retailing and housing are increasingly located in inaccessible places for non-car owners, such as on the edge of towns and cities. Public transport networks often do not adequately serve these locations. In many towns and cities during off-peak periods (early morning and late evening) buses can be infrequent, and in rural areas access to bus services is also limited (SEU, 2003) (Chapter 7). Typically, the reduction of social exclusion through improved public transport is treated as a general policy aim at local government level. This aim is made more difficult to achieve following deregulation due to a lack of control over public transport operators with regard to the price and quantity of public transport; however, there are policies for concessionary travel (e.g. for senior citizens and the disabled) and the buying in of socially necessary public transport services. This solution is tightly constrained by budget limits, and because such services are not allowed to compete with commercial services by diluting their use. In the UK, the implementation of Quality Partnership arrangements has increased frequencies and improved related infrastructure on some key urban corridors (Davison and Knowles, 2006), although no Quality Contracts have yet been established. In the USA, transport programmes have been introduced that address the problem of the 'reverse commute' (Cervero, 2004).

Bus policies

Public transport network coverage is a key public policy issue. Commercialization of local bus services has resulted in the development of 'Metro'-style high-frequency corridors in urban areas, because these are seen as essential by operators for future

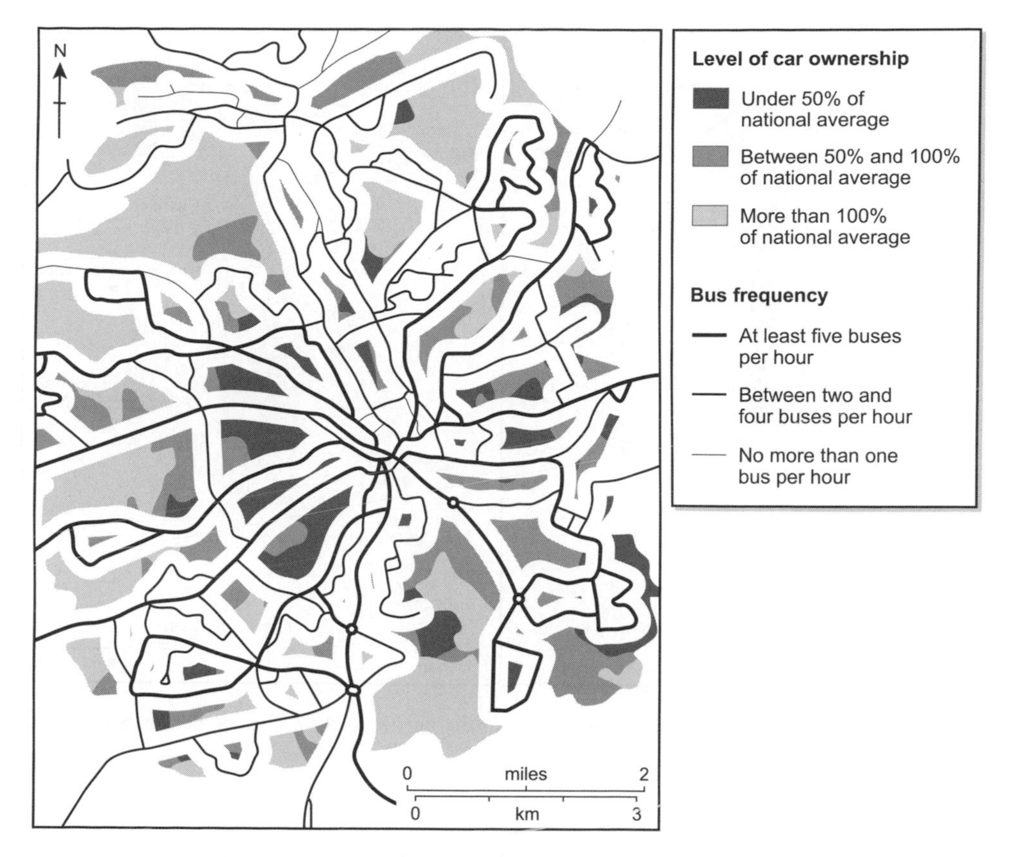

Figure 4.2. Distance from bus networks in Bradford by level of car ownership. Source: Friends of the Earth, 2001.

business growth. A consequence of this trend has been a movement away from the provision of the socially necessary services which are often in or adjacent to areas with high proportions of public sector housing. A study of public transport network change and levels of car ownership in Bradford (Friends of the Earth, 2001) shows that large areas with the lowest car ownership figures were over 200 metres from the city's high-frequency bus routes (Figure 4.2).

Work in Belfast has shown the importance of proximity to bus networks in affecting accessibility (Wu and Hine, 2003). The impact of different timetable schedules on network coverage can be clearly seen in Figure 4.3. The morning peak provides more frequent bus services with higher levels of accessibility on key corridors in the city, although there are areas located in the outer suburbs where accessibility is extremely poor in the morning peak. The network in the period between morning and evening peaks shrinks further with an associated decline in accessibility.

In the UK, an established method of improving access to bus services is through a general or targeted subsidy. The Transport Act (1985), which deregulated bus services throughout the country apart from London and Northern Ireland, heralded

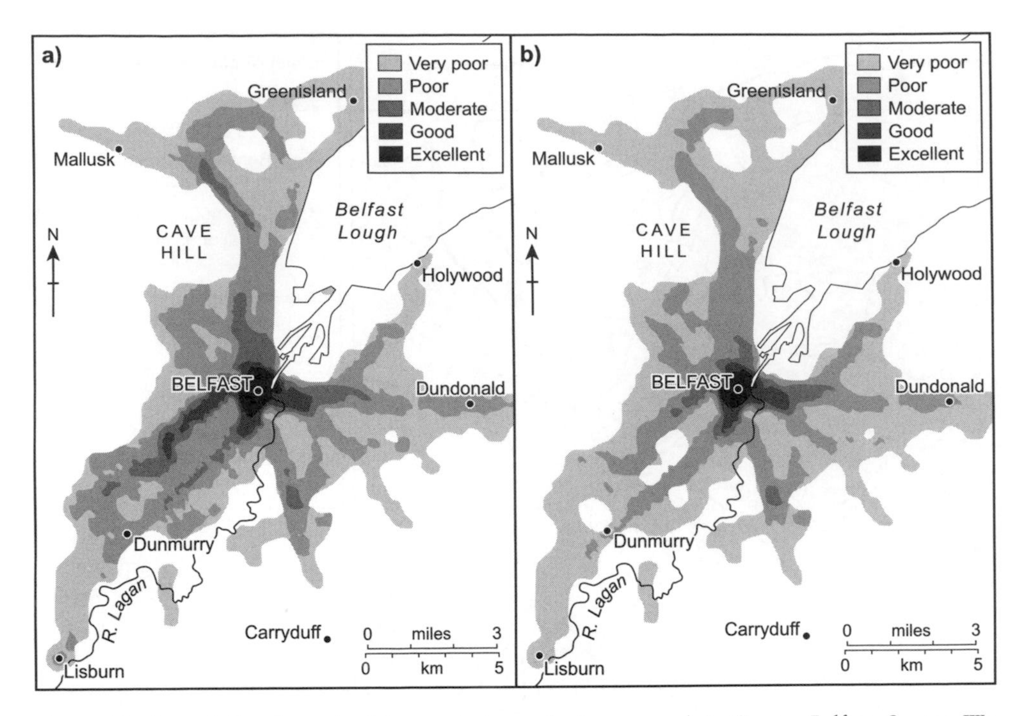

Figure 4.3. Accessibility determined by proximity to the bus network in Greater Belfast. Source: Wu and Hine, 2002.

the end of low fares policies because passenger transport authorities and local authorities were forbidden from subsidizing bus services except those which are unprofitable but deemed socially necessary. Donald and Pickup (1991) found that in Merseyside fare increases following deregulation were the main cause of reduced use. Interventions that produce a general fare subsidy are a positive step for low income groups (Hine & Mitchell, 2001; although see Grayling, 2001). Concessionary travel is commonly granted to senior citizens, the disabled, children under 16 and students aged up to 18 years in full-time education. Research indicates that these schemes encourage travel – those with concessions travel more often and further (Bonsall & Dunkerley, 1997; O'Reilly, 1989, 1990). Unfortunately, they convey less benefit to those who live in (typically rural) areas with low service levels. It is also possible for such schemes to be exclusionary if they do not cover other needy people such as the unemployed (Grayling, 2001).

Improving the physical accessibility of transport vehicles is another aspect of bus policy. The Disability Discrimination Act (1995), which legislates for mainstream public transport to become accessible to the disabled, when combined with local transport strategies and quality partnerships will ensure a fleet of accessible buses. The adaptation of street infrastructure, including bus boarders and raised platforms, upgrading of bus shelters, and the enforcement of parking controls in bus lanes and around bus stops, is also an important component of this approach (Evans and Smyth, 1997; York and Balcombe, 1997).

Specialist services

Specialist services are typically provided by the voluntary sector. These services include voluntary car schemes (DETR, 1999), 'shopmobility' facilities for those with walking difficulties (Gant, 2002), group hire bus services and dial a ride services (or 'Demand Responsive Transport' (DRT) services). The objective of DRT services was originally to provide a demand-responsive service serving low-density suburban areas. As a concept they have been around in the UK since the early 1970s. Initial experiments found that 'dial a ride' services were expensive and failed to cater for dispersed trip patterns (Oxley, 1977; DETR, 1999), although they remain an accepted method of delivering services to the elderly and disabled. The efficiency of DRT systems has been improved by computerized scheduling packages that in effect provide the operator of services with a reservation system for services (Brake *et al.*, 2004). A variety of DRT services is now offered by community transport operators (SEU, 2003) (Box 4.1) and such schemes can improve the quality of life for older people in a variety of ways (Ling & Mannion, 1995; see also Chapter 7).

Another specialist transport concept is the 'service route' developed in Scandinavia (Box 4.2). This is concerned with bringing the bus service closer to the residents (Stahl, 1992) and as such represents a move away from traditional route planning, which is based on radial routes coming into a city or town centre. The service route is a regular route network, but the route is based on where greater proportions of elderly and disabled people live, and on important destinations such as health centres, hospitals and shops (Evans and Smyth, 1997; Stahl, 1992). These schemes have been implemented in Denmark, Finland, Norway, Holland, Canada and the USA. Evidence indicates that service routes are cost effective (Stahl, 1992; Stahl and Brundell-Freij, 1995).

Box 4.1. Community Transport Association – North Walsham, Norfolk. Source: SEU, 2003.

In rural Norfolk the local community transport association offers a number of services to local residents. These include:

- Hospital Medi-Bus: This runs to and from Norfolk and Norwich University Hospital up to three times daily. Patients, escorts and visitors are picked up from home and guaranteed arrival at times that coincide with clinics.
- Dial-a-Medi-Ride: Volunteer drivers take registered users to doctors, dentists, opticians, etc. Passengers are charged according to distance the travelled.
- Group Transport: Provides transport for frail and elderly people wishing to attend group meetings, lunch clubs and similar social events. Excursions are often arranged to local places.

Box 4.2. Flexline, Gothenburg, Sweden. Source: SEU, 2003.

Flexline is a demand-responsive bus service with drop-off points at shopping centres, hospitals and other important destinations for elderly and disabled people.

- Services consist of small, fully accessible buses, which depart at half-hourly intervals from the end stops and collect passengers from designated meeting points within the service area.
- Meeting points are generously distributed within the area served so that at least 90 per cent of the residents live within 150 metres of a meeting point.
- Journeys must be booked at least 15 minutes before the bus is scheduled to leave. Times are confirmed 15 minutes prior to arrival at the meeting point through an automated call-back function once a computer has determined the optimum route.
- Users perceive an improvement in their mobility and activity as a result of the service.

Taxis

Taxis are the most flexible transport service (Beuret, 1994, 1995) and are a popular alternative to other modes in addressing the problems of the transport disadvantaged. Two forms of taxi operation have developed: taxis that are run through voluntary driver schemes and taxis operated by commercial firms. Voluntary car schemes have been concerned with transporting people for social services, health and education purposes, and this role has now expanded to shopping and leisure-based trips (DETR, 1999). These schemes have been effective although funding and volunteer resources dictate their availability, which is restricted according to specific eligibility criteria. Commercial taxis can be on average between five and seven times more expensive than other modes per passenger mile, but to combat such a high cost to the user, taxi card schemes exist as a subsidy mechanism (Trench & Lister, 1994).

Beuret (1995) outlined a number of shortcomings with taxis, including problems accessing vehicles for those with disabilities and wheelchairs, although a number of companies are pioneering wheelchair-accessible vehicles. It is also the case that those groups using taxis the most tend to be on lower incomes, although there has been a growth in their usage by the general population for leisure purposes. A number of authors have commented on other features of taxi systems where improvements can be made to facilitate easier use, including: accessible taxi ranks located near key facilities, use of black cab design vehicles rather than conventional vehicles, and 'smart card' technology for ease of use and also so that a record of concessions used can be maintained.

Conclusion

It is clear that transport has an important role to play in promoting accessibility, addressing social exclusion and thus enhancing social justice. By enabling or improving people's access to welfare goods such as health care, education, employment, social interaction, professional services and retail purchasing, transport can reduce disadvantage, particularly when this is due to location. Transport takes its place alongside other approaches to social exclusion and social justice that seek to reduce poverty.

As is apparent from the brief analysis in this chapter, however, there are difficulties. Integrating public policies on transport provision with other appropriate public policies such as those affecting the provision of healthcare and education is a challenge. So too is their integration with the private sector, which provides most public transport services on a commercial basis, along with other essentials such as retailing and professional services. A further challenge is the integration of public policy on transport with policies on poverty and similar, 'structurally based' factors affecting social exclusion.

There are also problems associated with the complexities of the notions of social exclusion and social justice. Engaging transport policy with goals of decreasing social exclusion and enhancing social justice is challenging. Different dimensions of space, time and cost need to be clearly understood, so as to appreciate what transport can and cannot do in addressing such goals. Only then can social justice be viewed as capable of influencing normative transport policy goals. None the less, the ideas of basic needs and well-being are powerful when the experiences of those on low incomes and in deprived communities are examined. Indeed, it is the examination of these transport experiences that should prompt closer examinations of the allocation of resources and the distributional consequences of policy decisions.

A final issue is the precise function of transport and transport policy in addressing social justice, in relation to other transport policy goals. In particular, trade-offs between the achievement of the three elements of sustainability – economic, environmental and social – are inevitable. For example, *within* the transport policy field itself, should the primary goal be to decrease social exclusion and enhance social justice, or to promote mode shift (for greater environmental sustainability and/or economic benefit) from car to bus, train, walking and cycling? In the wider context, how should a balance be achieved between the provision of better public transport opportunities through greater supply, and the possible environmental impacts where these are lightly used?

At these points of interaction with other policy fields, it is at the very least useful to recognise the contribution that transport can make to enhancing social justice, so that this role can be taken account of in forming both our view of transport, and also policy frameworks which address wider concerns.

Note

1 In the *National Travel Survey* this is measured as households within a 13-minute walk of an hourly bus service.

5 Transport Governance and Ownership

Jon Shaw, Richard Knowles and Iain Docherty

The preceding three chapters have been concerned with the relationships between transport, geography and the economy, the environment and society. During the course of their discussions, the authors of these chapters identified various transport-related issues, and considered policy responses developed by different governmental institutions seeking to shape the outcome of transport activities. Before moving on in Part 2 of this book to consider transport flows and spaces in more detail, it is necessary to dwell upon why and how the state – that is, those people and institutions determining the policy and regulatory framework of a particular territory – is involved in governing and owning transport networks and services.

As discussed in Chapter 1, transport has long been recognised as fundamental to the functioning of societies and economies, but the extent to which any given polity attempts to influence transport activities will vary and can depend upon a range of factors from the practical (such as the availability of finance) to the more abstract (such as changing political philosophies). The nature and scope of state intervention can also produce a range of different outcomes affecting the geography of not only transport but also economic, environmental and social activities. The purpose of this chapter is to explore the rationale for state involvement in transport, and in so doing to explain how trends in the governance and ownership of transport systems and operations – and their associated geographies – have changed over time.

Governance, transport and geography

Governments all over the world are involved in transport by providing, or regulating the provision of, infrastructure and services. Particular interventions into a transport market are determined by the policies developed and adopted to suit prevailing circumstances. Black (2003: 200) defines transport policy as 'the position that some level of government takes on a particular transport issue', and Tolley and Turton

(1995) refer to the process of regulating and controlling the provision of transport. Rodrigue *et al.* (2006) suggest that, in fact, transport policy can be both a public and a private sector endeavour – private companies will themselves have policies to govern their operations – but that governments are frequently the most involved in making and administering policies because many components of the transport system are owned or managed by the state.

Policies for transport are derived not just in relation to particular modes, but also to different spatial scales which themselves usually reflect hierarchies of government institutions. European Union (EU) regulations, for example, will apply supranationally; other policies will be national or regional in scope. Local authorities will generally be concerned with cities, towns or neighbourhoods. The originating body/bodies and relevant spatial scale of a policy will depend on the arrangements for government in existence in the territory to which the policy applies. In the USA, for example, transport policies might flow from federal, state, county and city governments, in addition to some multi-jurisdictional districts in metropolitan areas. As we discuss later in the chapter, the recent shift in many countries from a traditional type of *government* (involving the formal institutions of the state such as ministries, departments, municipalities and so on) towards a more complex system of *governance* (a more inclusive notion acknowledging that public policy is delivered by a range of organizations, from the state through to the private and voluntary sectors) can result in more joint working between layers of government and other stakeholders.

It is not surprising that changes in transport policy can have significant impacts upon the transport geography of the area in which they apply. It might, for example, alter a rail network by providing for the construction of new high-speed lines (such as the *Train à Grande Vitesse* (TGV) network in France – Figure 8.1) or the closure of a relatively little-used passenger route (such as the Amtrak service from Denver to Seattle in the USA). Passenger and freight flows between the affected city pairings will also change. Shifts in transport policy will not just alter the geography of transport activities: depending upon the type of change involved, the economic, social and/or environmental geographies of an area will also be affected. The TGV reduces greenhouse emissions by abstracting passengers from the airlines (an environmental impact at the global scale), and by reducing journey times influences the (re)location of businesses and residential developments (economic and social impacts on a local scale, and potential environmental benefits at the local and global scales if car usage is reduced as a result of more walking and public transport usage) (Chapter 8). Likewise, non-transport policies such as a rise in interest rates or the general level of taxation might impact upon the geography of transport activity within a given area by, say, cutting the demand for rail or air services and prompting a reduction in the levels of service provided to certain destinations.

Why is the state involved in the transport sector?

Transport is often described as a universal good, since most human activities depend on it to some extent. Markets depend on raw materials and finished products being

Figure 5.1. The Interstate system in the USA. After Rodrigue *et al.*, 2006.

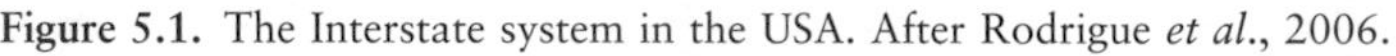

transported from producers to consumers (Chapter 2), and the myriad social and economic interactions we undertake in our daily lives – from visiting friends, to travelling to school, work, business meetings and the shops – require movement between places (Chapter 4). Managing this complex pattern of mobility and the infrastructure and services that it requires has long been at the heart of why governments seek to involve themselves in transport. A better understanding has also been gained of the need to manage transport's 'negative externalities', or the 'events [such as car pollution] that result in significant disbenefits for persons who had little or no role in the decision making that led to those events [such as a commuter deciding to drive to work rather than walking]' (Black, 2003: 199).

A further factor, and one which has recently increased in prominence following terrorist activity in Western countries, is the influence of military and national security considerations. In addition to facilities supporting, for example, air force and army bases, large-scale, ostensibly civilian networks such as the Interstate system in the USA have been developed with national security in mind (Figure 5.1) (Rodrigue *et al.*, 2006; Wood and Johnson, 1996). Equally, decommissioned military installations, especially airports, are often reclassified for use as civilian transport hubs (Behnen, 2004).

Excluding military imperatives, at the most basic level, the state is involved in the regulation and management of transport activities because the conditions rarely exist for an entirely – or even largely – free market in transport to function. Transport is 'peculiar' in a number of ways. To illustrate:

- the distribution of the benefits transport provides is very complex, and individuals, groups and companies derive many different outcomes, some of which are paid for directly by users (e.g. road tolls), some of which are not (bus subsidies);
- the costs of transport's externalities are seldom borne by those who produce them (the damage associated with, for example, vehicle emissions – poor health, global warming and so on – are not usually factored in to transport costs or are ascribed to the wrong parties);
- there has developed in many quarters a sociopolitical belief in the 'right to mobility', where everyone should be free to move around as they choose regardless of the consequences; and
- there are natural monopolies, particularly in infrastructure, which means it is cheaper and more efficient to provide one national network of roads or railway tracks than to have two or more networks competing with each other.

Throughout much of the twentieth century, the narrative of transport governance was dominated by economic and social imperatives. On the one hand was the need to regulate natural monopoly and address market failure within transport systems, and on the other was a desire to promote social equity in the transport opportunities available to individuals. These economic and social imperatives are, of course, interlinked.

Monopoly in infrastructure provision

The street is a good example of how certain aspects of transport activity tend towards natural monopoly. Streets are not just something which provide for mobility, they are also spaces that enable the exchange and sharing of goods and knowledge on which civic, economic and cultural life depends (Jacobs, 1968). It makes no sense to have competing sets of streets; as a public good, the benefit of the street is maximized when it is part of a single, coherent geography to which access is open to all. Conceiving of the street in this way brings the role of governance squarely into focus, since access to streets needs to be managed if the levels of security, social welfare, environmental protection and economic interaction demanded by liberal societies are to be realized.

Economic theory advanced nearly a century ago emphasizes the potential of carefully targeted state intervention to maximize a range of benefits to society unlikely to be expressed through private profit in the market (see Schumpeter, 1909). In transport terms this means that public provision of key infrastructure and services can be more efficient and equitable than the market alternative. Indeed, it is often essential, since the cost and risks involved mean private companies are frequently reluctant to provide them without some form of government underwriting. State intervention can also avoid wasteful competition and/or duplication of assets, such as separately owned, parallel streets and railway lines; this can also ensure that good value is obtained from the limited capital resources available within an economy if the planning and implementation of transport investment is carefully managed (Panigiua, 2004).

The power of this argument is reflected in the fact that it became orthodoxy for governments of all political persuasions for over 50 years. Moving away from the model of competing private sector railway companies (many of whom had gone bankrupt or were forced to merge – Figures 5.2a and b) and eliminating private toll highways had obvious appeal to the political Left (see Webb and Webb, 1963). Yet it was also attractive to many on the Right. Innovations such as expressways and passenger aviation provided the opportunity for the state to seek a step change in the efficiency of the market economy by transforming the level of mobility available to individuals and firms so that trade would increase substantially and competition would spread (Foster, 1981; Glaeser, 2004; Yago, 1984).

Social equity

The second key goal of transport governance has been to intervene in the provision of transport infrastructure and services so that a range of social policy goals can be achieved. Controls and systems of subsidy disbursement built up from the nineteenth century in order to promote customer, employee and vehicle safety, and to prevent the emergence of significant social and financial disparities between areas and groups (Foster, 1992; Leyshon, 1992). Examples of regulatory interventions are those designed to: ensure affordable rather than monopoly fares; guarantee concessionary fares for the young, old and disabled; promote a high quality of service provision; and stipulate the quantity of services provided to ensure (so far

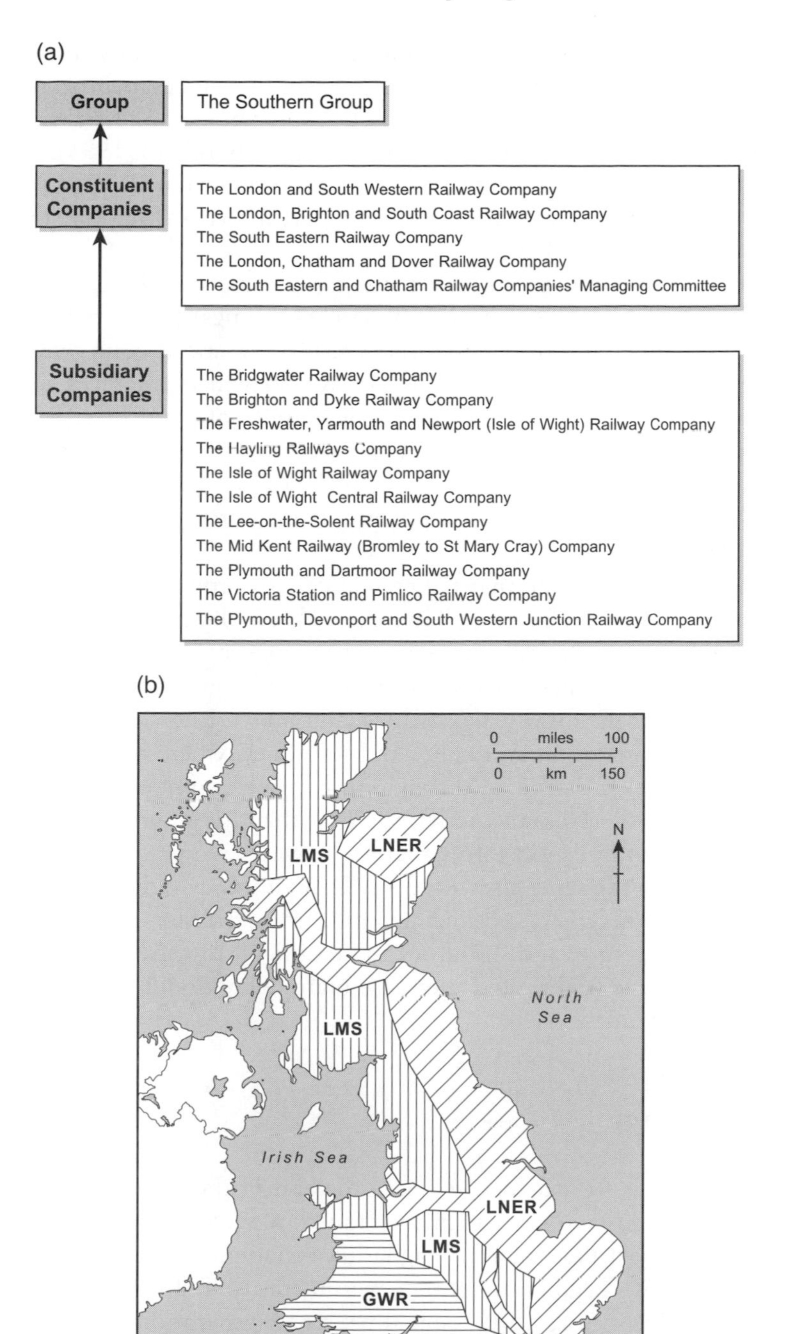

Figure 5.2. (a) Railway companies merged by Act of Parliament to form one of the 'big four' in Great Britain in 1923. After Shaw *et al.*, 1998. (b) The operating territories of the 'big four' railway companies, 1923–1947. These companies were combined and nationalized as British Railways in 1948. After Shaw *et al.*, 1998.

as was reasonable) a comprehensive transport network, especially in thinly populated areas where market failure was likely to occur. Social democratic and socialist governments alike nationalized various land transport modes, and air transport was publicly owned from an early stage in most countries (Graham, 1995). Even in the USA, public sector investment in transit systems began around 100 years ago, and by 1940 a substantial minority of public transport networks were in state or local government control (Neff, 1998).

A significant element in the notion of transport equity is the development and support of public transport to allow (among other things) for trips to be made by individuals in disadvantaged neighbourhoods. Overall social equity can be maximized through the provision of a transport system which enhances for as many people as possible the accessibility of a wide range of services (Buchan, 1992; Farrington and Farrington, 2005). Widening access to life opportunities for disadvantaged groups can also serve to promote broader economic benefits by, for example, increasing the mobility of a potentially neglected part of the labour force (Houston, 2001).

It is also important to note that the capital cost of providing any new transport initiative should preferably be at the lowest social cost. Black (2003) notes that this social cost will include the cost of the negative externalities associated with the transport development. These externalities were until relatively recently thought of as being primarily at the local scale, such as land take for new infrastructure developments, adverse impacts on air quality and severance, where a new piece of transport infrastructure is constructed through an existing community to the potential detriment of social or economic activity within that community. In addition, the building of a piece of transport infrastructure between two places reduces the available resources to do the same elsewhere. Investing in roads reduces the potential to enhance the railway network, subsidizing public transport fares reduces the capacity to improve infrastructure, and so on; it can be difficult to get the decision between infrastructure projects 'right' (and there are many who would argue that the state is generally a poor judge).

The neoliberal age

Compelling as it was for decades, towards the end of the twentieth century wide-ranging public sector involvement in the transport sector was undermined as an approach to transport governance. Stagnating economies and a sharp increase in oil prices in the 1970s began prompting governments to find ways of cutting spending across the public sector. At the same time, a new generation of national leaders, including Margaret Thatcher and Ronald Reagan, saw political and philosophical merit in reducing state expenditure and the size of the public sector more generally. Administrations in countries such as the UK, the USA and New Zealand started to adopt policies informed by neoliberal thinkers such as Friedman (1962), Hayek (1960, 1973), Niskanen (1973) and Tiebout (1956).

As a political and economic philosophy, neoliberalism countered prevailing assumptions underpinning the 'welfarist' economies of much of the developed world by stressing the need to reduce state intervention in the economy and promote market forces which encourage competition, enterprise and individual self-reliance (Harvey, 2005). Neoliberal ideas developed and spread through a range of institutional and political networks throughout the globe (Peck, 2001), including supra- and international organizations such as the European Union (EU) (although this retains a strong social regulationist agenda), the International Monetary Fund (IMF) and the World Bank, and were adopted by governments in Central and Eastern Europe following the collapse of the Berlin Wall. Peck and Tickell (2002: 380) suggest that neoliberalism has provided 'a kind of operating framework or "ideological software" for competitive globalisation, inspiring and imposing far-reaching programmes of state restructuring and rescaling across a wide range of national and local contexts'.

The transport sector has been affected significantly by this shift in political philosophy. Notwithstanding its peculiarities, from the neoliberal perspective transport is the same 'as any other good, subject to market forces and the rigours of competition' (Sutton, 1988: 132). The private car became seen as a symbol of freedom and progress – occupants can within reason go anywhere at any time – and governments deemed sustained increases in personal mobility facilitated by the car essential for economic growth (Meyer & Gomez-Ibanez, 1981). Large road building programmes were often implemented to facilitate increased traffic volumes (see Rodrigue *et al.*, 2006; Shaw & Walton, 2001). In respect of those modes in competition with the car, successive governments championed a range of reforms. Key among these was the removal of competition restrictions – the process of deregulation – in an attempt to open up a market for services previously provided only by the public sector. Downsizing, administrative fragmentation and in many cases outright privatization of state-owned transport undertakings were also actively pursued (Boxes 5.1 and 5.2). Such measures – grounded in the theory of contestable markets (Baumol, 1982) which emphasizes the importance of free entry (or even the *threat* of free entry) into a transport market – were deemed necessary to increase the efficiency and quality of services (whilst reducing the perceived oversupply of those services), improve

Box 5.1. Bus deregulation and privatization in the UK. Sources: DfT, 2006b; Kilvington & Cross, 1986; Knowles, 2006a; Preston, 2003a.

From 1930 to the 1980s bus and coach services were heavily regulated, and passenger numbers declined sharply as car ownership and use grew. Most rural – and an increasing number of urban – bus services became unprofitable and required subsidies to continue operating; nationally, bus subsidies rose from £10 million in 1972 to £520 million in 1982. The Transport Act (1980) abolished fares controls on all bus and coach services, breaking up geographically contiguous co-ordination and integration schemes. It also deregulated long-distance express coach services so that operators would decide routes and frequencies. There was an immediate growth in coach use as private operators competing with the state-owned National Express drove down fares, improved the quality of coaches and increased frequencies; many of these benefits have remained over time.

The 1980 Act also established trial areas to deregulate local bus services by scrapping route licences and letting the market determine the pattern of routes and frequency of services. The perceived success of a trial in Hereford led to the Transport Act (1985) which deregulated local bus services throughout Great Britain apart from in London (Figure 5.3). Ministers did realize that a system of completely commercial bus services in the provinces was likely to leave some suburban and rural areas, and indeed some urban areas during off-peak periods, without buses. As such, local authorities were allowed to specify and seek competitive tenders to provide 'socially necessary' services. In London, groups of routes were franchised to private companies, and the process of competitive tendering provided an incentive for bidders to cut costs. Finally, the legislation also led to the break-up and privatization of the state-owned National Bus Company and Scottish Bus Group, whilst Passenger Transport Authorities in the provincial conurbations and most councils were persuaded to sell off their bus companies.

The twin policies of bus deregulation and privatization have led to unexpected and geographically uneven results. By 1996, although over 90 per cent of bus services were run by private companies, mergers, takeovers and buyouts had led to the existence of only five major bus operators – First, Stagecoach, Arriva, Go-Ahead and National Express. 'On the road' competition between these companies has proved to be rare and short-lived, and uncontrolled monopolies have been established in many route corridors and districts. By 2004, market dominance by a single company had occurred in many counties and in all metropolitan counties and districts, and near-monopoly conditions – i.e. over two-thirds market share – had developed in 23 out of 36 metropolitan districts.

Figure 5.3. 'Bus wars' in central Manchester: vehicles from rival companies block the passage of a Metrolink tram. D. Hennigan.

Although London's franchised bus system experienced a 55 per cent increase in passenger numbers between 1985/86 and 2004/05, patronage of the deregulated bus system in England's provincial conurbations declined by 48 per cent in the same period. Similarly, patronage in English non-metropolitan areas, Scotland and Wales fell over the same period by 26, 34 and 31 per cent respectively. While heavy subsidies have kept fares in London relatively low, they rose sharply in the rest of the country: since 1986, fares in metropolitan areas have increased by 86 per cent and in shire counties by 37 per cent in real terms. Coupled with the fact that the overall cost of motoring has remained broadly unchanged since 1980, it is perhaps not surprising that patronage levels outside of London have continued to decline (Table 5.1). Bus deregulation and privatization have been very successful at cutting operating costs – mainly through lower wage rates for drivers, staff cuts and productivity increases – but the price of this success is considerable given the continuing decline in service and patronage levels in provincial areas.

Ironically, ministers from the current Labour administration, faced with charges of having done little to promote bus use since taking over from the Conservatives in 1997, have turned the stark geographical difference in bus patronage to their advantage. A 12 per cent ten-year *national* growth target for bus and light rail usage by 2010/11 will be easily achieved as the increase in passengers using London's franchised buses greatly outweighs the decline in other areas.

Table 5.1. Changes in Local Bus Passenger Journeys by Area 1985/86–2004/05 (million bus passenger kilometres). Source: Department for Transport (DfT), 2006b.

Area	*(a) 1985/86*	*(b) 2000/01*	*(c) 2004/05*	*% change (a) to c)*	*% change (b) to c)*
London	1152	1347	1782	+54.7	+32.3
England metropolitan[1]	2068	1166	1083	–47.6	–7.1
England shire[2]	1588	1247	1167	–26.5	–6.4
England total	*4807*	*3761*	*4032*	*–16.1*	+7.2
Scotland	671	443	465	–34.0	+5.0
Wales	163	116	113	–30.7	–2.6
Great Britain	*5641*	*4312*	*4609*	*–18.3*	+6.9

Notes:

1 Provincial conurbations of Greater Manchester, Merseyside, South Yorkshire, Tyne and Wear, West Midlands and West Yorkshire, centred respectively around the cities of Manchester, Liverpool, Sheffield, Newcastle, Birmingham and Leeds.

2 Non-metropolitan counties ranging from large cities to rural areas.

Box 5.2. Rail privatization. Sources: European Bank of Reconstruction and Development, 2001; Fullerton, 1990; Imashiro, 1997; Knowles, 2004; Wolmar, 2005.

Although some countries, notably the USA, developed and retained a network of private sector railways, state ownership of railways became widespread. Reasons vary from creating a national railway network in countries like Norway, where the private sector saw little potential, to nationalizing private sector railways for ideological reasons in countries like France and China, or in the late twentieth century to retain an unprofitable national network of passenger services. When road transport began to replace railways as the dominant mode of land transport in the mid-twentieth century, many railways required public subsidies or faced closure. Three different models of railway privatization have emerged: regional division and divestiture; outright privatization of national infrastructure and operation; and vertical separation of track and train ownership.

Regional division and divestiture

Commercial railway traffic can still make profits within loss-making state railway companies. Swedish Railways (SJ) was the first to recognise this when it divided its passenger railway operations into commercial and social railway networks in 1963. More recently the private sector has won tenders to operate part of the network. Japan began the process of privatizing the heavily subsidized Japanese National Railways (JNR) in 1987 with a regional division into six commercial passenger railway companies and one freight railway. Long-term debt was diverted into the JNR Settlement Corporation. The three profitable passenger companies with high-speed Shinkansen and commuter lines on the main island of Honshu – JNR West, East and Central – were first privatized in the 1990s. They had become attractive commercial propositions after downsizing and major productivity gains, and the removal of long-term debt. The other three passenger companies have a smaller customer base and less profit potential as they serve respectively the less populous islands of Hokkaido, Kyushu and Shikoku.

Network privatization

Outright privatization of national rail infrastructure and operation is uncommon because of limited opportunities to operate a whole rail network profitably. New Zealand Rail was privatized in 1993 for NZ$328 million to a consortium led by Wisconsin Central Transportation Corp., renamed Tranz Rail in 1995, and sold on to Toll Holdings of Australia in 2001. Inter-city passenger services ceased on all but four routes, track maintenance declined and much freight traffic was lost to road transport. After public and political dissatisfaction with the outcomes of rail privatization, the rail infrastructure was renationalized in 2004 and now trades as ONTRACK, although rail operations remain private and passenger services have recently been cut back further. In 2002, Estonia was the first former communist country in Eastern Europe to privatize its national railway operation. Supported by loans from the European Bank of Reconstruction and Development, the impact of Estonian Railways' privatization will have a substantial demonstration effect elsewhere in the region.

Vertical separation of track and train operation

Sweden pioneered the separation of track and train operations in 1988 to make the costs of railway operation more transparent. Banverket (the national railway infrastructure authority) remains in state ownership, but charges operators marginal cost for access to the network and a fixed charge for each wagon. Denmark followed Sweden in creating a state-owned track infrastructure company separate from its railway operating company. The European Commission then made this separation a policy recommendation for all European Union members.

The most significant example of this type of privatization has been in the UK. Great Britain's passenger and freight rail businesses were privatized between 1994 and 1997. A rail infrastructure company, Railtrack, was created in 1994 and privatized in 1996, and rail freight operations were split into seven businesses before being sold off. 25 Train Operating Companies (TOCs) were franchised by competitive tender in 1996 and 1997 to reduce government subsidies and maximize premium payments (Figure 5.4). Passenger services were specified at levels close to those previously operated by British Rail, and most ticket prices were regulated. Although designed to reduce taxpayer subsidies, this immensely complex structure more than doubled the total rail subsidy from £1.07 billion in 1993/94 to £2.19 billion in 1996/97, and debt write-off and privatization costs totalled £3.13 billion. Cost reduction and better marketing of services was meant to deliver year-on-year reductions in subsidies, but various developments, including the collapse of Railtrack and its subsequent replacement by the not for profit Network Rail, have led to further increases state support, to over £5 billion in 2006/07. A major success since the sell-off has been the unpredicted growth in rail patronage – the network is now being used more than at any time since the 1950s – although the influence of privatization is difficult to determine because demand for rail travel has also been boosted by continuous national economic growth since 1992.

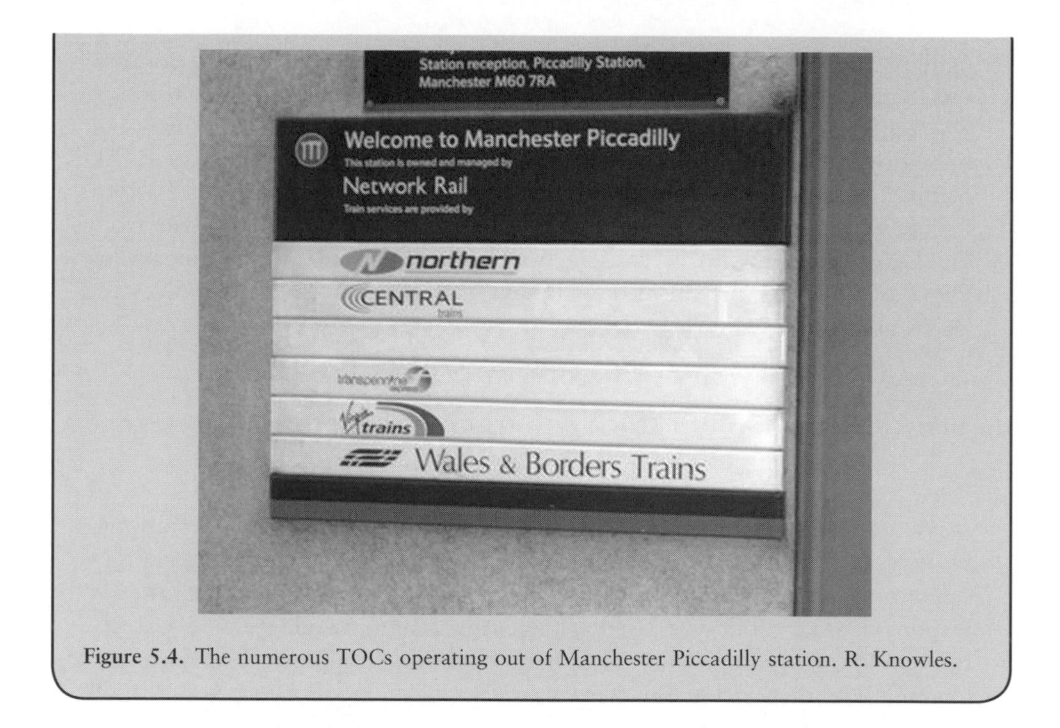

Figure 5.4. The numerous TOCs operating out of Manchester Piccadilly station. R. Knowles.

transport providers' responsiveness to changing customer demand (often a function of rapidly increasing car ownership and use), and reduce the need for public sector ownership, management and subsidy (Knowles, 2006a; Savage, 1985). The procurement and financing of infrastructure also became more dependent on the private sector as governments looked to reduce their spending and debt liabilities.

In continental European states, the 'welfare' economic model generally retained more influence, and the restructuring and liberalization of the transport sector only really gathered pace in the 1990s. Compared to what is (inaccurately) regarded as the 'Anglo-Saxon' world (Goodwin, 1997a), political discourse was far less influenced by faith in deregulation as the means to economic efficiency. Instead, reform proceeded at a slower pace, the objective being to deliver a true European common market for major transport services such as road haulage, aviation and the railways, whose needs dominated policy shifts (Gather, 1998). In some countries, this process was accompanied by the privatization of state-owned companies in an effort to strengthen their competitiveness as well as to limit public subsidy.

There is little doubt that some of these market reforms succeeded in realizing their intended cost savings, improvements in process efficiency and reductions in bureaucracy. Equally, others 'did not bring about most of the positive effects forecast', in terms of improving the scope and quality of transport services (Pucher and Lefevre, 1996). In many instances, reform was sufficient to alter significantly the pattern and nature of service provision (Boxes 5.1 and 5.2). This brought clear advantages, but there were also adverse effects such as service contraction, reductions in public transport patronage and increasing traffic congestion (Bell & Cloke,

1990; Commission for Integrated Transport (CfIT), 2001a; Goodwin, 1999; Knowles and Hall, 1998; Preston, 2003a; Wolmar 2005).

Redefining the role of the state

Just as the welfarist model of wide-ranging state provision of transport services was superseded by neoliberal ideas, so policies in the first decades of the twenty-first century seem likely to have a different emphasis to those in the 1980s and 1990s. In many countries, something of a retreat from – or at least a redefinition of – neoliberalism is evident, as politicians seek to re-engage the state with many areas of public policy. Governments have begun to adopt a more interventionist stance towards transport infrastructure and service provision. In the Anglo-Saxon world, where the pace of the neoliberal project was greatest, politicians are reflecting more on the experiences of other countries where reform proceeded more slowly, and where the desire to achieve economic efficiency never attained pre-eminent status.

Hollowing out and filling in

In one sense, the bold rhetoric of neoliberalism reducing the role of the state was always misleading. It is true that a 'hollowing out' (Jessop, 2004) of the state's existing functions was taking place, as privatization and deregulation significantly curtailed the public provision of services. Yet as Charlton and Gibb (1998a: 85) point out, despite the 'undeniable prominence of liberalised transport systems', there was 'an apparent contradiction between the broad policy support for deregulation and privatisation and the continued, and even enhanced, regulatory environment affecting many transport systems'. This observation refers to that fact that, in reality, it was never possible to privatize or deregulate many transport markets to the point where the state could simply withdraw from them. This is not least for the reasons discussed earlier in this chapter – monopolies could not be broken up, consumers had to be protected, social policy goals remained important – and progress towards neoliberal goals had to be tempered by a degree of *realpolitik*.

In other words, the process of 'filling in' – the introduction of new and different state functions to replace old ones being discontinued – was always in evidence. In the transport sector (as elsewhere), this has often been expressed by the introduction of contractual agreements – as a substitute for the command-and-control management style possible under direct ownership – to ensure that a range of outcomes specified by the state is actually delivered by the private sector. The decision to protect unprofitable rural bus services (Box 5.1) is one example. Also in the British bus sector, a regime of Quality Contracts is being investigated to address some of the ongoing shortcomings in service provision (Davison and Knowles, 2006). The privatization of Britain's railways has involved the tight regulation of a monopoly infrastructure owner and a plethora of contractual stipulations to which the passenger train operators have to adhere (Shaw, 2000) (Box 5.2). In large part this was to guard against the significant network contraction – effectively a complete recasting of Britain's railway geography – that would have ensued if private sector operators had sought to maximize revenue by closing unprofitable routes. The near collapse of the privatized rail industry in 2000/01 following a fatal crash at Hatfield, near London, saw ministers rescue the bankrupt infrastructure company and provided a stark illustration of the state's continuing role as guarantor of the last resort in many areas of transport endeavour, even in a supposedly privatized system of provision (Wolmar, 2005).

Governments are facing the reality that while market-based transport policies work well in some sectors, they function less effectively in others and in some cases tend not to function at all. The same applies in terms of the geographical implications of such policies: at the simplest level, the market will provide services in areas where they are profitable, but without appropriate regulation will precipitate their withdrawal or downscaling where money cannot be made. All of this means that privatization and deregulation should not be interpreted as a simple reduction of state intervention and control. Rather, they are part of a broader process of state restructuring which incorporates elements of both deregulation and re-regulation as established modes of organization are dismantled and new ones established (Peck, 2001).

Public and private interdependence

At the same time as this process of 'filling in' of the state's functions is becoming better understood, parallel developments generating increased interaction between the state and private transport interests are coming increasingly to the fore. At least three important trends are apparent here: the 'crisis of mobility' brought about by worsening congestion; the recognition of the need to pursue environmental policy goals through intervention in transport markets; and the general shift to a more complex system of governance.

The growth of car (and air) traffic levels in recent decades has left a legacy of severe congestion in much of the developed world (see CfIT, 2001a and Chapters 2 and 3). Despite significant road construction programmes, governments (or private sector holders of building concessions) are unable to build themselves out of congestion. Partly this is because the financial, political and practical limitations on building new infrastructure are real – roads are expensive, attract a lot of opposition and fill up very quickly after they are built (the phenomenon of 'induced traffic') – but politicians from the Right also realized that continuing to accommodate such vast increases in mobility was increasingly paradoxical as it implied very high levels of government spending.

The resulting crisis of mobility has strongly influenced the state's re-engagement in the regulation of the transport sector. Essentially, some form of demand management needs to be put in place once the level of congestion reaches a certain threshold, otherwise the costs of this congestion jeopardizes economic efficiency by creating delay and uncertainty in the whole transport system (Glaister, 2004; Thomson, 1977a). The London Congestion Charge was identified in Chapter 3 as a good working example of demand management – congestion has fallen by 26–30 per cent following the introduction of a daily charge (Transport for London (TfL), 2006) – and the UK government has indicated support for (though not committed itself to) a nationwide charging scheme when technology permits by around 2015 (Shaw *et al.*, 2006).

The second significant factor is the scale of transport's contribution to global environmental damage (Chapter 3). A critical introductory passage from the Brundtland Report noted the need for concerted government action because of 'a growing realization . . . that it is impossible to separate economic development issues from environment issues . . . and environmental degradation can undermine economic development' (World Commission on Environment and Development, 1987: 3). Not only has an extensive literature developed in the field of environmental economics that attempts to formalize and quantify some of these relationships (see, for example, Baumol & Oates, 1988; Kapp, 1988; Maddison *et al.*, 1996; Pearce *et al.*, 1989; Wicke, 1991), but also the case for government intervention in the transport market is much stronger since the international scale of transport-related pollution and emissions has become apparent (European Conference of Ministers of Transport (ECMT), 1990; Goodwin, 1999; Greene and Wegener, 1997). States and other government bodies are widely seen as the only institutions with the power to make any meaningful impact on such strategic processes, even if ultimately they rely on market mechanisms such as road user charging – derived from the work of environmental economists among others – to ascribe the costs of transport more appropriately to users (CfIT, 2006).

The recent recognition by many governments of the importance of environmental as well as economic and social policy aspirations underscores the contemporary 'three-legged stool' conception of sustainability, that seeks to balance these often competing demands (Docherty, 2003). It is also interesting to note that the emergence of congestion and pollution as significant multi-scalar issues repositions the state firmly in the role of arbitrator between public and private interests, as envisaged by those economists who identified, nearly 100 years ago, the potential for carefully targeted state intervention to benefit society more than perfect competition in all sectors of the economy (Schumpeter, 1909). In order to address congestion and pollution, it is becoming necessary for the state to take a leading role in managing individual mobility.

The third factor to consider is changes to the structures of governance witnessed in many countries as state institutions have sought to adapt to changing socio-economic contexts. Existing states have ceded some of their sovereign powers 'above' to supra-national institutions (such as the EU) and other international treaties and organizations, and in some countries there have also been challenges from 'below' in the form of nationalist desires for increased regional autonomy and discretion in policy making. The establishment of devolved governments within the UK in Scotland, Wales and Northern Ireland illustrates this well (Figure 5.5). Such *spatial* hollowing out – powers transferring up or down to institutions with different jurisdictional territories – combines with the processes of functional hollowing out and filling in to produce complex, multilayered systems of governance often involving stakeholders from the voluntary and private sectors (Table 5.2). Understanding how this impacts upon transport regulation and control – as in other areas of public

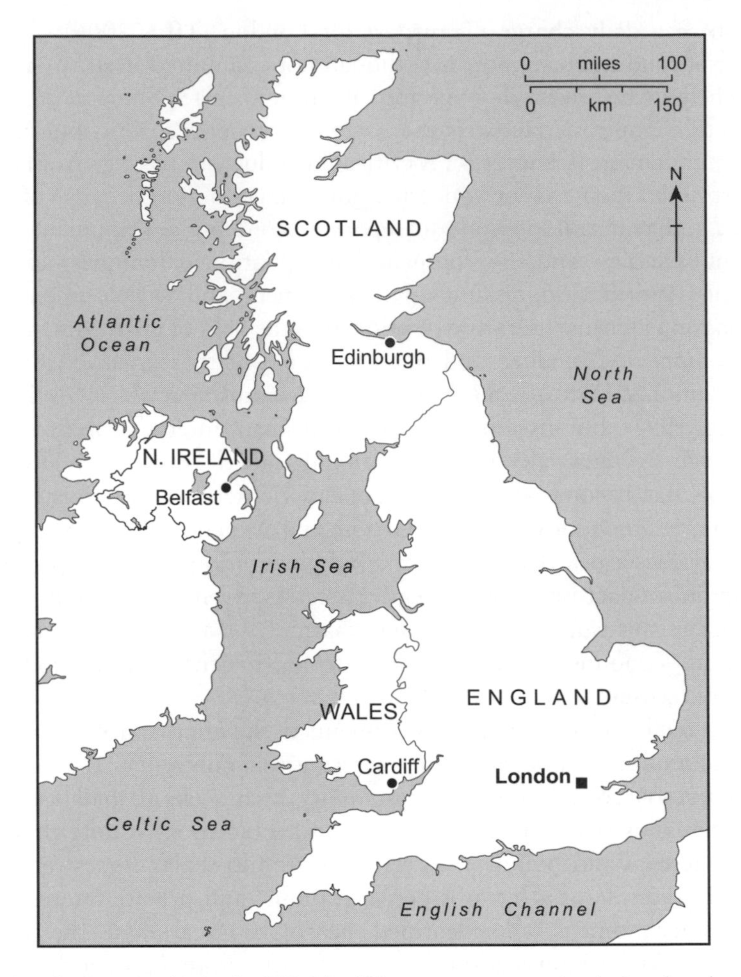

Figure 5.5. Devolved territories in the UK. The UK government remains in London, but power over many domestic matters has been transferred by Acts of the UK Parliament to the Scottish Parliament in Edinburgh, the Welsh Assembly in Cardiff and (when not suspended due to local squabbles) the Northern Ireland Assembly in Belfast.

Table 5.2. The devolved transport sector policy framework in the UK: *de facto* devolved and reserved matters. Source: After Smyth, 2003.

	Scotland	*Wales*	*Northern Ireland*
Road	Totally	Limited	Totally
Rail	Mostly	Limited	Totally
Bus	Totally	Limited	Totally
Air	Limited[1]	None[1]	None[1]
Sea (ferry)	Substantial	None	None

Note:

1 It is possible for each of the devolved administrations to influence aviation policy through various secondary mechanisms such as the application of their planning powers.

policy – can be tricky, and requires an analysis of 'the changing forms and mechanisms . . . in and through which' the state seeks to achieve its goals (Jessop, 1990: 203) (Chapters 6–10).

Jessop's (1997) work on the processes underlying the changing character of state reorganization is useful in explaining how the regulation of transport might begin to look in a 'post' neoliberal world. The first such process, denationalization, encapsulates the apparent spatial hollowing out of transport regulation. From 'above' comes, for example, the EU's drive for the harmonization of transport markets between states in areas such as air traffic control and open access to rail markets. From below, sub-national governments take on responsibility for the specification and franchising of transport activities such as local rail services in France, Germany and Sweden. The second process, destatization, reflects the increasing role played by non-state actors in regulation and service management, such as the control of some regional airports in Europe by Chambers of Commerce or wide-ranging public–private partnership consortia. Finally, the third process, the inter-state transfer of policy, is clearly evident across the transport sector, be it in the lessons learned from the privatization of British Rail (generally regarded as an overly complex and highly costly experiment), or the success of the German *Verkehrsverbund* system of fares management operating in major metropolitan areas such as Munich, Hamburg and the Rhein-Ruhr region (Pucher and Kurth, 2002).

In making sense of these developments, states are increasingly seeking to follow what Giddens (2000) characterizes as a 'third way', between neoliberal and welfarist thinking, by combining elements of market discipline in transport procurement and delivery with stronger state regulation for economic, social and environmental reasons. By adapting their transport strategies to respond to the 'changing connections and inter-relations between social, political and cultural factors' that characterize an internationalizing and dynamic political economy (Painter, 1995: 276), states are both reacting to new threats such as widespread congestion and the environmental crisis, and addressing less immediate but equally important trends such as ageing demographic profiles that place new demands on systems of mobility and connectivity. They also attempt to replicate the old consistency of state provision of services with complex regulatory systems based on contractual relationships. And in doing all of this they routinely – or at least frequently – share or cede/delegate responsibility to a far greater range of public, private and voluntary sector stakeholders than has historically been the case. Although the neoliberal experiment was designed to facilitate a shift to the market provision of transport services, the need for state intervention in the system, albeit in a different form, has very clearly reasserted itself.

Conclusion

This chapter has explored transport governance and ownership from a geographical perspective. It is important to understand the role of the state in transport because territorial governments are heavily involved in the management and ownership of transport activities; the strategies and policies they adopt over time will shape the development of transport infrastructure and services in a variety of ways at a variety of spatial scales. In addition to military and national security considerations, taken as a whole state interventions in the transport sector seek to promote economic development and efficiency whilst pursuing social objectives such as widespread mobility and accessibility, and addressing environmental imperatives like global warming and local air quality.

The level and scope of state intervention in transport activity will depend upon a range of factors. The finances available to politicians at any given time are clearly important, but more abstract factors such as political ideologies must also be taken into account. A major refocusing of state involvement in the transport sector was attempted in response to the rising influence of neoliberal ideology, although in reality the extent to which states were able to withdraw from the transport sector is debatable; in many territories direct ownership and operation of components within the transport market was replaced by management through contract and other regulatory apparatus. The neoliberal era certainly impacted upon the geography of transport activity, but the integrity of many potentially threatened networks and services was broadly retained as a result of ongoing state intervention in the transport market. Transport systems around the world are now witnessing something of a resurgence in state intervention arising from what might be described as a 'crisis of mobility', heightened concern about the global environmental impacts of transport, and increasing complexities within the state sector.

Part 2 | Transport Flows and Spaces

6 | Connected Cities

Iain Docherty, Genevieve Giuliano
and Donald Houston

This chapter reviews the links between urban transport systems and the functioning of contemporary cities in the developed world. It opens with the fundamental question, 'what is a city?' and goes on to review the critical roles played by transport in the ongoing processes of city building and renewal. The key concepts of mobility and accessibility, and of economic development and social justice, underline the importance of efficient and effective transport to the contemporary paradigm of city competitiveness. Yet with car travel now accounting for nearly 85 per cent of all motorized person kilometres in the 'EU 15', and an even greater proportion in the USA (Eurostat, 2006a; Bureau of Transportation Statistics, 2006a), rising congestion and environmental pollution increasingly constrain the functioning of key urban processes such as the labour market and reduce the attractiveness of the city for residents and investors alike. The chapter concludes by examining the key policy choices facing urban transport planners and decision makers, using examples from major cities across the world to highlight the potential for conflict between the often competing strategic objectives of mobility, accessibility and sustainability.

Transport's role in defining the city

We all have an intuitive understanding of the idea of what a 'city' is, and how it operates. Most of these understandings are based on the assumption that cities are in some way bigger and have a higher density of activity than elsewhere. The classic definition of the city is therefore as a location where the processes of economic, social and cultural activities are concentrated in space: a node of production, exchange and interaction within and between these functions. Dense movements of people, goods and information are required for cities to fulfil these roles, and as such a city's transportation system is crucial to its success, since an efficient urban transportation system can significantly boost the quality of life and investment opportunities in an urban area (Banister & Berechman, 2000).

But these ideas of scale and density of economic activity represent only one set of defining characteristics for the city. Urban hierarchies have also been defined on the basis of the economic, administrative and other functions of different settlements. The notion that cities might have distinctive place characteristics in addition to being larger, concentrated, more diverse and faster-changing spaces of economic activity is long-standing. Louis Wirth's classic essay of 1938, *Urbanism as a Way of Life*, identified the existence of a distinctive urban social order. Recognising the importance of size, density and diversity as basic characteristics of the city, Wirth proposed that there were nevertheless a number of attributes that made cities distinct from other places in terms of how their societies were organized, how people interacted in them, and how their identity and quality of life developed. Following on from this, Ferdinand Tönnies (1955) noted how the much greater level of choice available to individuals in cities, whether for work, education, leisure or other activities, generated a culture of 'association' that was very different to the much more rigid rural model of 'community' in which individuals' interactions with others were more closely regulated and controlled by long-standing social norms. These themes remain evident in the urban development literature, from Jane Jacobs' characterization of the vitality of great cities (1961) to Richard Florida's conceptualization of the urban 'creative class' (2005).

For transport, the importance of these two inter-related perspectives of *space* – the territory across which economic processes are played out – and *place* – bounded locations of activity with a specific social identity (see Tuan, 1977) – is that each implies a different set of objectives and priorities for the development and management of the urban transport system. As a centre of economic production and (especially) exchange, the city clearly requires a transport system capable of efficiently moving both goods and people (Burtenshaw *et al.*, 1991; Button and Gillingwater, 1986). But urban transport systems also influence the city as a place of social and cultural activity. Hanson (2004) neatly highlights the critical distinction between the physical mobility of goods and the workforce in the economic sphere, and the accessibility of urban functions and services to individuals implied by the notion of the city as a place of social association. White and Senior (1983: 1) identify the latter role of transport in urban life as fundamental, since 'transport creates the utilities of place' that together form the concept of urbanism as envisaged by Wirth. The need for urban transport is therefore derived from the wide range of activities people undertake as part of their daily lives (Table 6.1).

This categorization of the needs for urban transport can be taken a stage further by linking it to two important contemporary policy concepts – namely economic development and social justice (the third 'pillar' of sustainability, the environment, is also critically important, but is arguably better addressed at spatial scales above that of the individual city[1]). The provision of transport infrastructure is clearly necessary for efficient operation of the urban economy, given spatial separation of activities and households. The transport infrastructure, together with the resources available to individuals, determines each individual's level of *mobility* (the capability for physical movement); the spatial arrangement of activities and households

Table 6.1. Classification of purposes of urban personal travel. Source: Daniels and Warnes, 1980.

Activity	*Journey classification*
ECONOMIC	
Earning a living	To and from work In course of work
Acquiring goods and services	To and from shops and outlets for personal services In course of shopping or personal business
SOCIAL	
Forming, developing and maintaining personal relations	To and from homes of friends and relatives To and from non-home rendezvous
EDUCATIONAL	To and from schools, colleges and evening institutes
RECREATIONAL AND LEISURE	To and from places of recreation and entertainment In the course of recreation: walks, rides
CULTURAL	To and from places of worship To and from places of non-leisure activity, including cultural and political meetings

determines *accessibility* (the availability of employment, educational, social, and cultural opportunities). Mobility and accessibility are closely related: the more accessible the urban environment, the fewer mobility resources are required to carry out daily activities. The more mobility resources are available, the greater one's level of accessibility becomes (Chapters 4 and 7).

Both mobility and accessibility are important for social justice. A key challenge emerging over the past few decades has been the level and rate of decentralization and deconcentration of metropolitan areas, especially in North America. Since even maintaining a given level of accessibility requires the supply of more mobility in these circumstances, coping with decentralization and deconcentration has become a major urban transport policy concern. This is complicated still further by the fact that some social groups do not enjoy the same level of mobility as others: for example, those who cannot afford a car, or who cannot drive due to disability, or those who find transport costs prohibitive. As we discuss later in more detail, low density, decentralized land use patterns often make public transport modes less competitive, leaving those who remain dependent on them potentially disadvantaged (Chapter 4).

This distinction between economic and social approaches to urban transport also affects the processes through which network investments are prioritized and appraised. This is because economic benefits are often easier to quantify than social benefits – for example, whilst it is relatively straightforward to calculate the extra productivity a business obtains when a new bypass opens making it quicker for employees to travel between two different production sites, it is much more difficult to estimate how much benefit an elderly person gains from a new bus service giving him or her access to a new community centre, no matter how important that benefit is to the individual's quality of life (Standing Advisory Committee on Trunk Road Assessment (SACTRA), 1999). Advances in GIS technology promise to make it easier to more accurately capture and measure changes in accessibility, and, at the same time, the underlying assumptions of transport economics remain under scrutiny, given the potential to exaggerate and double count claimed economic efficiencies of new transport construction (Docherty *et al.*, 2007).

Transport and city building

The changing physical form of cities is directly related to the transport technology available at any point in time. As transport has made it possible for people and goods to travel further, at greater speed and at less cost, cities have both grown and developed distinctive land use structures reflecting changes in these transport patterns. Before around 1840, cities were usually modest in size, and structured around a basic system of roads for the use of pedestrians and horse-drawn vehicles. The dominant form was that of 'foot cities', dense places with little distinction between workplace and home, in which walking was the overwhelmingly dominant form of mobility.

The progressive motorization of transport seen from the mid-nineteenth century had profound impacts on cities, 'their internal structure, and the supply, demand, efficiency, speed and opportunities for movement within them' (Daniels and Warnes, 1980: 4). The technological shift from foot cities to 'tracked cities' made possible by the horse tramway enabled the first significant separation of homes and workplaces. The boom in mainline railway construction seen at the same time also made it possible for wealthier people to build new suburban communities at some distance from the congested, polluted city centres. Indeed, the railway companies often played a positive role in this process, offering free season tickets to 'people who agreed to build houses at any undeveloped location on the railway' (Thomas, 1971: 209). While horse tramways and railways provided the first means for separating home and work, it was the widespread and rapid growth of electric street tramways at the end of the nineteenth century that changed the geography of cities by making affordable commuting readily available to the industrial workforce, many of whom moved to the new high-density 'streetcar suburbs' built within convenient travelling time of city centres (Moorhouse, 1988; Vance, 1991; Ward, 1964).

In the larger and more congested cities, the sheer physical size of the built-up area encouraged the development of fully segregated railway systems in addition to the tram in the latter years of the nineteenth century. Cities such as London, Paris, Chicago and New York developed extensive rail transport networks. Electric traction was again the technology breakthrough; electric trains had better performance characteristics, and were less dangerous and polluting than the steam engines they replaced. Spurred on by the success of the Metropolitan Railway and later the London Underground system in supporting the creation of new garden suburbs (nicknamed 'Metro-Land' – Figure 6.1), Berlin became the first city in the world to separate its suburban railway operations from longer-distance mainline traffic with

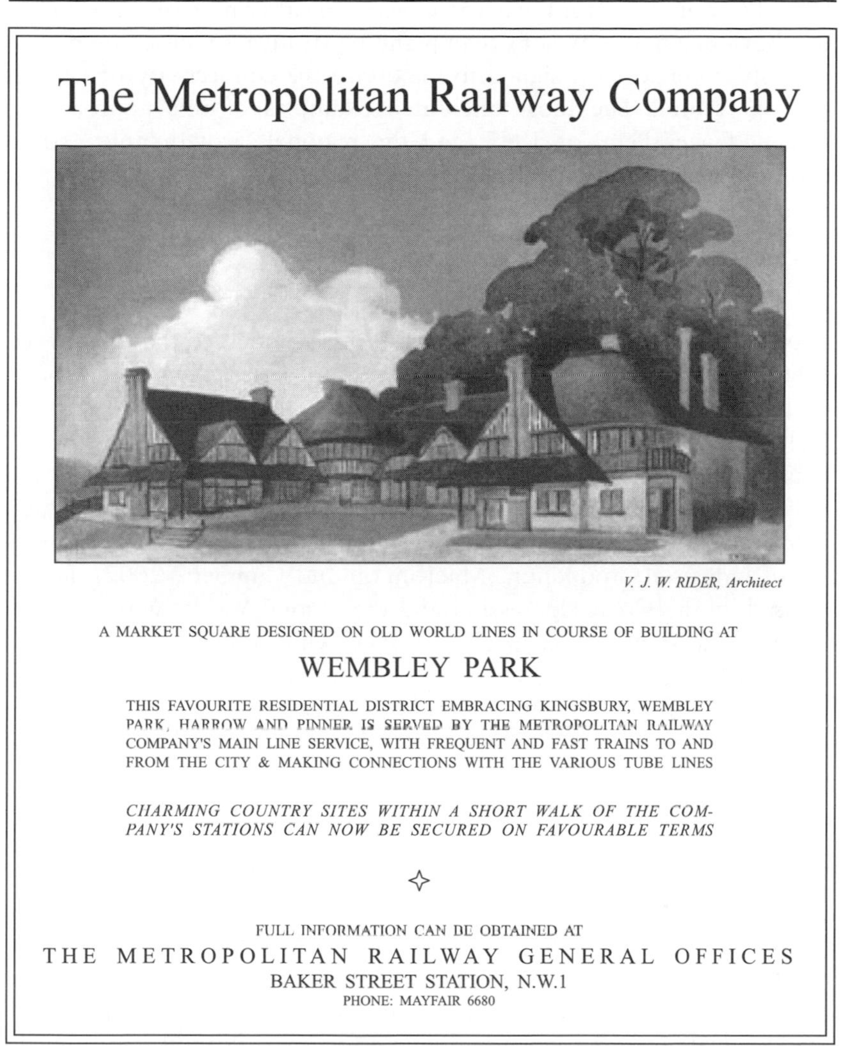

Figure 6.1. Metro-Land advertising poster, 1919.

the creation of the electrified S-Bahn rapid rail network in the early 1930s (Schreck *et al.*, 1979).

The step change in accessibility generated by these rail transit systems profoundly altered the urban geography of the cities in which they developed. The position of the city centre as the most accessible location was reinforced, with city cores becoming the preferred location not only for economic activities, but also for tertiary education, museums, public parks and other activities such as the new mass spectator sports. One of the main innovations of the growing public transport of this era was therefore the social integration achieved by opening up the wide variety of city services and activities to a broader social spectrum. And just as the radial tramways had influenced an earlier phase of growth in the nineteenth century, renewed focus on rail rapid transit in several European cities in the aftermath of the Second World War gave birth to a new wave of urban planning strategies focused on high-density public transport corridors designed to maximize the connectivity of new suburbs with the city centre. The most celebrated examples of this approach are the Copenhagen 'Finger Plan' of 1947, and the regional 'growth poles' strategy of the Ile-de-France region around Paris from the 1960s.

The love affair with the car

From about 1920, another urban transport narrative emerged that quickly came to eclipse the previous phases of the foot and tracked cities: the shift to the 'rubber city' of the bus, lorry and particularly the private car, which brought immense structural, social and economic change to almost every city in the world. In the USA, mass ownership of cars took hold almost overnight, reflecting the immense success of Henry Ford's Model T, a product that became synonymous not just with a transport revolution, but also with an even more profound revolution in the very structure of industrial production (MacKinnon and Cumbers, 2007). Despite setbacks caused by the Great Depression and the Second World War, the long-term trend of car ownership was set firmly upwards, with public transport use peaking in the early 1920s as a result. Over the next few decades, the level of personal mobility was transformed for those that could access a car, with many public transport systems in the USA (which were often privately owned) entering steep decline after the war and facing bankruptcy by the 1960s.

In Europe, mass car ownership gathered pace through the 1950s and 1960s. Even today, rates of car ownership in Europe trail the USA significantly – 472 cars per 1,000 population in EU25 in 2004 compared with 759 per 1,000 in the USA in 2003 (Eurostat, 2004). Despite these differences – which can be attributed in part to the influence of different public policy approaches on issues such as fuel and vehicle taxes, driver licensing policy and the importance afforded to new road construction – the end result in both contexts was to generate a previously unimaginable level of individual choice of when and where to travel. No longer confined to the clearly delineated corridors of the tracked city, travel patterns became much more complex, as trunk flows of movement to and from major urban centres along radial routes were increasingly supplemented and even replaced by a complex web of circumferential and tangential trips as homes became ever more separated from workplaces and preferred places of consumption.

In response to the steeply rising demand for car travel, many cities embarked on major road-building programmes, although these were often to prove difficult in the older, historic cities of Europe and the east coast of the USA, whose dense structures of relatively narrow streets would require complete rebuilding to accommodate the emerging level of private road traffic. Some cities in the UK, such as London, Birmingham, Liverpool, Manchester and Glasgow, attempted to build major highways through their city centres, but these were abandoned either before construction or midway through the implementation phase following protests over

Figure 6.2. Traffic in Wuhan, China, November 2004. G. Giuliano.

'motorway madness' destroying the urban environment (Starkie, 1972) and the capital rationing that followed the energy crisis of the mid-1970s. No such problems were apparent in the USA, where highways were seen as a way of reducing congestion and linking the cores of cities together. Some quite ambitious urban road building therefore took place, especially in the more densely developed regions, with perhaps the most celebrated examples being the elevated, iron structure of Interstate 95 through central Boston and the Hudson and FDR Parkways on either side of Manhattan Island.[2] Some cities in continental Europe followed the same example, with the Périphérique ring road around Paris and the A6 and A7 Autoroutes along the Lyon Quays in France, and the elevated Autostrade of Genoa and Naples in Italy, being important examples.

One of the enduring legacies of these different approaches is the varying degree to which the morphology of cities has developed in different parts of the world. In the USA, ambitious road construction enabled major cities to sprawl across huge areas, taking advantage of the new high-capacity parkways and freeways: as a result, many have extremely low residential densities, typically one-quarter that of older European cities of comparable size (Murphy, 1974; Newman and Kenworthy, 1999). Today, however, rapidly developing cities in China and India are following a different trajectory (Figure 6.2). Car ownership is growing very fast, but the transport infrastructure to support this level of car-based mobility is not being supplied. As a result, congestion is so severe that the suitability of these high-density cities for a high modal share for cars is rapidly being called into question (Lee, 2006).

Personal freedom or car dependence? The urban transport 'problem'

> For personal and family use, for the movement of people in mass, and for use in business, commerce and industry, the motor vehicle has become indispensable. (Ministry of Transport, 1963: 10)

> It is our contention that the urban crises which manifest themselves in so many ways have at least one common root. This is the increasing reliance on the automobile. In every urban area, the automobile has become the only means of transportation by which every part of the region can be reached. Wherever the automobile is *the* mode of travel, there access to transportation is distributed very unevenly between individuals. This is probably the greatest social fault of the automobile. (Schaeffer & Sclar, 1975: 103, original emphasis)

As these two starkly contrasting quotes demonstrate, the transformation in personal freedom and increased choice over where to live, work and spend leisure time brought about by mass car ownership cannot be underestimated. Overall, populations now have more mobility than ever before, and the demand for yet more expansion of personal mobility, either in terms of surface travel or increasingly for aviation, remains strong. But as the two citations also demonstrate, this freedom of choice for some has come at great cost for others: indeed, the negative externalities of car use have come to dominate the urban transport policy debate – at least in Europe – for the last 30 years (see Thomson, 1977b, for example, for an early analysis). These externalities include congestion, impacts on the environment and human health from poor air quality, and reduced accessibility and mobility for those who do not have access to a private vehicle. In the USA, where car ownership is higher and the 'transport-deprived' (Tolley & Turton, 1995) population much smaller, the reaction against the externalities of the private car has been significantly more muted, although still vociferous in certain locations.

As noted earlier, the near-universal adoption of the car and truck transformed the geography of cities, enabling economic activities to (re)locate in lower-cost areas, often in the suburbs. As people have become wealthier, they have tended to choose to move further from the urban core where land prices are cheaper, and larger family homes are more affordable. Over time, as populations became more dispersed, employment and other services such as education, retailing and leisure moved from the urban core to suburban locations in order to service these new demands. Reinforcing each other over several decades, these trends in land use and transport have resulted in a situation of widespread 'car dependence', in which many people now require (very) high levels of mobility simply to maintain the basic fabric of their lifestyles, such as travelling to work.

Table 6.2. Perspectives of the role urban transport. Developed from Docherty 1999.

Urban Perspective	City as economic space	City as social place
Policy objective	Economic development	Social justice
Prime function of urban transport	To maximise mobility of labour and capital	To maximise individual accessibility to urban amenities
Evaluation method	Quantitative analysis	Quantitative and qualitative analysis

Low densities and dispersed travel patterns are ideal for the car, but often incompatible with public transport, with its less flexible route structure. Thus the combination of rising car ownership and changing land use patterns has both reduced the demand for public transport and reduced its ability to compete effectively with the door-to-door, available-on-demand flexibility of the car. Writing in the mid-1990s, Pucher and Lefevre (1996: 136) noted that these processes had become so entrenched that,

> separating people from their cars is a well-nigh impossible task. It would require a considerable increase in public transport investment and the implementation of policies which favour public transport and alternative modes, neither of which seems very likely in the future.

As a consequence, those who cannot afford to drive, or for whom age, ill health or other factors make driving impossible, are doubly disadvantaged as public transport service quality and availability decline (Hanson, 2004; Torrance, 1992) (Chapter 4). Research by the UK government's Social Exclusion Unit has identified a number of other groups who are 'transport deprived' in urban areas, including lone parents, children, the elderly and those with disabilities (Social Exclusion Unit, 1998). In some extreme contexts, such as the semi-derelict urban environments of some of the USA's inner cities, even the very poor become 'auto dependent', because the car – even if it is shared between several people or is hired in the form of a taxi – is literally the *only* way to access employment and other basic activities (Deka, 2004; Giuliano, 2005).

The depth of social exclusion resulting from highly polarized levels of accessibility and mobility is particularly marked in the USA for several reasons (Tables 6.2 and 6.3). First, American metropolitan areas are spatially segmented by race as well as income, with poor and minority populations often concentrated in inner cities as a result of discriminatory housing and land use policies in suburban communities. Secondly, the sheer extent of decentralization and dispersion in most American cities has substantially eroded the relative accessibility of central locations to the extent that they are not really 'central' any more. Thirdly, the quality and flexibility of public transport systems in the USA can be exceptionally poor, exacerbating the perceived difference in amenity between these networks and the car. Finally, the limited nature of social policy and welfare support for the poor (especially compared with much of Europe) makes the divide between rich and poor particularly visible, resilient and extreme.

Table 6.3. Poverty and trip making in the USA, 1995. Source: Giuliano, 2005.

	Trips		*Distance (miles)*		*Time (minutes)*	
	Mean	*Median*	*Mean*	*Median*	*Mean*	*Median*
Poor	3.1	2	18.1	6	47.3	30
Not poor	4.0	4	30.9	20	61.1	50

Transit use and daily travel distance / time in the USA, 1995. Source: Giuliano, 2005.

		Regular user	*Occasional user*	*Not a user*
Daily travel Distance	Poor	14.0	15.3	18.9
	Not poor	23.9	31.9	31.2
Daily Travel Time	Poor	65.6	50.7	45.7
	Not poor	70.8	71.9	62.3
N = 39,693				

This combination of employment decentralization and residential segregation of poor, ethnic minority and other vulnerable groups has been summarized in what has become known as the spatial mismatch hypothesis. The argument is that the higher unemployment rates associated with these groups of people are explained in large part by their relative lack of mobility, which makes it difficult (and expensive) to access decentralized jobs (Kain, 1968). Many studies have been conducted to test the spatial mismatch hypothesis, and although the results remained mixed, there is still something of a consensus that it does help explain labour market patterns in many cities (Holzer, 1991; Houston, 2001; Ihlandfeldt & Sjoquist, 1998; Kain, 1992).

Despite the range of authors adopting normative viewpoints against urban deconcentration, there are voices that point to the positive aspects of decentralized land use patterns. Some have made the rather obvious yet important observation that deconcentrated urban forms developed in response to clear market demand for particular residential and commercial environments (Gordon & Richardson, 1996). Others identify larger trends in information and communications technology (ICT), preferences for low density and natural amenities, or the decreased value of agglomeration economies over time (Kotkin, 2000; Lang, 2003). In reality, development in many cities is a continuing mix of both inner-city redensification and continued suburban and exurbanization, although some of the most recent research has begun to focus again on the potential benefits of increased density and agglomeration (Graham, 2006).

In addition, feminist geographers have challenged common assumptions on the costs and benefits of urban dispersal as being overtly 'masculine' in their origin (Dobbs, 2005; Law, 1999; see also Hall, 2004a). This is because the traditional urban structure of discrete land use blocks connected by a transport system privileging the journey to work is argued to be less relevant to the needs of women than to those of men. For example, while suburbs can be argued to have developed as refuges for the male breadwinner from the hustle and bustle of the masculine city centre workplace, the suburban home is also the non-paid workplace of many women. Many city transport systems might provide good radial access between city centre and suburbs, thus meeting the needs of men, but they cater (much) less well for shorter and non-radial trips that women are more likely to make as part of their complex mix of work, childcare and other activities (Watson, 1999). Thus despite growing female labour market participation and a declining (although continuing) gender gap in personal mobility, the ability to drive cars and access to them, the question of 'whose city?' transport policies are intended to sustain remains very real (Hjorthol 2000; Knowles, 2006a; Polk 2004).

Tackling congestion

It is now over 40 years since two seminal documents were published by the UK government, crystallizing the debate about the impact of road traffic in that country and – in time – much of Europe and beyond. The first, *Traffic in Towns*, better known as the 'Buchanan Report' (Ministry of Transport, 1963), made the simple yet important observation that severe congestion was the inevitable outcome of the failure to deliver a greatly increased supply of road space to match the voracious appetite for car travel. Buchanan laid out the policy choice in stark terms: governments could either try to 'predict and provide' – that is, build sufficient new roadspace to deal with the forecast increase in car traffic – or find alternatives to unrestricted car-based mobility. In many cities, with their compact, historic form and high amenity values, predict and provide was never a serious option and so the total amount of road space in the city remained very limited in many instances.

The second document was the Smeed Report on road pricing issued a year later (Ministry of Transport, 1964). Smeed applied detailed technical modelling to rearticulate the philosophical dilemma underlying road use, which had been apparent (at least in theoretical terms) since Schumpeter's seminal work on welfare economics and the notion of the 'public good' was applied to transport problems in the early twentieth century (Docherty *et al.*, 2004; Pigou, 1932; Schumpeter, 1909). Smeed's argument can be summarized thus: each driver, by using roads in congested conditions such as peak times, imposes delays on everybody else using the road, and these delays have costs to others not paid by the individual drivers themselves. Therefore many journeys are made for which the marginal benefits to each driver are less than the marginal costs to society. Multiplied across the large number of such journeys, this means that very significant resources are wasted, primarily through congestion, but also through secondary factors such as diminished economic productivity in the city. Or as it has been elegantly summarized, traffic congestion is a daily battle between individual liberty and the wider common good in which neither wins (Goodwin, 1999).

One solution to this problem is congestion pricing: imposing a charge on each trip equivalent to the additional delay costs imposed on other travellers (Chapter 3). Such a charge would cause some travellers (those for whom the marginal benefits are less than the charge) to choose other modes or times, or to forego the trip (Santos

& Bhakar, 2006). Congestion pricing has many benefits: (i) it reduces total costs to society; (ii) it improves travel conditions on the road for everyone, including bus users; and (iii) it generates large revenue streams that could be used for investment in public transport alternatives. This in theory encourages further modal shift away from the car, improving economic efficiency by reducing congestion further still. At the same time, social inclusion is improved by providing better transport options for disadvantaged groups with low levels of car ownership:

> since further extension of the road infrastructure to meet growing demand for car use is not everywhere possible for urban planning and financial reasons, nor desirable from environmental, energy and often social policy standpoints, the only remaining transport policy option is to swing modal split in favour of public transport by investment and/or pricing policy measures. (Organisation for Economic Cooperation and Development (OECD), 1979: 149)

As well as the problems of congestion and car dependence, transport infrastructure itself can blight certain communities. For example, in the USA an unintended consequence of State and Interstate highways planned and built to connect cities was that these highways provided access to cheap land within reasonable commuting distance of employment concentrations. Not only did such roads sometimes physically split coherent communities, they also depleted the general amenity and aesthetic value of many cityscapes. Furthermore, the majority of the new commuters on these urban motorways and freeways were travelling from middle-class suburbs to white-collar jobs in city centres, and so the benefits went to one section of society while the costs fell on another. In some places, it even became a recognised planning objective to build further new roads to eliminate the 'blighted' neighbourhoods created by earlier road development schemes.

Transport in the contemporary city

It is widely believed that transport development plays a vital role in enhancing economic growth and competitiveness. Cities compete with one another for economic growth, and reducing transport cost is one strategy for becoming more competitive, although recent research on firm location suggests that many factors are important. In addition to the traditional criteria such as labour force quality, land price and availability, and regulatory environment, quality of life factors such as local amenities, arts and entertainment and a diverse population are significant for the so-called 'creative class' (Florida, 2005). The challenge for cities in the era of the competitiveness paradigm is therefore commonly framed in terms of improving their 'asset offer' so that they become more attractive places for people to live, work and invest. In other words, the attractiveness of the city is determined by the quality of its business environment and the quality of life it offers its citizens, combined with the standard of its environment. Places with the 'right' mix of these assets are the most competitive cities, which are most likely to attract more individuals and firms to locate there (Begg, 2001).

In recent years, substantial research into the importance of transport as one of these critical assets has been undertaken. In the mid-1990s, the UK government investigated the evidence for how transport investment promoted economic growth, noting that there are a number of important mechanisms through which transport improvements could, in principle, improve economic performance. These include: reorganization or rationalization of production, distribution and land use; extension of labour market catchment areas; increases in output resulting from lower costs of production; stimulation of inward investment; unlocking previously inaccessible sites for development; and a 'catalytic' effect whereby triggering growth through the elimination of a significant transport constraint unlocks further growth (SACTRA, 1994).

This corresponds with Banister and Berechman's (2000) assertion that cities with poor-quality transport are at a competitive disadvantage when compared with those with high-quality transport infrastructure. This is especially true for those cities where the existing level of infrastructure provision was poor enough to generate clear constraints on the functioning of the market, particularly congestion and unreliable journey times. Relieving problematic bottlenecks or improving critical interchanges can release constraints and start a virtuous cycle of growth, provided policy attends to the other factors necessary to maintain the city as a location attractive to outside investors (Porter and Ketels, 2004) (Box 6.1).

Box 6.1. Dublin, Republic of Ireland. Sources: Crafts & Leunig, 2005; Dublin Transportation Office, 2001).

The Irish economy experienced rapid levels of growth during the 1990s. Although growth rates are now more modest, Dublin remains near the top of the European city growth league table. There has been a significant level of debate around which economic factors explain Dublin's growth, with low corporate taxation levels, stable public finances, the adoption of a partnership approach with industry and policies designed to foster competition being highlighted.

Notably, the provision of physical transport infrastructure is not widely seen as having played a lead role in Dublin's and Ireland's spectacular economic growth, with taxation and education policies regarded as more important. But the lack of a reliable, modern transport infrastructure in Dublin is now regarded by both government and private sector as a serious impediment to *future* economic growth. This is because congestion is both restricting the mobility of people and goods, and exerting an increasing negative impact on quality of life in the city.

The Irish government, through the Dublin Transportation Office (DTO), is therefore pursuing an ambitious programme of transport infrastructure construction in Dublin over the next decade, worth around €20 billion. Its strategy for 2000–2016, *A Platform for Change*, puts transport in the context of the broader objectives for the city and region. The starting point for the development of the DTO Strategy was to ask the question: 'what type of city and region do we wish to live, work and relax in?' In broad terms, the Vision for Dublin, sees the Greater Dublin Area as:

- A City and Region which embraces the principles of sustainability.
- Encompassing a leading European City, proud of its heritage and looking to the future.
- Having at its heart the National Capital, seat of government and national centres of excellence.
- A strong, competitive, dynamic and sustainable Region.
- A Living City and Region, on a human scale, accessible to all and providing a good quality of life for its citizens.

Key infrastructure improvements proposed in the plan include:

- The completion and upgrading of the Dublin 'C-Ring' motorway (M50) and other roads and traffic management improvements, including the construction of the Dublin Port Tunnel to give a direct motorway link between key commercial areas on the waterfront, Dublin Airport and the national motorway system (€5.8 bn);
- Upgrading of the DART electric commuter rail service in the Dublin area to provide increased capacity and frequency (€5.6 bn);
- The LUAS light rail system (5 on street lines) and Dublin Metro (three fully segregated lines including a link to Dublin Airport) to provide an all-new, high-quality fixed public transport system for the city (€7.2 bn).

Many cities have therefore invested significantly in their transport systems over recent decades in the pursuit of improved competitiveness. In the USA, the primary purpose of investment in new public transport systems from the 1960s to the 1980s was inner-city revitalization and economic development, although many of these were not successful. Equally, the re-emergence in Europe from about the 1970s of large-scale urban public transport investment in fixed-track systems was a response to increasing problems of congestion in city centres. But many more recent schemes focus more on the potential wider benefits for city competitiveness and quality of life that can be achieved by directing expenditure to schemes that make the city a more attractive place in which to work, live and invest. In other words, the belief that transport infrastructure investment brings substantial competitiveness benefits is already commonplace across European cities, with some American cities such as Denver and Portland making renewed attempts to regenerate their downtown areas through public transport investment.

Excellent transport infrastructure is not in itself enough to guarantee economic success, however: there are many examples of cities that have invested significantly in their transport systems, but where the economy has failed to perform as well as locations that have invested less in their transport infrastructure. Economic success depends on numerous factors, many of which are outside the control of individual cities. Moreover, the relationship between transport infrastructure investment and economic growth is highly complex and therefore difficult to predict (Giuliano, 2004).

Balancing economic, environmental and social demands in the twenty-first century city

The key challenge facing policy makers at the start of the twenty-first century as they pursue improved competitiveness remains how to address the insatiable demand for more road traffic growth. Rising traffic levels create environmental problems at both the urban and global scale, damaging local air quality in cities and contributing to climate change through carbon emissions. In addition, congestion is now a very significant constraint on the functioning of the economy, a barrier to future growth, and a constraint on quality of life in many cities. Equally, the renewed drive towards minimizing social exclusion – especially in Europe – suggests renewed attention on public transport, and its role in ensuring accessibility to jobs and services for people without a car. Further, these two dimensions must be balanced against important questions about how the impact of our urban transport systems on the environment at the global scale will intensify if current measures designed to target climate change are seen to be insufficient (see Chapter 3). As a result, many cities are looking seriously at car restraint and the strongly pro-public transport policies of the postwar period in an attempt to turn the clock back to an earlier time. A notable example of this is the proposal by authorities in Edinburgh to apply the principles of the Copenhagen Finger Plan to their city in an attempt to channel growth along high-density rail transport corridors (City of Edinburgh Council, 2005; see also Knowles (2006a) for a review of some of the more resilient models of urban growth and their relationship to transport).

In short, although it is played out in different contexts in different locations with varying attitudes to the future of the ubiquity of the car, one question dominates the future of urban transport geographies. How do we maintain the very real economic and social benefits of mobility, whilst solving the equally apparent problems caused by mobility? Part of the answer to this policy conundrum is undoubtedly to rediscover and renew the concept of accessibility as the key to urban transport, encouraging land use patterns that make more activities available within a given area, and in so doing promote public transport solutions and non-motorized travel such as walking and cycling. In large part this explains the recent emergence of the concept of connectivity, which seeks to emphasize the importance of the density of economic activity as a means to focus finite motorized mobility resources on those trips and societal processes that most benefit from it.

Several examples demonstrate that this is (at least in part) readily achievable, with the revitalization of many British and continental European cities through determined city centre regeneration and repopulation, and the development of high-density new towns such as those in The Netherlands. In the USA, changing land use patterns and increasing public transport use is far more difficult. With the notable exception of Portland (Box 6.2), 'New Urbanist' development has few

Box 6.2. Portland, USA. Sources: Bae, 2001; City of Portland, 2004, 2006; Cox, 2001; Jun, 2004; Patterson, 1999; Nelson & Moore, 1993; Richardson & Gordon, 2001; Staley *et al.*, 1999.

Portland has long been a model for champions of normative planning ideals such as Smart Growth, New Urbanism, and New Regionalism. Deeply rooted in environmental protectionism, Portland has implemented several policy interventions aimed at reducing urban sprawl and promoting compact urban form, increasing the use of public transport, and achieving a strong and vibrant city core. Perhaps the most widely discussed and debated policy intervention in Portland has been its Urban Growth Boundary (UGB).

In 1972, the City of Portland began developing a Downtown Plan designed to revitalize the city centre through coordinated land use and transportation policies. The plan included strong landscape and urban design elements, supported by investments in local infrastructure, and extensive housing rehabilitation in older neighbourhoods. The Downtown Plan was supported by a decision of the Oregon State Legislature a year later to mandate all city and county governments to prepare and adopt comprehensive plans and land use regulations including a UGB that contained enough land to meet the region's need for the next 20 years. Key policy tools applied include rigorous land use zoning, free travel on all public transport trips that start or end in the central 'Transit Mall' area, and a restriction on the number of parking spaces permitted for new development, based on type of development and proximity to transit, but no minimum requirement.

Studies on the results of the Portland strategy are mixed, the city's experience being a good example of the tension between European-style top-down planning interventions and strong North American free market forces. Some argue that the UGB has indeed reduced urban sprawl and increased density, despite the fact that urban densities remain below target levels. Others argue that Portland's development trends are not very different from other American metropolitan areas such as Los Angeles, Atlanta, or trends in general in all other comparable metropolitan areas. One claim is that the major effect of Portland's UGB has been displacement of growth to adjacent Clark County in Washington State, which has lower house prices and local taxes. That is, the UGB has actually supported urban deconcentration and more car-based long-distance commuting.

claimed successes, although this has not deterred some city governments – even in highly car-dependent cities such as Atlanta and Dallas – from investing heavily in new fixed public transport systems.

Improving social equity in cities will also require a sophisticated combination of policies and investments. Inner urban revitalization, mixed use development, and densification tend to make economic activities and services more accessible to people without the need for motorized transport, with this kind of planning also likely to capture the greatest possible modal share for public transport services. More and better public transport services increase mobility for those who do not have access to cars, but also offer an attractive option for those who do have a choice of how to travel. In this way, overall urban social equity in the city can be maximized through the provision of a transport system that enables easier access to the full range of personal needs. This is important because,

> Policy makers in all societies have values or objectives that inform the making of social judgements and hence guide the making of social decisions. For most societies, these are likely to include attaining an efficient use of scarce resources *and* the promotion of an equitable or just distribution of those resources. (Le Grand, 1991: 1; emphasis added)

Indeed, Buchan (1992) includes the promotion of transport equity as a key aspiration within a comprehensive statement of scarce objectives for the enhancement of urban quality of life, which also incorporates a range of economic and environmental objectives (Table 6.4).

Finally, the question of how to support the third 'pillar' of urban transport sustainability – the critically important aspect of the environment – is perhaps the most challenging dilemma facing transport engineers and planners. Improved technology, such as 'ecocars' running on hydrogen fuel cells (Banister, 2000), would help enormously, if only to create their own form of 'clean' congestion. Shorter-term 'technological fixes' to reduce the environmental impact of vehicles include more fuel-efficient engines, hybrid cars, and stricter emissions standards for private cars, as have been introduced by the State of California and the European Commission. As discussed in Chapter 3, pricing instruments such as higher parking charges and restrictions, additional taxes (especially on the largest private vehicles), and urban congestion pricing hold particular promise, as illustrated by several parking charge studies in the USA (Shoup, 2005), as well as the experience of road pricing in Singapore, London and Norway (Goh, 2002; Ieromonachou *et al.*, 2006; Larsen, 1995; Richardson *et al.*, 2004; Santos and Bhakar, 2006; Transport for London, 2006). But with the increasing dominance of environmental considerations across the wider discourse of how our transport systems – and related infrastructures such as energy supply – should adapt to the real and potential threats of climate change, it is likely that decisions on how transport should respond will be taken at the larger spatial scales over which these processes operate, and at which national and supra-national governmental institutions can act.

Table 6.4. Quality of life objectives for transport. Source: Buchan, 1992.

Accessibility
- To encourage and provide a transport system which will give people access to workplaces, to shops and public buildings, to industry and commerce, to facilities like doctors' surgeries, to centres for recreation and entertainment, to other goods and services, and to one another;
- To co-ordinate transport planning with land use and economic development planning with the aim of minimising the overall need to travel.

Environment
- To protect and enhance the quality of the environment as an objective in its own right and not merely to minimise the damage resulting from transport developments;
- To set quality standards which should apply throughout and not just in certain areas;
- To ensure that transport policies contribute towards reducing environmental damage nationally and worldwide;
- To set clear constraints which prevent the destruction of irreplaceable environmental assets.

Economic Development
- To create patterns of transport infrastructure which support sustainable economic development at the local and national level;
- To encourage research, innovation and technological process both in manufacturing for, and in the operation of, transport industries;
- To regulate the transport industry itself to provide reasonable pay and conditions and fulfilment at work.

Fairness and Choice
- To improve freedom of choice of destination and mode for everyone, tackling the inequalities that currently exist;
- To ensure that a major part of benefits from the design and operation of the transport system are distributed to those who are most in need of them.

Safety and Security
- To reduce the risks and fear of personal injury, assault and harassment on all modes of transport (including walking);
- To reduce the number, risk and level of severity of road traffic accidents for drivers, passengers, pedestrians and cyclists.

Energy and Efficiency
- To meet the accessibility requirements of residents, visitors, industry and commerce at the lowest resource cost;
- To minimise consumption of non-renewable sources of energy;
- To reduce congestion, and encourage transport efficiency.

Accountability
- To give people an unequivocal right to participate in the transport planning process;
- To establish mechanisms for people to exert influence through local democratic means over the decision-making process for transport schemes;
- To set up mechanisms by which users can directly influence the quality of service offered by the transport providers.

Flexibility
- To make the system responsive to changes in the external constraints operating on the transport system, and to new understanding about its impacts.

Notes

1 This is not, of course, to exclude genuinely local impacts – such as air pollution – which are often addressed by actions at the urban scale.
2 The I-95 freeway through central Boston was relocated underground in the multi-billion dollar 'Big Dig' project in the 1990s and early 2000s to eliminate the severe dislocation of the waterfront from downtown caused by the earlier road.

7 | Geographies of Rural Transport

David Gray, John Farrington and Andreas Kagermeier

In this chapter we consider geographies of rural transport. Planning and providing transport in rural areas present particular challenges, a function of often large geographic areas and low population densities. While many of the transport themes discussed elsewhere in this book (car use, accessibility and mobility constraints) are present in the countryside, they often manifest themselves in different ways. Issues and problems can be magnified or discounted in the rural context, journey makers are often preoccupied by uniquely rural constraints, while policy makers face different challenges and draw upon different solutions compared to those working in large towns and cities. And although we discuss transport in the rural context, we also demonstrate that a simple urban–rural distinction is itself unhelpful; in considering transport – and rural society and economy more generally – it is important to consider the 'rural' as an aggregate of heterogeneous 'rurals'. In so doing, we draw upon a typology of rural areas which groups localities by geographical and transport characteristics.

The motor car has a central place in rural life, not only as a means of transport, but also culturally and as a focal point for political dissension. We argue that it is first necessary to understand the role of the car in modern rural life before we can understand issues such as mobility, access, inclusion and sustainability and the policies required to underpin them. People living in rural communities are often limited in regard to how, where and when they can travel, and we consider mobility and accessibility in terms of opportunity and constraint, identifying social groups that are especially vulnerable. We finish by considering the range of interventions available to rural transport policy makers, including demand-responsive transport, accessibility approaches, capacity building and virtual mobility, as well as highlighting the continued importance of informal lift giving.

Different rurals

The recognition that areas are 'rural' rather than 'urban' is an important theme in geography (e.g. Philip & Shucksmith, 2003; Woods, 2005) and the definition of what is meant by 'rural' is a contested issue which has produced an extensive literature of its own (see Cloke *et al.*, 2005). The previous chapter noted that economic, social and cultural activities are concentrated in cities. In spatial terms the opposite is true in the countryside where people, activity and interaction are dispersed. Rurality can consequently be defined by some combination of low population density and distance from concentrations of urban activity. In contrast to large towns and cities, these localities are less able to support a wide range of services (including health, education, retail, other high-order services and public transport). Chapter 6 also emphasized the diversity of activity *within* cities, as places. Rural geographers are also concerned with economic, social and cultural diversity, but it is the differences (and similarities) *between* areas (which help define and characterize rural 'communities of place') that are the prime concern. To this end, it is first of all useful to identify a meaningful classification of rurality.

One approach to defining rural areas – at least in transport terms – is to use journey to work areas to indicate urban functionality, and to take rural areas as existing in the periphery beyond. Such an approach can be useful in recognising the functional and economic links between rural and urban areas. Nevertheless, a simple travel to work definition quickly runs into difficulties. In Southern Germany, for example, almost the entire area would be incorporated in journey to work areas, including large areas of agriculture, woodland and other pockets of 'countryside'. Similarly in Southern England, most of the area would be defined as urban using this method, and yet large parts of the London commuter belt – Kent, Suffolk, Surrey, etc. – are regarded firmly as rural in character.

Systems of definition that recognise degrees of rurality are therefore useful. Not only do they accord with general perceptions that some areas are more truly (or 'deeply') rural than others, but they reflect some of the constraints affecting the provision of transport and other services. A slightly more sophisticated classification of rurality is thus employed in the Scottish Household Survey (Scottish Executive, 2006b), which uses a sixfold classification of urban and rural areas: large urban areas, other urban areas, accessible small towns, remote small towns, accessible rural areas and remote rural areas. This type of classification distinguishes between urban and rural areas, while acknowledging the relationship between urban centre and rural hinterland and the hierarchy governing urban centres of different sizes. While it might be unfashionable for geographers to cite Christaller's (1933) seminal work on Central Place Theory, the fact that settlements of different size have different spheres of influence and can or cannot support services of differing market thresholds is pertinent for those looking at both the spatial coverage and viability of public transport networks, and wider access to a range of services (Christaller 1933, 1966) (Figure 2.1).

The hinterlands surrounding major centres of population and services can often support 'thick' conventional bus services. Consequently, those without a car are not unduly constrained in their ability to access jobs, services, education and training, health care, recreation and social networks. A more remote locality not served by a sizeable service centre presents greater challenges. Even if there is sufficient demand for a bus service, services are often infrequent (especially in the evening and at weekends) and journey distances and travel times are lengthy, limiting the extent to which they can be used for travelling to work or for off-peak travel. In

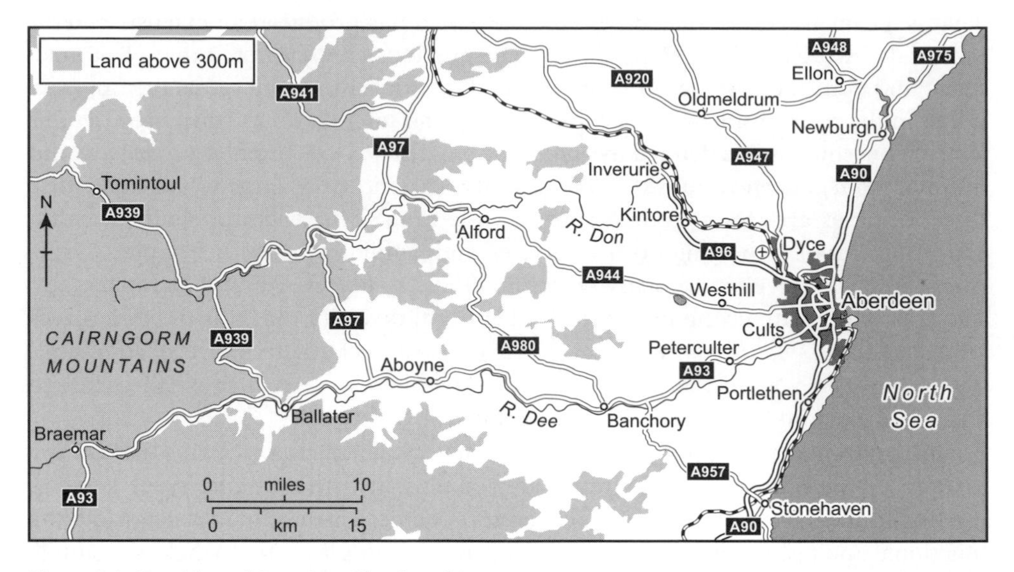

Figure 7.1. Deeside and Donside, Aberdeenshire.

the most remote communities, there is often insufficient demand to support any service at all.

Space, place and population distribution rarely conform to neat geometric models, however, either geographically or over time. The Scottish Household Survey classification is similarly limited. Take Aberdeenshire in Northern Scotland, for example; specifically, the countryside around the River Dee and the River Don (Figure 7.1). Much of the upper Dee valley is hemmed in by mountains, so the rural population is largely nucleated into discrete settlements located along the main road into Aberdeen. The road runs through Ballater, Banchory and Aboyne, key secondary service centres, and there are no other competing urban centres of any significance nearby. Because these key nodes are connected along one route and the majority of households are situated relatively close to the road, the rural population in this area is well served by an hourly bus service. In contrast, while there is also an hourly bus service between Aberdeen and Alford, the main secondary service centre for upper Donside, the rural population is dispersed across a wider area and a far lower proportion of households is located within walking distance of the bus route. Inverurie also competes with Aberdeen and Alford as a destination of choice. Thus, while Deeside and Donside are only several miles apart, their transport geographies are quite different.

Any rural definition must recognise geographical and functional differences (in terms of land use, the local economy, employment, recreation and tourism) between rural areas. It is also important to acknowledge that the extent to which transport is regarded as a key social concern, and the nature of the dominant rural transport issue, can take quite different forms in different places. For some car owners in proximate suburbs, transport 'problems' equate to rising traffic levels, lengthening journey times and an awareness that fuel prices can increase from time to time. In some market towns it may be escalating congestion, coupled with lack of parking and social severance (as people on the wrong side of a busy road find themselves cut off from the rest of the town), that are the pressing concerns. In other towns it is the *lack* of traffic that is causing problems, as consumers in their historical hinterland take advantage of road improvements or a bypass to miss out the market town in favour of shopping in an edge of town shopping development at a larger urban centre. A good example of this process is Dingwall in the Scottish Highlands, where the high street is in crisis, local shops having lost out to a large out of town development 14 miles (23 km) away in Inverness.

For others without regular access to a car and without public transport, a lack of alternatives to the car and rising motoring costs present fundamental challenges to their ability to participate fully in society. In remote localities, these can be salient and highly politicized issues, key concerns that are perceived to be intrinsically linked with the viability of communities themselves. In short, while rural transport is of little concern for some, for others it is *the* main local issue, a focal point that plays a part in defining and sustaining a sense of individual and community identity (Farrington *et al.*, 1998).

In taking account of the need to consider the heterogeneity of the rural transport 'experience' and the limitations of other rural/urban classifications, Gray (2004) developed a typology comprising eight rural transport types based on population density, geographical conditions, dominant transport concerns and the relationship between transport and social exclusion (Table 7.1). This typology comprises two peri-urban areas (A1 and A2), three market town/hinterland types (B1, B2 and B3) – where the main preoccupation is local accessibility – and three remote rural types (C1, C2 and C3). In one of these 'C' types, a rural tourist 'honey-pot', policy makers not only have to contend with many of the problems of isolation, but also with unsustainable levels of traffic at peak times (Chapter 12). The remaining two are characterized by the classic problems of remoteness: car dependence, lack of access to services, shops and health care and high fuel costs. In these communities, there is a fundamental relationship between transport constraint and social exclusion (Chapter 4).

While it is difficult to estimate, somewhere around 90 per cent of the rural population is contained in rural typologies A1 to B3, although the problems associated with the remote areas (C1 to C3) often dominate debates on rural transport. It is also the case that all transport issues are probably experienced within each typology area to some degree (and that some areas can fall into more than one category). Nevertheless, different issues dominate in different localities and 'the most salient transport problems and concerns facing a community shape the views of the public, politicians, and local media, mediating how rural transport policy is received and contested at local level' (Gray, 2004: 180).

The car and modern rural life

Western societies are increasingly dependent on the car (ESRC Transport Unit, 1995) and before going on to consider issues such as access and mobility in more

Table 7.1. Rural transport typology. Source: Gray, 2004.

Type	*Characteristics*	*Car ownership*	*Households within 13 mins. of hourly bus service*	*Proportion of journeys made by car*	*Key transport issues include*	*Contribution of transport to social exclusion*
A1*	Rural areas close to a major conurbation (for example)	75–85%	Up to 80%	65–75%	■ Increasing traffic levels ■ Social severance ■ Deteriorating condition of non-trunk road network ■ Road safety and congestion ■ Local impact of park and ride schemes	Partial/secondary
A2	rural locality surrounding a freestanding city					
B1	Retail and service provision of a smaller market town is increasingly overshadowed by a larger urban centre.	75–85%	Up to 60%	65–75%	■ Decline of retail and service provision in market town and smaller villages ■ Increased bypass traffic on minor roads and trunk routes ■ Falling demand for non-trunk public transport services ■ Bus services do not go where people want to travel ■ Accessibility of households removed from main public transport routes ■ Cost of public transport fares and subsidies on non-trunk routes	Partial/significant
B2	Market town servicing a dispersed rural population, making effective public transport difficult to provide	75–85%	Up to 50%	70–75%	■ Increasing traffic levels and lack of parking in town centre ■ Decline of local shop, service and health provision in smaller villages ■ Coverage and frequency of bus network ■ Falling demand for existing bus services ■ Increasing cost of public transport fares and subsidies ■ Accessibility of households removed from main public transport routes ■ Low income; proportion of weekly expenditure accounted for by transport	Significant

B3	Market town servicing a rural population dispersed in a linear fashion along main routes (e.g a valley) making public transport easier to provide.	75–85%	Up to 60%	70–75%	■ Increasing traffic levels and lack of parking in town centre ■ Decline in local shop, service and health provision in smaller villages ■ Falling demand for off peak public transport. ■ Increasing cost of public transport fares and subsidies ■ Low income; proportion of weekly expenditure accounted for by transport	Significant
C1	A remote 'honey pot' or 'tourist' location. Problems of isolation combined with rising traffic levels and congestion	80–90%	Under 35%	70–80%	■ Increasing traffic levels, congestion and parking difficulties, especially at tourist spots ■ Deteriorating condition of non-trunk road network ■ Decline of shop, service provision outside tourist areas ■ Non-tourist bus services; Lack of demand & increasing cost of fares & subsidies ■ High fuel prices ■ Low income; proportion of weekly expenditure accounted for by transport ■ Accessibility & mobility of non-car owners	Substantial
C2	An isolated village or villages	85–95%	Under 20%	75–85%	■ Decline or absence of local shop, service and health provision in smaller villages ■ Frequency of public transport service; lack of demand for existing services ■ High fuel prices and proximity of filling stations ■ High cost of public transport fares and subsidies ■ Low income; proportion of weekly expenditure accounted for by transport ■ Accessibility & mobility of non-car owners	Substantial
C3†	Extremely isolated settlement or households well removed from main roads and/ or bus routes	90–100%	0%	80–90%	■ Absence of local shop, service and health provision ■ Absence of alternative to the car ■ High fuel prices and absence of filling stations ■ Lack of demand for public transport ■ High journey distance and cost ■ Low income; proportion of weekly expenditure accounted for by transport ■ Accessibility and mobility of all occupants	Fundamental

Note:
*Example area: Rural Kent – relatively close to Greater London.
†Example area: Northwest Sutherland in the Scottish Highlands.

detail, it is important to consider the role of the car in the countryside. It is difficult to overstate the importance of the motor car in shaping and underpinning rural life in the early twenty-first century. The car has changed how rural communities function, created new patterns of migration, and transformed the way those living in the countryside lead their lives.

The car has allowed people to live ever further away from where they work, and has enabled retirees and other migrants to move to remoter rural areas in search of a better quality of life, facilitating a reversal of the centuries-old drift of people from the countryside to towns and cities. While Western society as a whole has become increasingly reliant on the car, it is often suggested that rural life depends on it. Certainly, in the 'C type' rural areas outlined above, access to a car is virtually a necessity. Rural car ownership rates are typically significantly higher than those in urban areas. In the most rural parts the UK and Western Europe, over 80 per cent of households own at least one car. Car ownership is even higher in Australia and the USA, where population densities are lower and economic conditions favourable. Around 94 per cent of rural households own at least one car (Nutley 1996, 2003), and in some parts of the USA ownership rates are approaching one per head of population. Rapidly increasing car ownership rates are expected in Eastern Europe, along with rural depopulation, as those countries' economies adjust to EU membership.

On the whole, the car has had a positive impact in the countryside, as it has bestowed opportunity and choice to rural households. Nevertheless, in collapsing space and time, mass car ownership has promoted wider societal changes in the way that people live and consume (Knowles, 2006a). The built environment of western countries has evolved to cater for a more *automobile* society, in turn shaping norms, expectations and patterns of behaviour. The car-owning rural 'majority' are now much less likely to work, shop, access services and even socialize locally than rural populations of fifty years ago, when rural dwellers had more limited spheres of activity, and the social and economic fabric of the countryside has been altered as a result. Rural shops, services, banking and health care are in decline. Car dependence coupled with other centralizing tendencies in the provision of schools, health care and benefits, have resulted in a continuing withdrawal of services often into fewer, larger units such as supermarkets, large secondary schools, and large hospitals (Boardman, 1998; Gray *et al.*, 2001; Knowles, 2006a).

Expanding spheres of activity have also undermined the social ties and networks – the social capital – that traditionally bound rural communities together (Urry 2004a; Gray *et al.*, 2006). This has been exacerbated by the loss of traditional arenas of social interaction: primary schools, banks, post offices, GPs' surgeries and public houses. Longer commuting distances, coupled with quality of life migration, have increased demand for rural housing, inflating property prices and squeezing low-income, local first-time buyers out of the market. And for many, the car is not merely a mode of transport, it is a very powerful cultural symbol. Where it is regarded as crucial to the viability of a community, its relationship with personal and cultural identity is even more fundamental and explicit. In this context, policies which raise the cost of, or threaten, car ownership are seen as a direct and personal attack on the sustainability of rural communities. In subsequently considering rural accessibility, many of the problems, constraints and barriers to progress are a direct result of the strong and ever-strengthening relationship between rural society and the car.

Accessibility and mobility

As discussed in Chapter 4, accessibility is essentially people's ability to reach certain important services, job opportunities, friends and family and recreational facilities, and increasingly it can also include the ability to access 'virtual' communities of interest via the Internet and other ICT technology. In the policy context, accessibility can be defined for a *place* (accessible, with certain conditions and constraints such as cost and travel time, to a certain number of people) or for *people* with their constrained access to places. Regarding access to services, it is assumed that people should have reasonable access to education, employment, health care, professional services, social, recreational and entertainment opportunities, and retail services (ranging from everyday convenience goods to higher-order goods such as furniture) (Farrington and Farrington, 2005). Mobility, the ability to move around between places, is an important aspect of accessibility but in the context of this discussion should be seen as a means to the broader end of making places and services 'get-at-able' (Moseley, 1979).

When considering mobility and access, it is useful to think in terms of opportunity and constraint. Members of a household living with access to at least two cars, located in a mid-sized rural town with a good range of shops and services – which is itself connected to a nearby city by good rail and bus links – will have a range of options regarding how, where, when and how far they travel. Many of their journeys can be carried out on foot or by public transport, access to a car is rarely a limiting factor, and they are unlikely to suffer from any access or mobility difficulties. In contrast, the members of a household with one and especially no cars living in a very isolated locality, are much more constrained in their travel choices. Some or all members of this second household may have significant and potentially insurmountable mobility and access difficulties.

As we have noted, the dominant trend in Western Europe and North America since the mid-twentieth century has been one of a gradual decline in the number and range of services available in rural areas. As in the example of Dingwall, rural shops cannot compete on price or variety of goods with large retail outlets which are designed to be easily accessible by car. Greater mobility has similarly undermined rural petrol stations, banks (which have also been acutely hit by the growth of internet banking) and rural post offices (again coupled with changing operational practices such as the government's withdrawal of pension payments in cash). Rationalization and cost cutting has also acted to centralize health care and educational provision (Figure 7.2).

In the UK, the Commission for Rural Communities (2005) reported a decline in the geographical availability of eight out of 10 services surveyed between 2000 and 2005, while the number of rural households located within 4 km of a bank or building society fell from 66 to 60 per cent over the same period. Similarly, the proportion

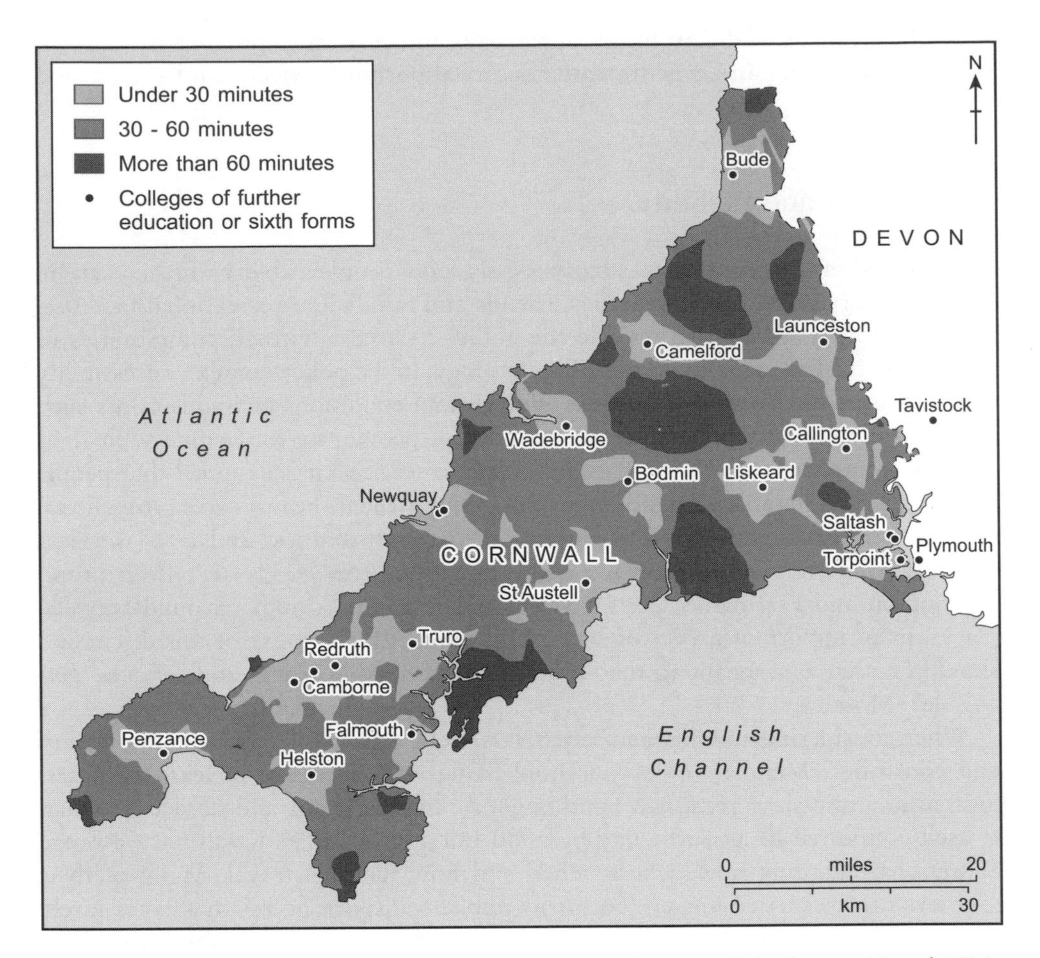

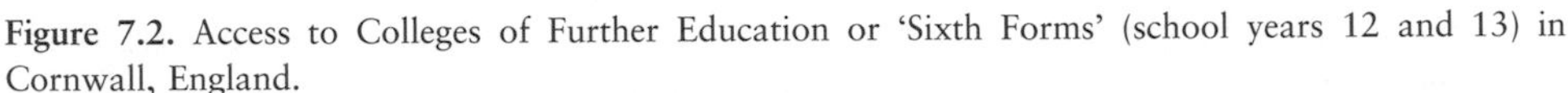
Figure 7.2. Access to Colleges of Further Education or 'Sixth Forms' (school years 12 and 13) in Cornwall, England.

living within 2 km of a post office declined from 90 to 85 per cent between 2000 and 2005 (Gray *et al.*, 2006). Until 1998, there was also a similar decline in public transport. The UK Commission for Integrated Transport (2001b) reported that nearly a fifth (18.5 per cent) of villages and towns in England with a population of up to 2,000 had a service level 'below subsistence' while the number of rural parishes without any bus service increased from 14 to 22 per cent between 1991 and 1997.

Where public transport struggles to provide an alternative, many (often those on low incomes) have little choice but to use local shops and services, where prices are invariably more expensive than in urban or out of town centres. Where such services have left a village altogether, people can have acute difficulty accessing them. Certain household members and – in aggregate – social groups are prone to experiencing constrained mobility and access, either because they are non-car owners (cannot afford to, or they have yet to learn or are too old/mobility impaired to drive) or have limited access to a household car at certain times of the day (typically because it is used by the main income-earner for the journey to work). Specifically, these groups include the elderly and disabled, young people, job seekers and those involved in child care.

Elderly and less able-bodied people – who make up the majority of rural non-car owning households – are especially vulnerable to mobility and access problems, particularly if they have difficulty in making local journeys on foot. People's needs for access to certain services, notably health care, increase with age, while the chances of retaining car-based mobility declines. This problem will potentially worsen as many western countries have an ageing rural population. Young people attend school during the school day, but often find difficulty in accessing other young people and related social activities during evenings, weekends and school holidays. Not yet able legally to drive, they aspire to greater mobility to provide this access (Kingham *et al*., 2004). School leavers and longer-term job seekers need mobility to attend job training, interviews and work itself. Public transport services do not always allow journeys to work to be made, and some employers are reluctant to employ people living in rural areas and dependent on public transport, since they perceive it as unreliable.

Those involved in child care while their partner is at work are often, but not always, women. In the past, only a minority of women held driving licences, although current trends suggest that gendering in this context may be declining as an increasing proportion of women are licence-holders. As this trend moves upwards through the age cohorts, gender equality in licence-holding is being approached. In 1975–6, 29 per cent of women held full driving licences in Great Britain, compared with 69 per cent of men. By 2002–3, 61 per cent of women and 81 per cent of men held full driving licences – still a marked difference, but a narrower gap (Hamilton *et al*., 2005). In a one-car household, however, it may still be the case that a male takes priority in car-use, so that a female is still not the primary car-user. The increasing number of women with driving licences and in paid employment is promoting the acquisition of second, and further, cars in households.

Rural transport policy

National philosophies

Different countries take quite divergent approaches to the prioritization and provision of rural transport. In general, public transport in low-density rural areas is not commercially viable. In most cases only between one-third and one-half of the costs are covered by revenue from tickets sold (Kagermeier, 2004). A major factor in determining whether the transport network can meet the needs of journey makers in rural communities is the attitude of planners, politicians and the state, and the willingness of these actors and agencies to intervene in the rural transport market to provide a higher transport service level than the market alone would support.

At one extreme, for example Australia and the USA, there is little or no state-subsidized rural public transport. With low population densities and 94 per cent of households owning a car, no rural transport 'problem' is perceived to exist (Nutley, 1996, 2003). Those without a car are to all intents and purposes politically 'invisible' (Nutley, 1996), although studies have recognised the existence of transport-deprived rural groups such as the elderly, young people below the legal driving age, the physically handicapped and people on low incomes (Briggs & McKelvey, 1975; Kidder, 1989; Nutley, 2003). At the other end of the spectrum, it is more common in some parts of Western Europe (where car ownership is lower and population densities are higher) to find a preparedness to support extensive rail, bus and ferry networks. In Switzerland and Austria – two countries characterized by alpine rural areas – providing mobility opportunities has traditionally been regarded as a central component of equity-based public policy. Even small villages and hamlets have high levels of public transport, and bus and rail transport are closely integrated.

In Germany, and particularly in the UK, networks are less comprehensive. While central resources are available to underpin the provision of rural public transport, much depends on the attitude at district or local level. In Germany, a basic network is guaranteed by public funding for school transport (€1.5 billion per annum) that can also be used by the wider public. The level of additional investment subsequently depends on individual districts and municipalities, and sharp differences in provision between districts can occur. Some districts subsidize an hourly service for most of the rural population, while in others school transport accounts for up to 80 per cent of public transport on offer while the district limits itself to financing two buses a day. In an ageing population with a declining number of school pupils, the relationship between school and public transport will have to be revisited, and there is an ongoing public debate in Germany about the extent to which mobility should be guaranteed for the entire country.

In the UK, bus services outside London and Northern Ireland are provided by companies as commercial operations in a deregulated environment, with local authorities subsidizing services that are not commercially viable. In rural areas typically all or most services are subsidized and local authorities receive support for rural transport services from a range of grant schemes offered by central and devolved government. Since 1998, grants worth £80 million a year have been paid out in England, and the proportion of UK rural households living within 13 minutes' walk of a bus stop increased by 5 per cent between 2002 and 2004 (Gray *et al.*, 2006). Any additional supported services must be funded by local authorities, however, and as in Germany some rural authorities are more willing to invest in transport than others.

Several Western European countries have similarly introduced market and privatization elements into public transport systems. The quality and coverage of public transport provision and accessibility seen in many countries has varied as a result, generally to the detriment of access in rural areas. Marell and Westin (2002), for example, demonstrate how fares rose in rural Sweden while efficiency decreased following the deregulation of taxicab services.

Public transport

The 'big bus' – the conventional bus seating 40 or more people, operating on scheduled services and routes – has traditionally been the mainstay of rural transport networks (Figure 7.3). While *most* rural networks are not commercially viable, it is entirely possible – depending on geographical circumstances – for rural communities to be well served by such services, especially in urban hinterlands where the population is aligned along a limited number of main routes. On the whole, 'rural to urban' journey makers travelling into major cities and conurbations tend to be well served (such as in the example of Deeside noted above). Since inter-urban public transport routes tend to cross rural areas, some of the latter benefit from improved services (greater frequency through longer time periods in the day/week). The increasing popularity of rail and bus Park and Ride has also enhanced access to towns and cities for those living in the neighbouring countryside.

For what can be termed 'rural to rural', or 'deep rural to urban' journey making, the provision of viable public transport represents more of a challenge. Particular problems are seen in the decline in, or absence of, early morning, evening and weekend services. Increasingly, flexible and/or demand-responsive transport (DRT) is seen as the answer in these areas. Such services have traditionally been provided by a range of agencies including health, education and social service authorities, as well as by voluntary organizations and even postal services. A review of DRT by the Scottish Executive identified four strategies for providing demand-responsive services (Table 7.2). Increasingly, DRT is being provided by community groups and also by municipal authorities. They are popular with passengers because they are more flexible and personal than conventional services and take closer account of the needs of those who use them (Brake *et al.*, 2004; Mageean and Nelson 2003), although they often require high per-passenger subsidy, at least in their early stages. Examples of successful rural DRT include the Wiltshire Wiggly Bus, which was set

Figure 7.3. Traditional rural 'big bus' service approaching Aberdeen. D. Gray.

Table 7.2. Demand-responsive transport: route options. Source: Scottish Executive 2006c.

Fixed routes	Service journey departing from an end stopping point (terminal) at prescribed times. This is effectively a regular bus route.
Semi-fixed routes	Depart from an end stopping point (terminal) at prescribed times. Stops at any fixed intermediate stopping points at prescribed times. Deviations to other stopping points upon request.
Flexible routes	Depart from an end stopping point (terminal) at prescribed times. The vehicle only calls at stopping points upon request.
Area-wide services	No fixed end or intermediate stopping points. No scheduled departure times from any stopping point. Limited by operational hours and area limit. Only calls upon request.

Figure 7.4. 'Publicar' service provided by Swiss Post. A. Kagermeier.

up in 1999 with a grant from the UK Government's Rural Bus Challenge Fund. The Wigglybus is a semi-fixed route operation used by nearly 4,000 passengers per month in south-west England comprising three circular core routes that link into conventional operation (Brake *et al.*, 2004). Other examples include the Rural LIFT community transport project serving West Cavan and Leitrim in Ireland (another semi-fixed route service) and the 'Publicar' service provided by Swiss Post, which is fully integrated into the national transport network (Figure 7.4).

Figure 7.5. Organized hitchhiking: the CARLOS project in Switzerland. A. Kagermeier.

The Rural LIFT service is a good example of the involvement of communities themselves in the provision of DRT. In Scotland, the Rural Community Transport Initiative assists community groups to apply for funding and set up schemes. Under Scottish legislation, communities can 'buy-in' specified services from transport operators, or purchase a community vehicle staffed by volunteer drivers. An even more flexible approach to providing public transport is the CARLOS project in Switzerland (CARLOS, 2005), which is essentially organized hitchhiking (Figure 7.5). In small villages, terminals are installed where a journey maker can enter a desired destination which is then shown on a display so that car drivers know where the hitchhikers want to go. For insurance purposes, hitchhikers must first register before using the service.

Accessibility-based solutions

Accessibility-based solutions essentially involve bringing services to people and/or optimizing public transport and available services through planning so as to maximize access. A well-established idea in rural accessibility is the mobilization of services by putting them into road vehicles. Mobile grocers, fishmongers, banks and bakers, for example, were once common sights in rural Britain, but the number of mobile services declined due to increasing car ownership, decreasing price competitiveness and, latterly, the cost of meeting health and safety regulations. Now only a few survive. By contrast, health care providers are increasingly looking at ways of bringing services to people to reduce patients' need to travel. Approaches being considered include giving 'practice nurses' more training and clinical responsibility to deliver care, and allocating a bigger role for pharmacists in helping patients to identify treatments for non-urgent conditions (see, for example, Kerr, 2005). In education, the Extended Schools Programme in the UK encourages schools to adopt more flexible opening hours with a view to provide child care, family, health and social care and other services to the wider community.

Another innovation in the UK is the emergence of Accessibility Auditing and Accessibility Planning. This approach is based on the recognition that people's ability to access services and opportunities is an outcome of a range of policy sectors, and that these must be 'joined up' in a holistic way to maximize access for individuals and communities (Social Exclusion Unit, 2003). Accessibility planning is now a statutory requirement for local authorities in England and Wales in the preparation of their Local Transport Plans. It is expected to shape local authorities' provision of public transport and, ultimately, their land use planning strategies. The UK Department for Transport (DfT) is working on definitions for mandatory Core Accessibility Indicators (to be supplemented by local ones), to include the fields of education, work, health care and shopping. Indicators will include the percentage of households without a car who are within 15 and 30 minutes of (i) a general medical practitioner, and (ii) a major shopping centre, by public transport (DfT, 2006c).

Capacity building

As well as maximizing accessibility, rural policy in the UK is also concerned with local capacity building, essentially encouraging and empowering local communities to make their own decisions and to take responsibility for local development, welfare, service provision and transport. This bottom-up approach relies on local knowledge of problems and local ownership of solutions. In the transport context, outcomes of this process might involve the identification of local mobility and access issues and an appropriate solution: community transport, the provision of better travel information, or the involvement of a commercial retailer in providing a transport service to a new supermarket in a market town. Communities can also tap into government funding to run local shops. The viability of schemes can be enhanced by locating different, but complementary services – such as petrol, shop and post office and even a pub – in the same premises (Department of Environment, Food and Rural Affairs (DEFRA), 2006). As well as building community capacity, such innovations also help to retain or create arenas for social interaction. Rural churches and faith groups also involve themselves in such community initiatives and are making a growing contribution in some rural areas.

ICT and virtual mobility

In recent years, academics and policy makers have been attracted to the idea of virtual mobility, which visualizes greater use of Internet and other technologies to allow interaction between people and services without incurring penalties of distance, travel time and travel cost (Banister and Stead, 2004) (Chapter 11). Household Internet access varies considerably, even in the developed world. In the USA, 55 per cent of households had Internet access in 2003 (US Census Bureau, 2005). In Europe, household Internet access in 2006 varied from 23 per cent in Greece to 80 per cent in the Netherlands, and averaged 52 per cent across the EU (Eurostat 2006b). Growth in household Internet access is generally rapid: for example, within the UK, 63 per cent of households had access in 2006, compared with 27 per cent in mid-2002 (Eurostat, 2006b; Office of National Statistics, 2002). This increasing household Internet access could be harnessed to the accessibility and social inclusion agenda by replacing the need for travel to places. Internet shopping and banking are already widespread, while health care by video consultation has been pioneered in Orkney. Moreover, a great deal of research has been carried out on the possibilities offered by teleworking at home. One example of successful teleworking is the Co-Op Travel Group's *Future Travel* call centre. The UK's largest virtual call centre employs 630 ABTA-certified travel agents, all working from home (Flexibility, 2006).

Developments in ICT and the Internet now mean that virtual *communities of interest* are often more important to people than *communities of place*. Yet with Internet access far from universal, commentators such as Skerratt and Warren (2004) argue that those who are traditionally vulnerable to access exclusion (such as the elderly) are also likely to be on the wrong side of the 'digital divide'. The increasing importance of ICT and virtual mobility might merely act to reinforce existing patterns of exclusion or act to create new ones.

The need to travel and 'informal' mobility

A contrasting view of the value of ICT in replacing the need for travel is seen in the ideas of 'new mobilities' (Sheller & Urry 2006). Urry (2002) dismisses the idea that technology can ever replace the need for travel, arguing that 'co-presence' will continue to be a fundamental requirement for maintaining social contact with family and friends and for recreation. In other words, the importance of 'being there' in person – either somewhere, with someone, or for some event – will remain undiminished, regardless of technological advance. In fact, Urry (2004a) argues that virtual social networks can eventually generate additional travel as spatially dispersed members meet up. It was noted above that a decline in services has meant that many places where people did meet face-to-face (for example, in shops and post offices) have been lost, and there is evidence that the popularity of demand-responsive community transport has as much to do with providing an arena for social interaction as it does with accessibility (Gray *et al.*, 2006). Groups of friends will book a ride at the same time in order for them to socialize. The journey is therefore as important as the services it accesses. The question of whether it is more important to maximize access or maximize mobility is emerging as a key debate among rural transport geographers.

Another emerging area of interest is the relationship between social networks (or social capital) and mobility, through informal lift giving. The primacy of the car for providing transport in rural areas is such that non-car owners make more journeys by this mode than they do by public transport (Commission for Integrated Transport, 2001b; Nutley, 2005). In other words, those without a car (either non-car owners or those who are without a vehicle for much of the day) rely more on lifts from family, friends, neighbours and others in the community than they do on buses or DRT. Rural social networks are thus crucial for providing mobility and access where public transport cannot provide an alternative. Consequently, lack of access to someone else's car is arguably more of a constraint than the absence of a bus service, and it raises the question of whether state funding of rural bus services is misplaced. Nevertheless, subsidizing the lift giver via the passenger (possibly using smart cards) would require a significant paradigm shift among policy makers, while there would also be substantial start-up costs for such an approach (Gray *et al.*, 2006).

Conclusion

The car has transformed Western rural geographies in the decades since the mid-twentieth century. It has brought great advantages to many, providing flexible mobility and high levels of accessibility to those able to use it. Declining real costs of ownership and use have brought these advantages to increasing numbers of rural households. A consequence has been continued decline in transport and other rural services, as people have preferred to use more centralized services, and as service providers have taken advantage of car-based mobility to centralize their provision. Unfortunately, those without access to car use have been far less able to take part in these changes, and have faced increasing problems in their lives. Constrained in their journey-making choices, they compensate by taking lifts in others' cars, or by getting others to interact with services for them (e.g. proxy shopping), but still experience lower levels of mobility and accessibility. Many in this group – the transport poor – locate in villages or small towns to overcome isolation and poverty of access, but if they prefer to remain in more remote rural areas, or are unable to move, they can experience isolation and social exclusion, especially when they have limited or no access to someone else's car.

A key assumption underpinning this chapter, however, is that we should not be seduced into thinking that the plight of those with limited access or mobility is necessarily the norm. Rural places – and the communities, households and individuals located within them – are heterogeneous. The concerns of those dwelling in a remote hamlet or a cottage located many miles from another house are not shared by those living in a metropolitan hinterland or a small town. Journey-making opportunities and constraints vary considerably from place to place, as do the nature and strength of feeling surrounding the resulting transport problems, and the solutions required to tackle them. In this context, we have demonstrated how rural transport policy making is evolving, highlighted in the latest policy thinking, and catalogued the wide range of approaches – public transport, accessibility, IT and even subsidizing the individual – available to assist those constrained by poor access and mobility.

In conclusion, therefore, we argue that the days of a 'one size fits all' – or, more accurately, a 'one size of bus fits all' – approach should be confined to history, and we have emphasized the need for decision makers at national, regional and local levels to consider a suite of possible transport and accessibility interventions, before identifying the solution which best fits local circumstances. In particular, we have noted the importance of the local voice and local capacities in finding interventions best suited to local geographies, societies and economies.

8 | Inter-Urban and Regional Transport

Clive Charlton and Tim Vowles

This chapter focuses on inter-urban and regional transport (IURT), which is undeniably complex and necessarily flexible, and as such eludes simple definition. IURT varies on several dimensions: distance and scale, socio-economic and political-territorial settings, transport modes and markets, and travel and trip motivations. Its bounds lie somewhat imprecisely between transport that is evidently local in terms of physical distance (commuting, intra-urban and intra-metropolitan movements and many rural journeys, typically with daily frequency) and long-range intercontinental and global movements. As such, much IURT is inherently trans-jurisdictional, not least in Europe. Transport infrastructure and many services cross intra- and international borders, so demanding administrative and political cooperation which is, understandably, sometimes difficult to achieve.

In the chapter we consider IURT in the context of the region before moving on to explore selected IURT themes and modes, namely high-speed rail (HSR) systems and Trans-European Transport Networks (TEN-T). The recent meteoric rise of Low Cost Carriers (LCCs) in the air transport sector is clearly of significance to IURT, although this issue is examined in Chapters 9 and 12. Discussion then turns to relating IURT to some current social science perspectives on transport – sustainability, social justice, mobilities and aspects of governance and regulation – since these offer geographers a rich enhancement of the tools at their disposal for analysing transport issues. A short conclusion brings the chapter to a close.

Transport and contested regions

Geographers have long debated the validity of simple notions of scale, especially assumptions about hierarchical and discrete categorization from the local, via regional and national to global (see, for example, Herod, 2003; Shepherd and McMaster, 2004). Understandings of 'the region' and 'regional' are ambiguous and contested, not only among Anglo-Saxon geographers, but also within the long-established French tradition (Claval, 1998). Whereas regions were interpreted as bounded territorial units, as depicted for example in the 'uniform' and 'functional' regions identified by Taaffe and Gauthier (1973), and persist as distinct administrative and political units, the notion of 'coherent regional economies is at best complex and problematic and at worst, downright misleading' (Cloke *et al.*, 2004: 49). More recent observers interpret both cities and regions within a network society of spaces of flows rather than spaces of places (Castells, 1996; Taylor, 2004). Thus movements and mobilities define space rather than static locations. Regions may be understood as intricate sets of connections, focused through urban nodes and often acting over long distances. They are also unbounded, discontinuous, and internally diverse, rather than homogeneous (Allen *et al.*, 1998). Reductions in the generalized costs of transport, in conjunction with enhanced communications technologies, have stretched these networks and disrupted the coherence and internal consistency of regions and cities.

In this context, inter-urban and regional transport might be taken as referring, typically, to journeys and trips of between 100 and 800 km, and in many situations will refer to domestic movements within countries. At the upper end of the spectrum, such bounds are clearly problematic, for example in the cases of North America, Australia and Russia, where many domestic flows exceed 2,000 km (although the United States (Bureau of Transportation Statistics, 2006b) classifies a long-distance trip as more than 50 miles from home to the furthest destination). Where political geographies are more fragmented, such as in Europe or Central America, IURT includes much short-range international transport.

A further complexity arises in that infrastructure and vehicles often form part of multi-scalar transport systems. There is often overlap between IURT and other scales of transport, not only in cities, but also in more peripheral locations. Most major European rail termini and their approaches handle both long-distance and suburban traffic, and similarly, limited access highways and orbital ring roads in major urban centres see a constant blending of very local traffic with freight and passenger trips travelling much longer distances. Boeing 777s and 747s from New York or Sydney sequentially share approach paths into London Heathrow with B737s and A320s from Amsterdam or Edinburgh. Such overlap between inter-urban and more local transport sub-systems could be viewed in some cases as advantageous in maximizing the utilization of infrastructure and facilitating interchange, but may often present problems of overcrowding and congestion on urban network links and at terminals. The latter point is confirmed in a recent report to the UK government on its overcrowded and ageing transport network (Eddington, 2006: 46):

> The most significant long-term problems for UK inter-urban travel . . . are the rising levels of road congestion and overcrowding on the rail network, particularly where inter-urban and commuting journeys compete to use the same networks. On the railways . . . [it] is principally on the approaches to major urban areas, where significant commuter flows compete for line space with inter-urban flows, that congestion and overcrowding is [sic] found.

Finally, within many individual trips, the identity of IURT cannot be isolated. The majority of journeys are inevitably multi-stage, and only partially 'inter-urban' or 'regional'. Most IURT trips involving rail or air are also necessarily multi-modal, with the initial and/or final stages by private car, taxi, bus, cycling or walking, with the great majority by car (Wolmar, 2006). The trajectories of many freight shipments are likely to be similarly multi-stage, with many consumer goods moved by 'local' and 'global' transport systems as well as 'mid-scale' IURT.

Increasing mobility and trip distance

The expansion of longer-distance passenger and freight transport reflects the impact of technological and organizational innovations in reducing generalized transport costs, changes in society and the development of increasingly internationalized economic systems. Recent increases in individual mobility have been expressed more in greater distances travelled rather than additional time spent travelling or a higher frequency of trips. Schafer and Victor (2000) show that time and money budgets are essentially stable over space and time, and mobility rises in approximate proportion to income. According to the UK *National Travel Survey 2005* (DfT, 2006d), the average annual distance travelled by individuals rose from 4,476 miles in 1972/73 to 7,208 miles in 2005 – an increase of around 60 per cent. Although this change is in part accounted for by 9 per cent more annual trips made per person, it is notable that average trip length rose by nearly half.

Similar trends are evident in both the EU as a whole and in the United States (Bureau of Transportation Statistics, 2006c; Eurostat, 2006c). The total length of passenger journeys in the EU rose by 14 per cent between 1995 and 2003 (Eurostat, 2006c), whilst in the United States annual distance travelled per car went up from 8,813 miles in 1980 to 12,242 miles in 2004 (Bureau of Transportation Statistics, 2006c). The length of freight trips has also risen. For example, whereas the average length of truck haul in the United States was 286 miles in 1975, it was 485 miles by 2001, paralleled by an increase in Class 1 railroad hauls from 541 to 859 miles (Bureau of Transportation Statistics, 2006d). Globally, production and consumption chains have become more complex and more international, a fundamental trend that both generates and is fostered by longer-distance freight transport and logistics systems (Chapter 10). Nevertheless, the generalized time-space convergence has been differential, with considerable spatial inequalities (Knowles, 2006a).

Increases in trip distance have both stimulated and been facilitated by enhanced IURT systems, especially in the road transport sector. The total length of motorways in what is now the EU 25 tripled between 1972 and 2002 (Eurostat, 2005), and the European stock of passenger cars increased by 38 per cent between 1990 and 2004, with the most rapid rise in Greece, Portugal and the recent accession states. Data from 2004 show a positive linear relation between the average daily distance covered per person by car and the number of passenger cars per 1,000 inhabitants (Eurostat, 2006c), although the propensity for medium- and longer-distance travel is differential, and is determined by gender, household composition and income, particularly in the case of work- or business-related trips (Limtanakool *et al.*, 2006a, 2006b). Patterns of IURT also reflect the geographical structure of a country's urban system and population density.

Indeed, geographical patterns and processes are fundamental for the understanding of IURT systems (Knowles, 2006a; Taaffe and Gauthier, 1973). There are major differences between the spatial patterns and associated IURT networks of, for example, the UK, the Netherlands and Japan, and those of Spain, Turkey, the USA or Canada, although simple distinctions are less reliable at the sub-national scale. Such comparisons are picked up by Eddington (2006c: 22) in dismissing high-speed rail investment in the UK:

> . . . the UK's economic geography means that the principal task of the UK transport system is not, in comparison to the needs of France or Spain, to put in place very high-speed networks to bring distant cities and regions closer together . . . Instead, because the UK's economic activity is . . . densely located in and around urban areas, domestic freight routes and international gateways, the greater task is to deal with the resulting density of transport demand.

With this context in mind, we now turn to discuss two key examples of IURT – HSR and the European Union's (EU's) TEN-T policy. These examples neatly demonstrate the complexities involved in coordinating IURT across a series of closely collaborating yet distinct international political jurisdictions.

High-speed rail

The first commercial HSR operations began in 1964, when the Japanese Shinkansen services began between Tokyo and Osaka (Murayama, 1994). The purpose-built Shinkansen HSR lines are reserved solely for fast passenger services, and overcome difficult terrain through heavy civil engineering. Japan has maintained its momentum in the operation of HSR services between major urban centres, with two new lines (Hokuriku and Kyushu Shinkansens) and an extension of the Tohoku Shinkansen to Hachinohe (Kitagawa, 2005). In 2006, four further Shinkansen lines were under construction, to integrate key peripheral cities into the network.

The French *Train à Grande Vitesse* (TGV) pioneered high-speed rail services in Europe and became a prestigious symbol of French technical prowess and an instrument of regional development strategy (Charlton and Gibb, 1998b) (Figure 8.1). The TGV concept is based on frequent high-quality services on new *Lignes à Grande Vitesse* (LGV) dedicated exclusively to high-speed trains and serving only a limited number of access points on the line itself. The absence of slower traffic, along with powerful electric traction and sophisticated train control systems allows line speeds up to 320 km/h on steeper gradients than normal, thus reducing construction costs. Three years after the first TGV services began between Paris and Lyon in 1981, rail traffic between the two cities had risen by 45 per cent, much of it diverted from air transport. The most recent addition to the network is the LGV Est européene, opened in 2007 from near Paris to Baudrecourt in Moselle (LGV Est européene, 2006). The new line, to be extended towards Strasbourg, links Lorraine

Figure 8.1. The TGV, Thalys and Eurostar networks.

and Alsace with the capital, as well as HSR connections with Luxembourg and Germany (Hughes, 2006).

France has maintained its place in the world's 'top three' in the *Railway Gazette*'s 'World Speed Survey' for the past 30 years (Taylor, 2005) and an ambitious Master Plan has set out a programme of new LGV projects into the twenty-first century. In addition to those new lines already completed (Figure 8.1), future extensions of the TGV network will include: the LGV Rhin-Rhône from Dijon to Mulhouse, which will speed travel between eastern France and southern Germany to the Mediterranean, forming a 'linchpin in a continent-wide network' (LGV Rhin-Rhône website, 2006: unpaginated); extensions of the TGV-Atlantique to Rennes and Bordeaux; connections from Bordeaux to the Spanish frontier and to Toulouse; extensions from the LGV Mediterranée westwards beyond Montpellier to the

Spanish high-speed system in Catalonia; and from Lyon through the Alps to Turin (*Railway Gazette*, 2004).

The German HSR system combines newly-constructed *Neubaustrecke* (NBS) and existing main lines adapted to high-speed standards. The first two NBS, Mannheim–Stuttgart and Hannover–Würzburg (opened in 1991), improved north–south rail communications in the former West Germany. When the Wolfsburg–Berlin NBS opened in 1998, however, it reflected the new spatial imperatives of reunification and the need for faster transport to the restored capital. The German HSR network has been further extended by the Frankfurt–Cologne NBS, which replaced the picturesque but tortuous route along the Rhine valley, and in 2006 a further north–south NBS from Nuremberg to Ingolstadt, for faster services into southern Bavaria.

HSR development in Italy has focused on improving internal intercity connections and, marketed as Eurostar Italia, is based on trains capable of operating at up to 300 km/h on purpose-built HSR routes, notably the strategic north–south *Direttissima* route connecting Naples, Rome and Florence, which is being extended to Bologna and Milan and Turin, with proposals to link Milan with Genoa and Venice, and the tilting-body *Pendolino* trains, which can negotiate curves at higher speeds without causing undue passenger discomfort (Haydock, 1995). The successful *Pendolino* technology has been exported to other European countries, including the Czech Republic, Finland, Germany and the UK.

Spain's impressive *Alta Velocidad Española* (AVE) high-speed system confirms the country's joining the European mainstream both in terms of technical and operational achievement and by its use of the standard European 1,435 mm gauge, rather than the broader Iberian variant. The first AVE line from Madrid to Seville has been followed by an extension to Malaga, the new trunk line from Madrid to Barcelona (with line speeds of up to 350 km/h), the line north from the capital via a base tunnel through the Sierra de Guadarrama, and the connection from Barcelona to the French frontier, due to open in 2009.

The most distinctly international components of the European HSR system are the 'Thalys' and Eurostar networks. The former links Paris, Brussels, Amsterdam and Cologne, using a dedicated fleet of high-speed trains based on the French TGV design, but adapted to operate with different national train control systems (Charlton and Gibb, 1998b). The opening in 1998 of the Belgian high-speed line from the French frontier into the suburbs of Brussels cut the journey between the two capitals to 1 hour 25 minutes, and boosted rail's share of traffic from air and road (Perren, 2006). Thalys services will accelerate on completion of the HSL Zuid high-speed line from Antwerp to the Netherlands, and investments in the Brussels–Aachen corridor (Railway-Technology.com, 2006a).

Eurostar services via the Channel Tunnel dominate the London–Paris and London–Brussels intercity passenger market, with 7.85 million passengers in 2006, 5.45 per cent more than in the previous year. Completion of the 'High Speed 1' link from the Channel Tunnel to the new Eurostar terminus at St Pancras in London in 2007 cut journey times by 25 minutes. This further increased rail's share of the market at the expense of air transport which is suffering delays from tighter security measures and a doubling in air passenger departure tax in response to fears about environmental impacts. Interchange at Brussels with Thalys services adds to the overall connectivity of HSR in Europe's economic core area.

High-speed rail, geography and IURT

The spectrum and scope of European high-speed rail services has clearly widened since its first decade. While high-speed operations are considered to be optimum for two- to three-hour journeys, some shorter-distance regional high-speed services operate, including Spain's *Avant* services from Madrid to Ciudad Real and Toledo (the latter a mere half hour sprint from the capital), and the transformed Svealand Line from Stockholm to Eskilstuna and Örebro (Fröidh, 2005). At the other extreme, there are now some surprisingly long high-speed services, such as the direct Eurostar services from London to Avignon and the *Thalys Neige* and *Thalys Soleil* seasonal trains from Brussels to southern France; the latter covers the 1,054 km to Marseilles in just 4 hours 31 minutes (Taylor, 2005). Once additional HSR connections have been completed from Barcelona to Perpignan, it is envisaged that through high-speed services will operate between Paris and Madrid.

The planning of HSR presents a number of dilemmas. High capital and operational costs mean operations are optimized between large cities, especially those with dense concentrations of passenger origins and destinations close to city centres, where most main termini are located, as is still the case in Europe. To optimize the advantages of high speeds and line capacity, the number of stations must be kept to a minimum, even though more stops widens access to HSR (Givoni, 2006). But as commercial and residential development disperses, the dominance of city centres as origins and destinations, and therefore the potential of HSR to capture a higher share of IURT, may be threatened. This challenge is apparent, for example, in the case of Madrid, where there is a rapid expansion and spatial restructuring to lower densities of both new residential developments and commercial activities (Gutiérrez and Garcia Palomares, 2007). The swathes of new developments around the capital are relatively isolated by distance and congestion from access to the high-speed trains that sweep across the horizon.

Nevertheless, a degree of adaptation to this type of spatial discordance is possible. The dislocation of the dynamic outer suburban zones of Madrid from Spain's AVE system is partly compensated by high-frequency suburban services that serve the central Atocha AVE station. There are similar interchanges in many European cities, including Berlin Hauptbahnhof and Paris HSR termini. Adjustment to changing spatial patterns can also occur via peri-urban HSR stations, as, for example, Massy (Paris) and Ebbsfleet on the British High Speed 1 Channel Tunnel link demonstrate. Other HSR stations serve both major city airports and wider metropolitan zones, such as at Paris Charles de Gaulle, Frankfurt Main and Cologne-Bonn airports. Still others have been located in relatively 'rural' locations to serve a more dispersed regional market with modal interchange to car or feeder bus. Spatial adjustment also operates in the reverse direction, whereby central city locations near HSR terminals stimulate 're-densification' by attracting substantial new commercial development, as is the case, for example, of Lyon Part Dieu, Lille Europe and Rotterdam Centraal (in anticipation of the completion of the HSL Zuid).

Transport policy for IURT, as at other scales, conventionally prioritizes speed and the fastest modes of transport that 'save' time, based on assumptions that travel time is 'wasted' time. Even modest reductions in journey times warrant substantial investment – for example, the €11 million viaduct that opened in 2006 to take high-speed tracks into Brussels South station saves just three minutes (Haydock, 2007). This orthodox perspective is challenged by many observers, however, despite the evident appeal of higher speeds, as confirmed by frequent examples of traffic increase following the introduction of faster services. Crozet (2005) points out that time saved is frequently 'spent' on travelling longer distances and more frequently, while Adam (2001) also questions how time 'saved' on journeys is used. The ability to undertake different activities while travelling means that travel time may often be used very productively (Jain and Lyons, 2005; Lyons, 2005; Lyons and Urry, 2005). Such observations might provide justification for IURT policy to place higher value on qualities other than very high speeds, such as cost, convenience and capacity, with corresponding allocation of scarce resources on, for example, more conventional, shorter-range IURT rail services.

What will be the future status of HSR from a wider global perspective? Outside Europe, this mode is being developed with impressive vigour in East Asia. Besides the continued extension of Japan's Shinkansen network (Kitagawa, 2005), new HSR trunk lines now operate in South Korea and Taiwan, and the 1,318 km Beijing–Shanghai high-speed line is under construction as part of China's very dynamic programme of IURT investment (*Railway Gazette*, 2005; Railway-Technology.com, 2006b). Elsewhere, air and road dominate the IURT market, especially in the United States, Canada and Australia. In these cases HSR is distinctly underdeveloped, within IURT rail systems that are overwhelmingly dominated by freight. India will be especially interesting over the next decade. Whereas rail has long been dominant in the IURT sector, road and especially low-cost air transport are now becoming serious challengers. It is open to question how far India will be able or willing to invest in the radical upgrading of its trunk rail routes demanded for HSR operations, especially given parallel pressures to satisfy other rail markets like high-density urban passenger and freight.

European transport policy and TEN-T

Enhanced IURT systems are seen as essential for the EU's objectives. Efficient transport allows the free movement of goods, people and services and thereby the achievement of the single internal market (Van Reeven, 2005). However, the EU's Common Transport Policy has had a slow and somewhat incomplete gestation. Despite the aspirations of the European Commission, there was relatively limited progress with meaningful legislation and action until the Maastricht Treaty came into force in 1993 (CEC, 2001b). In the 1980s, the emphasis in European transport policy moved towards liberalization of transport markets, with the intention of opening them to competition (Jensen, 2005). This fundamental policy goal matched the wider drive towards the completion of the competitive Single European Market. *Keep Europe Moving*, the mid-term review of the European Commission's 2001

Transport White Paper (CEC, 2006b: 3), restates the EU's core concerns in the development of its common transport policy: 'to help provide Europeans with efficient, effective transportation systems that offer a high level of mobility to people and businesses throughout the Union and protect the environment, ensure energy security, promote minimum labour standards for the sector and protect the passenger and the citizen', as well as innovation in pursuit of these two policies.

Although the core EU transport policy objectives are regarded as stable, the context within which they operate has evolved. Particularly important is the enlargement of the EU to encompass 27 member states. Different levels of development, in general and in terms of transport, implies divergence in priorities. In the wealthier states, sustainability issues such as pollution and congestion are prominent, whereas accessibility is a more pressing deficiency in recently acceded states. Shifts in the European transport context are also reflected in rapid technological innovation (for instance, in the development of intelligent and more environmentally benign transport systems), changes in the organization of the transport industry (e.g. consolidation and greater international competition) and the need to meet international environmental commitments such as those within the Kyoto protocol as well as new energy imperatives.

The Maastricht Treaty identified Trans European Networks as fundamental for the creation of a competitive single internal market, reinforcing economic and social cohesion and ensuring that the EU's development is balanced and sustainable (CEC, 2006c). This demands the interconnection and interoperability of national networks and access to these networks by potential operators. 'Projects of common interest' have been supported via specific TEN funds, the Structural and Cohesion Funds and European Investment Bank (EIB) loans. The TEN concept embraces energy and telecommunications interconnections as well as transport (TEN-T) projects. Community guidelines for the development of the TEN-T cover roads, railways, inland waterways, airports, seaports, inland ports and international traffic management systems, with the emphasis on longer-distance IURT-scale connections, and the improved cross-border interaction. Two notable examples of international multimodal TEN-T projects are the Øresund fixed link between Denmark and Sweden, which opened in 2000, and the Fehmarn Belt project which would enhance connections from Germany to Scandinavia (Hansen, 2005; Knowles, 2000, 2006b; Matthiessen, 2000). By 2020, when traffic between member states is expected to have doubled, the TEN-T is intended to include 94,000 km of railways, including around 20,000 km of high-speed lines (speeds of at least 200 km/h) and 89,500 km of roads (CEC, 2005b) (Figure 8.2; Box 8.1).

Member states have prime responsibility for developing and funding the TEN-T, but progress has been slower than intended, so 30 key transnational axes were prioritized (CEC, 2003, 2005b). Their cost had increased to €252 billion by 2005 (CEC, 2005) whilst the whole trans-European network will cost €600 billion. The European Commission has proposed increasing its share of the TEN-T budget, especially to support cross-border projects, using structural and cohesion funds, and EIB loans to leverage national public funding. Supplementary funding sources include European loan guarantees to support public–private partnerships (used, for example, for construction of the 8.2 km HSR tunnel under the Pyrenees on the

Figure 8.2. Selected Trans-European Transport Network priority projects.

Box 8.1. Examples of key TEN-T corridor projects.

- Berlin–Verona/Milan–Bologna–Naples–Messina–Palermo railway axis, intended to speed passenger and freight traffic between northern Europe and Italy. The most ambitious and costly component of this axis is the 56 km Brenner base tunnel, with a target completion date in 2015. This project would, it is claimed, reduce congestion and bring important environmental benefits to the 'ecologically sensitive Alpine region'. At its lower end the route improvements, including a bridge over the Straits of Messina, would provide better connections with the peripheral southern regions of Italy.
- High-speed railway axis of south-west Europe: three new standard European-gauge high-speed railway lines connecting major cities on the Iberian peninsula, with links to the French high-speed rail network. Substantial stretches of this TEN-T package are already complete or under construction, notably the high-speed line between Madrid and Barcelona. When fully operational, this line will cut the rail journey time between the two largest Spanish cities from 6 hours 50 minutes to 2 hours 25 minutes. This TEN-T package reflects pressure to strengthen further the connectivity of the more peripheral south-western regions of Europe, especially Portugal.
- 'Rail Baltica' axis Warsaw–Kaunas–Riga–Tallinn–Helsinki: a major upgrading and renewal of the north–south rail network linking Estonia, Latvia, Lithuania and Poland, to make it more efficient, speedy and interoperable with the rest of the European network. This is one of a number of TEN-T priority projects that reflect the impact on transport policy of the EU's recently admitted Convergence states.

Figueras–Perpignan route), plus road user charging in the form of tunnel and bridge tolls and the *Eurovignette* scheme for distance-based charges on lorries to help defray infrastructure costs (CEC, 2006c; Euractiv.com, 2006).

The advance of the TEN-T project has confirmed the need for appropriate international arrangements for the governance and management of cross-border operations. The French rail infrastructure manager, Réseau Ferré de France (RFF), has formed partnerships with equivalent agencies in Spain, Italy and Germany to co-ordinate the major investments required for the further development of key European corridors, such as Hendaye (France)–Vitoria (Spain), the Lyon–Turin trunk route and that between Paris and Frankfurt, which will be the first international rail link to use the European Rail Traffic Management System (ERTMS) (RFF, 2006). Whilst the European Commission has pursued the TEN-T network as fundamental to its core purposes, the strategy has been accused of favouring the EU's economic growth and development imperatives over declared concerns for a more sustainable future. The pressure group CEE Bankwatch claims that some of the proposals to build new motorways through strategic corridors as part of the TEN-T network are misguided and undesirable on environmental, social and economic grounds (Stefanova, 2006). In comparison with the support being offered to trunk road projects, it is claimed that rail alternatives are being neglected.

Perspectives on IURT development

Whilst well-established approaches to transport geography remain strong, perspectives on transport issues have diversified in recent years, in tune with those adopted in cognate disciplines. Alongside the long-established analyses of changing transport systems, networks, technologies and operations, and transport's relationships with spatial patterns and processes, complementary themes and concerns have emerged. These include concerns for sustainability, the 'new mobilities' paradigm, and the relationships between transport and social exclusion and justice, and these perspectives offer some valuable insights on recent developments in IURT.

Sustainability

It is widely recognised that current patterns of transport activity are not sustainable (Black, 1998, 2003; Steg and Gifford, 2005; Whitelegg and Haq, 2003; Knowles, 2006a). The sustainability of the IURT sector is clearly questionable, given the prominence of both distance and speed. The continued expansion of longer distance road networks and traffic is a fundamental sustainability challenge. The EC's White Paper on Transport (CEC, 2001b: 10) claims that 'the trans-European transport network itself suffers increasingly from chronic congestion: some 10% of the road network is affected daily by traffic jams', while 16 of the EU's major airports suffered delays of at least 15 minutes on over 30 per cent of flights. Overall these delays accounted for an extra 1.9 billion litres of fuel consumption (see Chapters 3 and 9 for more detailed discussions on air transport and sustainability). The future pattern of IURT faces other sustainability challenges. Hall *et al.* (2006: 1406) suggest that looming constraints on the cheap availability of oil will result in 'a strong reality check for cheap transportation and its spatial structure', so that 'fallacies such as the death of distance' will be exposed, hence reversing the dominant trend of relative liberation of economic activity from the spatial constraints imposed by transport costs. Knowles (2006a) suggests that dearer fuel costs, alongside traffic congestion, could reverse the long-established pattern of time-space convergence.

The close relationship between transport improvements and efficiency, and economic development and prosperity is well-established (Chapter 2). For example, Eddington (2006c: 14) identifies seven 'micro-drivers' through which transport impacts on the economy: 'increasing business efficiency, investment and innovation, improving the functioning of agglomerations and labour markets, increasing competition, increasing trade, and attracting globally mobile resources'. But acute awareness that such economic benefits are accompanied by external costs, especially on the environment, have made 'decoupling' of the negative environmental impacts of transport from economic growth a key policy goal (Organisation for Economic Cooperation and Development (OECD), 2006a; Stead and Banister, 2006). A significant reduction of emissions, as well as road congestion, requires the large-scale implementation of road pricing, coupled with an improvement of the quality and accessibility of rail, to encourage modal shift from road and air. Eddington (2006c: 19) accepts that transport users 'should pay the full costs of their travel, whether those are the costs of congestion or environmental damage' and lends weight to a move towards system-wide road pricing. With the exception of motorway tolls in countries such as the United States, France, Spain, Italy and Mexico (Farrell, 1999), road user charging at the IURT scale still lags behind urban and suburban policy developments, as represented, notably, by the Congestion Charge in London (Chapter 3). And the constraints on IURT activity that may be necessary for a more sustainable future will not be easy for governments to impose (Shaw and Thomas, 2006). The potential for resistance to major initiatives intended to restrict private road traffic was revealed by nearly 1.8 million signatures to an online petition protesting at UK government proposals for compulsory fitting of in-car tracking systems as a step towards universal road user charging (Porter, 2007).

Mobilities

Conventional evaluation of passenger flows has tended to aggregate travellers as uniform actors subject to forces that are principally economic, but recent transport related research in the social sciences interrogates the complex and changing motivations and conditions that determine mobility and travel patterns (Sheller and Urry, 2006). This approach highlights the constitution and spatial disposition of social networks, which have been extending from relatively simple, localised communities to more diffuse, complex patterns of social interaction and interdependence. Mobility patterns reflect social changes: evolving family and friendship ties and obligations, and fluid household structures in which the nuclear family is less dominant are important. Despite communicative technologies such as email and mobile phones, direct face to face 'co-presence' has retained its power in social networks (Larsen *et al.*, 2006a; 2006b; Urry, 2002, 2003). Besides the imperatives of social networks, the 'need to be elsewhere' is also explained by factors such as employment patterns and leisure. The organization of work in contemporary post-industrial societies is more fluid and shorter term than in the past, and stronger drives to self-fulfilment and the collection of social capital through tourist trips also provoke more frequent and longer distance IURT movement.

This 'new mobilities paradigm' can offer useful perspectives on the expanding demand for IURT and strategies to contain it. The spatial dispersal of social networks adds to the expanding demand for travel, at least in terms of total distances travelled, if not the trip frequency. While IURT-length personal trips per capita are less frequent than local-scale trips, their significance within personal and professional networks may be especially high for the building and maintenance of social capital (Larsen *et al.*, 2006a, 2006b). More dispersed, travel-inducing social networks have emerged in a context of social change and cheaper and faster IURT mobility (Cass *et al.*, 2005). Future policies that seek to reverse this mobility, in deference to new sustainability and economic imperatives, will inevitably have important social consequences that may not be readily incorporated into conventional approaches to transport planning. While some IURT trips may increasingly be regarded by critics as unnecessary and/or wasteful, many are personally essential because 'networked relationships are conducted at-a-distance, so encountering, visiting and seeing networks members face-to-face is crucial' (Urry, 2003: 162).

Social justice

The increases in mobility afforded by transport developments have not been experienced uniformly either in spatial or social terms (Chapter 4). Yet interpretations of transport exclusion tend to emphasize access to services required over short distances on a regular, even daily basis. How far are notions of transport and social justice applicable to the longer distances associated with IURT? In many respects, the issues are similar across spatial scales. Low-income individuals and households, women, the disabled, the elderly and young are often excluded from some kinds of long-distance travel. Evidence from the USA reveals a gender imbalance in the number of annual trips over 50 miles; in the 25–54 years age group, urban males averaged 13.0 trips in 2001 in comparison with only 8.69 by urban females (Bureau of Transportation Statistics, 2006e). Among the over-75 population, there was, unsurprisingly, less long-distance trip-taking (urban males: 4.92, urban females 3.23). On both sides of the Atlantic, the proportion of elderly people in the population is rising, with the 'older' elderly increasing fastest. This group is less able to undertake longer journeys by private car or air and is more dependent on surface public transport. As social networks widen spatially, face-to-face meetings with significant contacts can only be achieved through longer-distance trips. In the context of IURT, social disadvantage results from exclusion, through immobility, from the maintenance of social networks and social capital 'at a distance' – i.e. the disadvantage that individuals, households, organisations and places may suffer through being excluded from IURT.

The impact of trends in IURT on mobility-based exclusion or inclusion is ambivalent. The costs of private motoring have fallen, while low-cost air services have reduced barriers to longer-distance travel between many urban centres, especially for certain market segments such as younger adults employed in major cities, and the active retired. Some surface passenger transport initiatives have kept down the cost of IURT travel – for example, the 'Megabus' inter-city coach services operated in Britain by Stagecoach. Even in wealthy economies, however, practical and financial factors exclude many from IURT-scale mobility, notably for those in households without cars. For example, the cost of open access rail fares in Great Britain has continued to rise relative to prevailing incomes and inflation on longer-distance routes (Barkham, 2007; Webster, 2007). In future, there will be awkward social challenges if operators and users are required to bear the full external costs of transport in the pursuit of more sustainable transport policies.

IURT systems, especially major trunk roads, but also high-speed railways and airports, also cause social disadvantage through the noise, pollution, and physical and visual disruption they inflict on adjacent communities (Social Exclusion Unit, 2003). Those affected by the intrusion of major transport investment may see few direct benefits, and may be caught in a double bind – their own longer-distance mobility constrained by poverty, yet subjected to the incessant noise of motorways and airports or the periodic passage of high-speed trains. In the case of the Brussels South high-speed rail interchange station, Albrechts and Coppens (2003) concluded that large-scale strategic transport investments and property development overrode less influential interests. The local economic and social fabric in the vicinity of the new station was disrupted, with many households forced to leave the neighbourhood; the 'space of places' that made up the district close to the new station was usurped by a 'space of flows'.

Governance and regulation

Although Chapter 5 explores the changing governance and regulation of transport in some detail, this theme has important implications for IURT. The dominant tendency has been for less intervention and regulation of transport systems by states, with more emphasis on free market processes, privatization and the dwindling of state monopolies and control (Nash, 2005). Deregulation of air transport has facilitated the surge of IURT-range low-cost air services in the United States, the EU, Australia and now India. Changes in passenger rail service ownership and governance has separated rail infrastructure from train operations in European countries such as Great Britain, Germany, Sweden, Denmark and the Netherlands and created competition *for* rail markets, if less *in* rail markets (CER, 2006; Jackson, 2005). While many devolved passenger services are local operations, the EU has pressed for open access for international, and, more recently, domestic rail freight operations (Nash, 2006).

The tendency to deregulation is not uniform, however. Docherty *et al.* (2004) indicate that rising concern about the environmental and social impacts of transport have made it necessary for states to re-engage in the delivery of transport policy, albeit within varied systems of governance that combine regulation with market influences. In Europe, especially, the international nature of many IURT systems, as well as the rising number of actors in an increasingly devolved political space, has made trans-jurisdictional co-ordination and intervention more desirable, not least in the planning, construction and operation of enhanced transport infrastructure, as represented by the TEN-T strategy (Ollivier-Trigalo, 2001).

Contemporary economic and social systems form complex, diffuse spatial structures consisting of interdependent nodes connected by multi-modal transport networks. The concept of the European megacorridor exemplifies this notion (Zonneveld and Trip, 2003); polynuclear, broadly linear spatial forms, in which transportation, economic development and urban development are strongly interrelated (Priemus and Zonneveld, 2003). In northwestern Europe, such megacorridors override national and regional boundaries, as for example the Randstad–Flemish Diamond and the Randstad–RheinRuhr megacorridors. Multiple nodes, administrations and transport modes demand innovative planning and governance (de Vries and Priemus, 2003). Romein *et al.* (2003: 211) note how 'infrastructure development in a megacorridor is a multi-dimensional affair, less and less tied to one particular administrative area and one single level of governance'.

Effective planning and co-ordination of complex IURT-scale transport networks presents a major challenge, as is evident from many examples. Progress towards a more competitive, efficient and inter-operable European international rail system demands enhanced collaborative arrangements between both public and private sector organizations. Giorgi and Schmidt (2005: 216) highlight how within Europe 'the co-ordination of policy towards sustainable mobility in the Alps will require establishing a supranational mode of governance . . .' that has so far proved elusive. In England, the multi-modal studies undertaken in the early twenty-first century were intended as a 'significant new approach to developing integrated transport strategies, examining how all modes can contribute to developing long-term solutions to some of the country's most severe transport problems' (AEA Technology, 2004: 1). However, Shaw *et al.* (2006: 576) concluded that the studies' outcome was largely 'approval of a host of new road schemes' rather than the genuinely integrated transport strategies envisaged at the outset.

Conclusion

It is clear that IURT encompasses a diverse set of transport systems that differ considerably in terms of geographical, socio-economic and political settings. It is embedded within a self-reinforcing cycle of technical and organizational improvement and increasing frequency, diversity and distance of travel and trips. This dynamic context for IURT brings a series of interwoven challenges. In many parts of the world, the expansion of IURT generates more trans-jurisdictional movements that demand new, better-integrated and collaborative modes of governance, as in the case of HSR operations and the development of the TEN-T system. Future evolution of IURT will be increasingly determined by rising concerns about sustainability, with profound questions on the validity and viability of continued transport growth and on the optimum mix of modes in different IURT markets. Policies aimed at securing more sustainable IURT are complicated by the need for transport policy makers – and transport geographers – to incorporate the insights provided by fresh interpretations of personal mobility patterns and social networks. Likewise, the implications of dominant trends in transport, including the pursuit of higher speeds, for social inclusion/exclusion must be understood.

The issues raised in this chapter provoke obvious further questions. Besides the need to interrogate modes that have not been addressed in any detail here – the supremely significant road passenger and freight modes, air transport and short-sea operations – it is also important to recognise the many other locations for IURT. Besides the distinctive cases of North America and Australia, where IURT has particular domestic and long-distance dimensions, the evolution of IURT in developing economies is especially critical. The growth of IURT in China and India poses immense challenges; in both cases, the demand for IURT-scale air travel and road transport appears to be rising exponentially, raising questions about the status of current public transport modes as well as profound implications in terms of sustainability and socio-economic and spatial differentials.

The planning and management of IURT systems will therefore require a holistic approach – long the hallmark of the geographer – in which social and environmental concerns are incorporated alongside the dominant economic and technical perspectives. At the same time, there should be sufficient flexibility and awareness of the specific socio-economic, political and territorial setting in which IURT operates, which is also a central concern of geographical enquiry.

9 | Global Air Transport

Brian Graham and Andrew R. Goetz

Although the world's air transport networks were largely pioneered prior to the Second World War, the origins of mass air travel – at least in the developed world – date back to no earlier than around 1960. While aggregate growth rates since then have been quite dramatic, they still possess two contradictory characteristics which, together, create a sense of volatility in the industry. The long-term underlying aggregate growth in demand for air transport has been driven by growing gross domestic product (GDP) per capita and disposable incomes, originally in developed countries and, latterly, in Asia. These financial trends have been accompanied by radical changes in the geopolitics of air transport, as government regulation and control have increasingly been replaced by an ethos of deregulation, liberalization, privatization and increased competition. Cumulatively, the result has been a step change in the supply and pricing of air transport. Simultaneously, however, these aggregate growth trends have been punctuated by sharp falls in demand caused by external events such as '9/11', the SARS outbreak in Asia in 2002–03, and the US-led invasion of Iraq in 2003.

When combined with sharply rising fuel costs in 2005–2006, these and other events caused the global airline industry to lose US$40.7 billion between 2001 and 2005 (International Air Transport Association (IATA), 2006), although the negative effects have been felt disproportionately by US carriers. Nevertheless, the aviation industry remains bullish about long-term growth, despite concerns about fuel supplies and costs, shortages of airport capacity in many key markets, and the negative environmental impacts of air transport. It is projected, for example, that international air traffic will grow at an average rate of 5.6 per cent between 2005 and 2009 (a rate equivalent to a doubling in demand every 12 years; IATA, 2006).

This chapter begins with a brief outline of global trends and patterns in air transport and an assessment of the current state of the industry. We then examine three interlinked dimensions to contemporary air transport: the impacts of globalization and liberalization on the geopolitics and structure of the industry; the resultant trends in air transport networks; and the interlinkages with sustainability issues. Following a brief consideration of the impact of these processes on airports, the latter part of the chapter summarizes the principal characteristics of the major air transport regions.

Global trends and patterns

Given the correlation between GDP, disposable income and demand for air transport, it is not surprising that virtually all the top-ranked countries for scheduled air traffic are located within North America, Europe and Asia-Pacific. The United States has by far the largest domestic market, accounting for over 35 per cent of all passengers worldwide (Rodrigue *et al.*, 2006) (Figure 9.1). The largest share of international scheduled traffic – around 30 per cent – is carried by European airlines. The dominant intercontinental traffic axes interlink Europe, North America, and the Asia-Pacific region. Consequently, at the global scale, the demand for, and provision of, air transport has a pronounced east–west bias, the basic network interconnecting some 20 or so of the world cities that serve as the gatekeepers of the global service economy (Zook and Brunn, 2006). Metropoles such as London, New York, Chicago, Tokyo, Singapore and Hong Kong constitute a set of commercial and financial nodes, joined by a series of linkages including virtual and physical transport flows. Inevitably, they have become the hubs for global airline networks. North–south routes into South America, Africa and the South-West

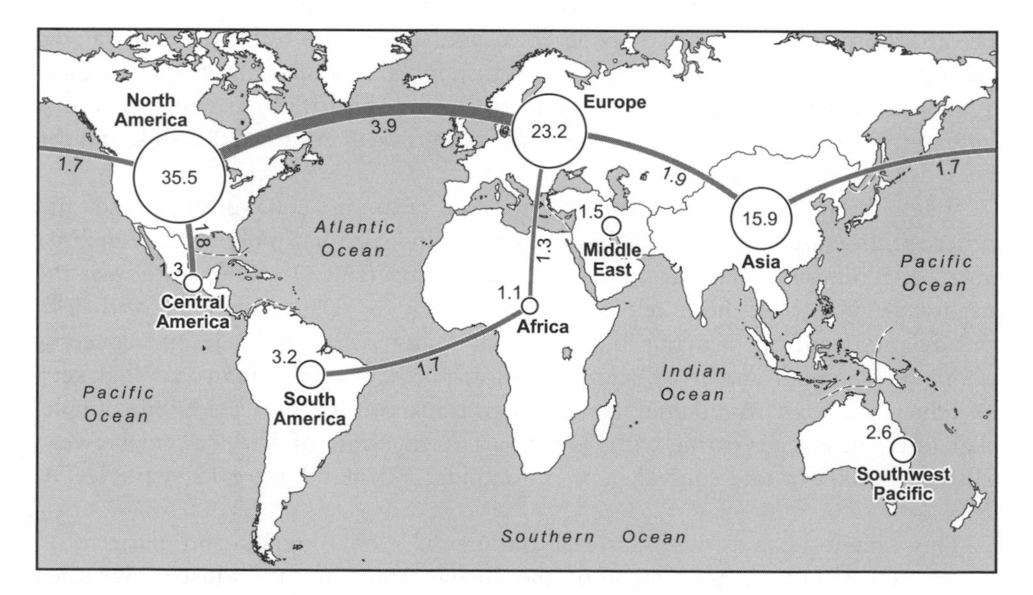

Figure 9.1. Principal international air traffic flows (IATA).

Table 9.1. World's top 30 airline groups by sales, 2005. Source: *Airline Business*, 2006a.

Rank	*Airline group*	*Country*	*Group revenues (US$million)*
1	Air France-KLM	France/Netherlands	26,063
2	Lufthansa Group	Germany	22,371
3	FedEx	USA	21,446
4	AMR Corporation (American)	USA	20,712
5	Japan Airlines Corporation	Japan	19,346
6	United Airlines	USA	17,379
7	Delta Air Lines	USA	16,191
8	British Airways	UK	15,122
9	Northwest Airlines	USA	12,286
10	ANA group	Japan	12,040
11	Continental Airlines	USA	11,208
12	US Airways Group	USA	10,610
13	Qantas	Australia	9,524
14	ACE/Air Canada	Canada	8,422
15	SAS Group	Denmark/Norway/Sweden	8,225
16	Singapore Airlines Group	Singapore	8,030
17	Southwest Airlines	USA	7,584
18	Korean Air	South Korea	7,424
19	Cathay Pacific	China	6,548
20	Emirates	UAE	6,281
21	Iberia Airlines	Spain	6,073
22	Alitalia	Italy	5,940
23	China Southern Airlines	China	4,682
24	Air China	China	4,681
25	United Postal Service	USA	4,105
26	Thai Airways	Thailand	4,056
27	Saudi Arabian Airlines	Saudi Arabia	3,901
28	Consorcio Aeromexico	Mexico	3,604
29	China Eastern Airlines	China	3,357
30	Virgin Group	UK	3,326

Pacific are effectively little more than capillaries, connecting Buenos Aires, Johannesburg and Sydney to the other world cities.

Not surprisingly, the global distribution of major airline operators reflects these trends. Of the world's top 30 scheduled air transport groups in 2005, ranked by revenue, eleven were North American (including Mexico), 10 Asian-Pacific, seven European and two were based in the Middle East (Table 9.1). Of the top six largest passenger carriers (ranked by revenue passenger kilometre – RPK), five were North American, reflecting the size of that internal market (Table 9.2). European, Asian-Pacific airlines and, increasingly, a handful of Middle Eastern carriers, are, however, much more dominant in terms of international passenger traffic, a function of the political fragmentation of their home regions and the restricted spatial extent of some domestic markets.

Table 9.2. World's top 20 passenger airlines, 2005. Source: *Airline Business*, 2006a.

Rank	*Airline*	*Country*	*Passenger traffic (mRPK)*
1	American Airlines	USA	222,412
2	Delta Air Lines	USA	193,006
3	Air France-KLM Group	France/Netherlands	189,253
4	United Airlines	USA	183,262
5	Northwest Airlines	USA	121,994
6	Continental Airlines	USA	114,659
7	British Airways	UK	111,859
8	Lufthansa	Germany	108,185
9	Japan Airlines Corporation	Japan	100,345
10	Southwest Airlines	USA	96,899
11	Qantas Airways	Australia	86,986
12	Singapore Airlines	Singapore	82,742
13	Air Canada	Canada	75,290
14	Cathay Pacific	China	65,110
15	US Airways	USA	62,582
16	Emirates	UAE	62,260
17	China Southern Airlines	China	61,923
18	ANA	Japan	58,949
19	Air China	China	52,543
20	Thai Airways	Thailand	49,930

Globalization, liberalization and sustainability

Recent academic research concerning air transport has been dominated by the effects of deregulation and, to a lesser extent, of globalization. The contemporary debate on transport and social change as a whole, however, is perhaps more concerned with the broad realm of sustainability which interconnects transport, the economy and development with environmental issues (Chapters 2, 3 and 13). It is not at all clear how globalization, liberalization, and sustainability are interrelated in the context of air transport (Goetz and Graham, 2004; Figure 9.2). One perspective maintains that globalization and liberalization strategies have rationalized the airline sector into a more efficient operation that enhances long-term sustainability. An alternative interpretation argues that globalization and liberalization have resulted in excessive air traffic growth and wasteful competition, thereby exacerbating the mode's negative social and environmental externalities which are incompatible with long-term sustainability. Air transport remains the fastest-growing cause of the emissions that lead to global warming, while internationally agreed mitigation controls on noise and atmospheric pollution are being more than offset by the growth in aviation (Intergovernmental Panel on Climate Change (IPCC), 1999; Upham *et al.*, 2003).

Globalization

The air transportation industry worldwide has been re-shaped dramatically since the step change introduced by the deregulation of the US internal aviation

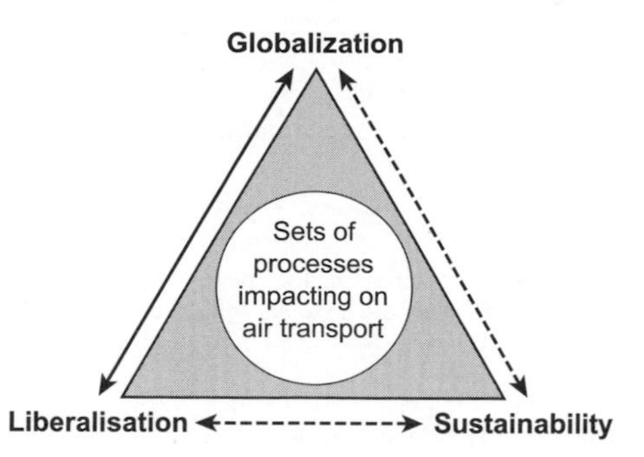

Figure 9.2. The globalization/liberalization/sustainability nexus.

market in 1978. Long-established regulatory regimes have been modified, and in some cases abolished, as a result of liberalization policies, resulting in mergers, acquisitions and/or strategic alliances among the largest carriers (Goetz, 2002; Graham, 1998). There has been a concomitant trend towards privatization, although many major air carriers still remain wholly or partially government-owned while there are also still significant constraints on foreign shareholdings in airlines. These changes are interconnected with globalization, which is fundamentally altering the volumes, patterns, directions, ownership, and control of air passenger and freight flows world-wide (Capineri and Leinbach, 2004). Thus, 'contemporary transportation systems can be better understood when viewed as simultaneously weaving together and shaping patterns of competitive advantage in production networks across multiple scales' (Bowen and Leinbach, 2006: 164).

Although it can be difficult to determine the direction of cause–effect relationships, globalization would simply not be possible without air transportation. Likewise, the airline industry would be much less significant without concomitant global expansion. For example, it is estimated that about 40 per cent of global freight trade by value (if only 2 per cent by weight) is moved by air (Bowen and Leinbach, 2006; Upham *et al.*, 2003). Airline freight operations are shared between integrators such as FedEx and UPS, both of which have global networks, and combination carriers which use dedicated freighters but also the considerable belly-hold freight capacity of wide-bodied passenger aircraft (Bowen, 2004).

The historical regulation of air transport does, however, still impose significant constraints on the sector's ability to respond to globalization. At the international scale, air service provision between countries was controlled historically by bilateral agreements, negotiated between pairs of governments. These governed the applicability of the nine so-called 'freedoms' of civil aviation (Figure 9.3). The basic principle of all bilaterals is reciprocity or equivalency, the agreements covering fares, capacity, frequency, number of carriers and routes flown. Since domestic airline deregulation in 1978, the US government has pursued a global policy – congruent with US national interests – to liberalize international bilaterals. Most recently, it has sought so-called 'open-skies' bilaterals, allowing unrestricted market entry and code-sharing alliances (in which one service is operated under the flight codes

FREEDOM	
FIRST	Right to overfly one country en-route to another.
SECOND	Right to make a technical stop in another country, for example, to re-fuel. There are no traffic rights to and from country A.
THIRD	Right to carry passengers from home country to another country.
FOURTH	Right to carry passengers to home country from another country.
FIFTH	Right to carry passengers between two countries by an airline of a third, on a route which originates or terminates in the home country.
SIXTH	Right to carry passengers between two countries by an airline of a third, using two routes which connect in the home country. May involve stop-over, change of aircraft, or altered flight codes.
SEVENTH	Right to carry passengers between two countries by an airline of a third, on a route outside its home country.
EIGHTH (Cabotage)	Right to carry passengers within a country by an airline of another. Route may have to originate within home country.
NINTH (Stand-Alone Cabotage)	Right to carry passengers within a country by an airline of another country.

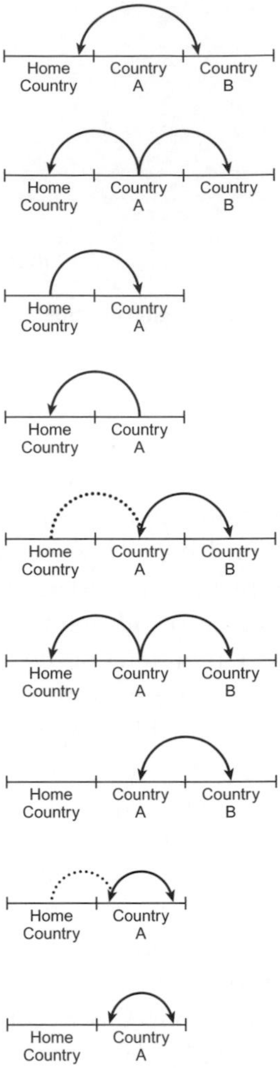

Figure 9.3. The air transport freedoms.

of two airlines). This version of open skies has been accompanied by the offer of anti-trust immunisation for various airline alliances and mergers. The logical outcome of full open skies is the replacement of bilateral with multilateral agreements, in which groups of like-minded countries permit any airline virtually unlimited access to any market within their boundaries. While this has occurred *within* regional markets such as the European Union (EU) and the North American Free Trade Area (NAFTA), the provision of both passenger and freight air transport between these blocs and many individual countries still remains constrained by bilaterals and continuing restrictions on foreign ownership of airlines (not least within the United States).

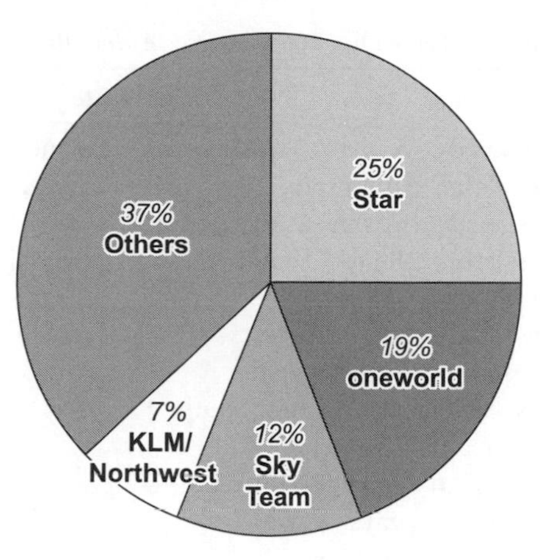

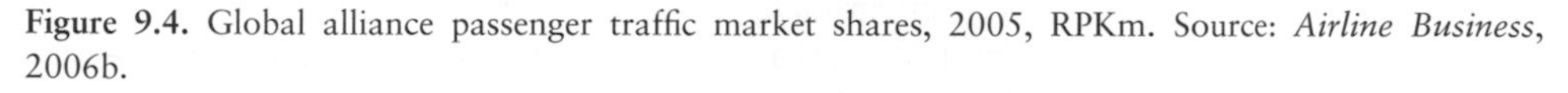

Figure 9.4. Global alliance passenger traffic market shares, 2005, RPKm. Source: *Airline Business*, 2006b.

The result is that no one airline could ever mount a global operation without recourse to partners. This has led to the creation of global airline alliances which, at one level, provide a means of circumventing at least some of the restrictions on international services. There are three principal groupings – Star Alliance, oneworld (*sic*) and Sky Team (Figure 9.4; Table 9.3). Each alliance is based on core members in the key air transport regions, supplemented by affiliate carriers in less strategic markets. To an extent, alliances reflect that the globalized world, paradoxically, still remains a bounded and sovereign space in which historical processes of localized economic development continue to influence the location of economic activity. Despite the revolution in air transport and other communications technologies, all economic activity is grounded in specific locations, 'both physical[ly], in the form of sunk costs, and less tangibl[ly] in the form of localised social relationships' (Dicken, 1998: 11). One consequence is noted by Zook and Brunn (2006) who adapt Goetz's 'pockets of pain' concept (2002) – those places that 'lost out' in US deregulation – to the global context. They argue that there are similar 'forgotten places', actively being forged as a result of processes of inclusion and exclusion in the global economy.

Liberalization

Globalization is partially explicable by economic shifts, which have encouraged free trade and increased competition achieved through world-wide processes of deregulation and liberalization and the removal of trade barriers. In this present context, deregulation involves the exposure of air transport to *laissez-faire*, or free-market, forces, achieved through the removal of most regulatory controls over pricing, while permitting carriers to enter and leave markets at will. Governments (and the European Commission) do still retain significant direct and indirect regulatory powers over air transport, which is why the term liberalization is often

Table 9.3. Airline global alliances, September 2004. Source: *Airline Business*, 2006b.

Alliance	*Member airlines*	*Notes*
Star Alliance	Adria Airways*, Air Canada, Air New Zealand, All Nippon Airlines, Asiana Airlines, Austrian Airlines Group, Blue1*, bmi British Midland, Croatia Airlines*, LOT Polish Airlines, Lufthansa, SAS Scandinavian Airlines, Singapore Airlines, South African, Spanair, Swiss, TAP Portugal, Thai Airways, United Airlines, US Airways, Varig	New members may include: Air China, Shanghai Airlines
oneworld	Aer Lingus, American Airlines, British Airways (BA), Cathay Pacific, Finnair, Iberia, LAN Airlines, Qantas	Crucially, oneworld lacks US anti-trust immunity for the relationship between its two most powerful members, BA and American Airlines, a factor that significantly disempowers the entire alliance. Aer Lingus is leaving oneworld. New members may include; Japan Airlines, Malev, Royal Jordanian
SkyTeam	Aeroflot Russian, Aeromexico, Air France–KLM Group, Alitalia, Continental Airlines, CSA Czech Airlines, Delta Air Lines, Korean Air, Northwest Airlines	New members may/will include: Air Europa*, Copa*, China Southern, Kenya Airways*, Middle East Airlines*, PGA-Portugalia*, Tarom*

Note:
* Associate or regional member.

preferred to deregulation. There is generally a time-lag between the implementation of domestic and international deregulation, the former being much easier to realize, either at the scale of the individual country or a trading bloc such as the EU.

First in North America, then in the EU and, now, elsewhere in the world, the most important outcome of liberalization has been the rise of the low-cost carrier (LCC), the chronology of airline market deregulation contributing to the uneven spread and speed of the low-cost model around the world (Francis *et al.*, 2006; Lawton, 2002). LCCs are exploiting the derived demand for air transport by selling mobility at low cost and therefore promoting behavioural changes in leisure and business traffic. The definition of a LCC is rather ambiguous as there are numerous product differentiations within the sector. All the airlines share, however, in a commitment to what Lawton (2003) terms the 'cult of cost reduction', a business model that offers low fares and strips out overall costs.

Table 9.4. Key principles of the low-cost carrier model. Source: Graham and Vowles, 2006.

- High-capacity seating;
- Minimum legal crew;
- Cabin service only at additional cost;
- Fast turn-rounds;
- On-board air stairs instead of airport air bridges;
- Operating procedures to minimize take-off thrust and braking on landing, congruent with runway length;
- Point-to-point traffic only;
- No freight;
- Advantageous rates from airport operators;
- Generally sectors of less than two hours to maximize aircraft utilization;
- Online booking to eradicate travel agent commission;
- Supplements for payment by credit card;
- Sophisticated websites with extensive information on destinations;
- One-size and type fleets (although some LCCs have compromised on this point).

Doganis (2001) identifies three critical areas in this process. First, labour unit costs (which account for between 25–35 per cent of total costs) have to be contained and the productivity of that labour increased. Secondly, airlines themselves have to implement e-commerce sales, ticketing and distribution (15–20 per cent of total costs). Finally, costs can be further reduced through operational and service changes including: reductions in cabin service (and staff); outsourcing of maintenance and ground handling; single-aircraft type; and enhanced aircraft utilization. Doganis estimates that a LCC can operate sustainably at 40–50 per cent of the unit cost of the average network carrier. Overall, LCCs have unit costs up to 60 per cent lower than network carriers achieved through a set of now relatively standardized operating practices (Table 9.4). Comparing Asian and European LCCs, O'Connell and Williams (2005) found that there are no major cross-cultural differences in passenger expectations or profiles. In a market strongly biased towards younger people, fares are the dominant factor whereas passengers on incumbent carriers are prepared also to value quality, reliability, flight schedules, frequent flyer programmes and comfort.

Crucially for mainline network (or 'legacy') carriers, the LCCs have not just changed airline ticket pricing but also consumer price expectations. The low-cost model was pioneered by Southwest Airlines in the United States and has been widely emulated by other North American carriers such as AirTran, JetBlue and WestJet, and in Europe by Ryanair and easyJet. Alamdari and Fagan (2005: 391) argue that adherence to an 'original' Southwest model based on cost leadership, 'could potentially ensure greater profitability', although they recognise that LCCs have to fine-tune the model to specific market conditions (not least being the need to appeal to business as well as leisure travellers). The low-cost model has also penetrated Asia-Pacific, pioneered by Virgin Blue in Australia and the Malaysian carrier, AirAsia. There have been a number of new start-ups, particularly in India, but it is, as yet, less than clear how the model will evolve in this region where air traffic between states is still generally controlled by restrictive bilaterals. Whichever the region, however, distinct advantages accrue to 'first movers' in that market.

Faced with this dramatic expansion of the low-cost sector, mainline carriers have been forced to refocus their operations in terms of costs and yields (the average revenue produced by each passenger-km or tonne-km carried). There has been a step change in the nature of the liberalized airline industry since around 1990: 'Cost reduction is no longer a short-term response to declining yields or falling load factors. It is a continued and permanent requirement if airlines are to be profitable' (Doganis, 2001: 222). The legacy carriers have responded to competition from LCCs by reducing labour costs (through layoffs, wage and benefit cutbacks, and shedding pension benefit plans), lowering their own fares for point-to-point services, reducing the standard of cabin service, and by promoting electronic booking on the Internet. Some have even established their own low-cost 'carrier-within-carrier' subsidiaries (Graham and Vowles, 2006; Morrell, 2005).

Network strategies

The impact of the LCCs has also altered the geography of air transport networks. The initial response by legacy carriers in the US market to the threat of start-up competition in the deregulated internal market after 1978 lay in the creation of hub-and-spoke networks. Strictly speaking, a hub is an integrated air transport interchange through which (normally) a single carrier operates synchronized banks – or waves – of flights (Figure 9.5). In these, the hub-arrival times of aircraft, originating from cities at the ends of numerous spokes, are co-ordinated into a short time period. After the minimum interval necessary to redistribute passengers and baggage, an equally large number of aircraft departs to the spoke cities. This pattern is repeated several times during the day (Dennis, 1994; Graham, 1995; Vowles, 2006).

Hub dominance has been regarded as a large incumbent's most effective defensive tactic in a liberalized market because, especially when combined with airport congestion and linked to an alliance strategy, it offers the real possibility of pre-empting

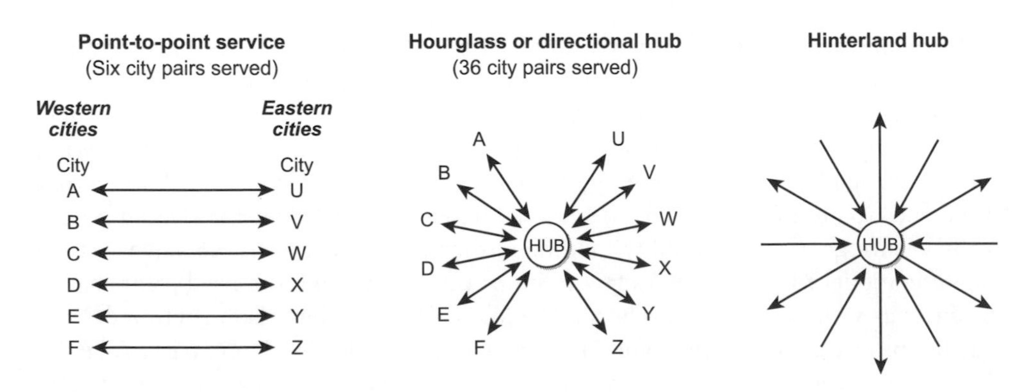

Figure 9.5. Hub-and-spoke model.

– or at least controlling – competition at a particular airport. Efficient hub operation is dependent upon available runway and terminal capacity to handle the peaks, combined with extensive feeder connections, often employing smaller aircraft operated by regional airlines. The US hub-and-spoke model, with its dominant carriers and dedicated terminals and gates, has not been replicated fully in the EU, largely because of existing restrictions on airport capacity and political fragmentation. The LCCs, conversely, following the model pioneered by Southwest in the United States and then by Ryanair in Europe, operate point-to-point networks incorporating a significant number of 'bases'. These are often but not necessarily located at secondary airports which offer lower operating costs and faster turn-around times to maximize aircraft utilization.

Despite the importance of hub concentration, the future strategic domination of the mega-hub has been questioned because of an apparently contradictory trend towards dispersal or 'fragmentation'. Liberalized bilaterals have allowed a proliferation of intercontinental routes, with smaller aircraft serving a much larger number of point-to-point city-pairs. This first occurred on the North Atlantic but is now increasingly apparent in other markets. To some extent, fragmentation is a reflection of the ways in which globalization encourages long-distance interaction, thereby elongating supply lines and demanding the use of smaller vehicles. Transport demand in general for both passengers and freight is becoming more customized and dispersed as global activity is dispersed away from the top-ranked global cities (O'Connor, 2003). Moreover, despite real-time information and communications technology (ICT), the continuing demand for face-to-face contact requires more low-density routings. Banister and Stead (2004) believe that there appears to be no limit to the relentless demand for air transport. Nor is ICT an effective alternative to air transport. Much leisure traffic is undertaken for its own sake while ICT can act as a facilitator and generator of traffic by providing greater knowledge and easier booking of air travel through the Internet (Giuliano and Gillespie, 2002; Janelle and Gillespie, 2004).

Airbus (A) and Boeing (B), the two companies that dominate global aircraft manufacturing, have diametrically opposed perspectives on hub-and-spoke versus fragmentation. Boeing, which is developing the carbon-titanium B787 due to enter service in 2008, favours point-to-point or one-stop connecting services over a single hub as alternatives to multi-sector journeys. This network strategy requires airlines to 'maintain or reduce airplane size to provide frequent, non-stop service' (Boeing, 2003: 11). Airbus, conversely, argues that 'in response to increasingly severe cost pressures, established airlines [as distinct from LCCs] will be driven even further to improve the efficiency of their route networks and to use low-unit-cost aircraft'. This will involve the replacement of point-to-point systems by 'lower-cost, lower-fare "hub" systems' (Airbus, 2002: 13 and 17). Thus the airframer has developed the A380 as a low-cost people carrier catering for the bulk of long-haul passengers concentrated in the world's major centres of population and being moved across hubs. It acknowledges that high-yield traffic will demand direct, frequent non-stop point-to-point flights. On this point, Boeing concurs but its fragmentation model sees large aircraft being flown only 'on dense routes by a limited number of airlines'. Instead, 'most growth in the world's airlines will manifest as increased frequencies, more nonstops, and new city pairs served by small- and intermediate-size airplanes' such as the B787 (Boeing, 2003: 14).

Sustainability

The consensus of the most recent study into the prospect of what some commentators regard as the 'oxymoron of sustainable aviation' is that, at best, the environmental sustainability of the air transport industry is in doubt, although, conversely, aviation is delivering social and economic goods (Upham *et al.*, 2003). Air transport policy making has been driven by the concern to introduce, implement and protect the competitive marketplace. Nevertheless, as is a characteristic of all transport modes, such policies do not encourage individual restraint in the use of environmental resources on the part of any one airline. Air travel can be viewed as another 'tragedy of the commons', the situation in which people believe that any individual sacrifice for the greater good (in this case, the environment) would have no value unless followed by all others (Shaw and Thomas, 2006).

The principal environmental sustainability externalities created by air transport are:

- Noise from aircraft engines, airframes and ground traffic;
- Atmospheric pollution, especially from the effects of contrails in the upper atmosphere, nitric oxide/nitrogen dioxide (collectively NO_x), and carbon dioxide (CO_2), the principal cause of global warming; conventional turbofan engines, no matter how refined, cannot achieve emissions levels that will allow a stabilisation of CO_2 concentration in the atmosphere (Åkerman, 2005);
- Terrestrial pollution at airports, both airside and landside including: water pollution from surface run-off; waste; congestion; and
- Rate of aviation fuel use exceeding the rate at which substitutes are being developed.

Technological improvements to reduce both noise and emissions are being offset by the escalating growth trends in air travel which has been expanding at nearly 2.5 times average economic growth rates since 1960. Moreover, traffic growth is increasing exponentially. For example, predictions for demand at UK airports range between 400–600 million passengers per annum (mppa) by 2030, compared to around 200 mppa in 2003 (DfT, 2003a). Again, an increasing percentage of the projected increase of global demand (in excess of 5 per cent per annum) is being satisfied by short-haul flights (Whitelegg and Cambridge, 2004). Air transport's principal external costs are not 'damage' as such but 'costs unaccounted for' (Wit *et al.*, 2002: 13). Noise and emissions costs are not internalized and are 'thus excessive in terms of what might be anticipated if a sustainable environment is to be attained' (Button, 2001: 70). Airlines pay no tax on kerosene while new aircraft in the EU are zero-rated for value-added tax (VAT) as are air tickets, although there are alternative taxes such as the United Kingdom's (UK) Air Passenger Duty (APD).

Public opposition to aviation, which is both situationally and culturally determined, tends to focus primarily on noise rather than emissions. Whereas modern aircraft are quieter than their predecessors, it is the volume of traffic – both airside and landside – that compounds public exposure to noise, particularly for residents in the hinterlands of major airports. In general terms, internationally negotiated and implemented noise controls have largely realized the potential returns from contemporary aircraft engine technology and future gains are most likely to come from advances in airframe design. It is now widely recognised, however, that the most serious sustainability impacts of air transport stem from atmospheric pollution at both global and local scales (Environmental Change Institute, 2006). Again, although technology has been successful in reducing atmospheric emissions per individual aircraft and passenger, the technological returns are diminishing and being offset by aviation's growth. If present growth trends continue, 'air travel will become one of the major sources of anthropogenic climate change by 2050' (Royal Commission on Environmental Pollution (RCEP), 2002b: 37).

If 'aviation is moving in an unsustainable direction due to absolute increases in environmental consumption and emissions' (Upham *et al.*, 2003: 16), the industry itself is pinning its hopes on emissions trading rather than fiscal measures to reduce demand. One justification is that sustainability links concerns with environmental carrying capacity to strategies for long-term economic development, social needs and equity. Such targets, especially when applied to more peripheral or disadvantaged regions, demand accessibility, while the access of such isolated areas to wider networks is a basic social equity objective. Firms also require accessibility to factors of production and markets. But the infrastructure created to enhance accessibility also encourages mobility, which is essentially a behavioural attribute, and moreover one easily manipulated by price. Arguably it is the provision and ready availability of cheap mobility, best exemplified here by the growth of LCCs, which provides the basic challenge to the environmental dimension of sustainability (Graham, 2003).

Airports in a globalizing age

Not surprisingly, given the global pre-eminence of US carriers and the size and spatial extent of the US domestic market, the principal US airports dominate the world rankings for passenger traffic (Table 9.5). Of the top 30 airports world-wide in 2005, ranked by total passengers handled, no less than 17 were located within the 48 contiguous states, compared to six in Asia-Pacific, six in Europe and one in Canada. The top ten airports for international traffic, however, are located in Europe or Asia. London remains by far the dominant international node in the global air transport network. Memphis, Tennessee, home hub of the world's largest cargo carrier, FedEx, leads the air freight rankings.

Airport congestion – both airside (runways and aprons) and landside (terminals and parking) – and mounting public hostility to the provision of additional capacity are among the most serious problems facing the liberalized air transport industry. Many airport operators have been fully or, more commonly, partially privatized but that strategy in itself, while leading to increased profits, is essentially irrelevant to the capacity issue (for a comprehensive survey of airport operations, see

Table 9.5. Top 30 airports, 2005. Source: *Airline Business*, 2006c.

Rank	*Airport*	*Total terminal passengers handled (m)*
1	Atlanta Hartsfield	85.9
2	Chicago O'Hare International	76.8
3	London Heathrow	67.9
4	Tokyo Haneda	63.3
5	Los Angeles International	61.5
6	Dallas/Fort Worth International	59.1
7	Paris Charles de Gaulle	53.6
8	Frankfurt International	52.2
9	Las Vegas McCarran	44.3
10	Amsterdam Schiphol	44.2
11	Denver International	43.3
12	Madrid Barajas	41.9
13	Phoenix Sky Harbor	41.2
14	Beijing Capital	41.0
15	New York JFK	40.6
16	Hong Kong Chek Lap Kok	40.3
17	Houston George Bush	39.7
18	Bangkok International	39.0
19	Minneapolis/St Paul International	37.6
20	Detroit Wayne County	36.4
21	Orlando International	33.9
22	San Francisco International	33.6
23	Newark Liberty	33.0
24	London Gatwick	32.8
25	Singapore Changi	32.4
26	Tokyo Narita	31.5
27	Philadelphia International	31.5
28	Miami International	31.0
29	Toronto Lester B. Pearson	29.9
30	Seattle/Tacoma International	29.2

A. Graham, 2001). Most major European airports, and a significant number on the United States and Asia-Pacific, are capacity-constrained, which effectively means that airline access is rationed. Yet, not least because of environmental objections, very few new airports or even runways are currently being built or even planned in Europe. In the United States, the controversial Denver International, which finally opened to traffic in 1995 following serious delays and escalating costs (largely incurred by the failure of a high-technology baggage system), remains the only important recent new airport (Dempsey *et al.*, 1997; Goetz and Szyliowicz, 1997). The most substantial airport developments are taking place in Asia and the Middle East (Dempsey, 2000). In Asia, these include: Kansai International, sited on an artificial island in Osaka Bay and opened in 1994; Macau (1995) and Chek Lap Kok (1998) in Hong Kong; Seoul Incheon (2001) and Bangkok Suvarnabhumi (2006). China is also radically enhancing and extending its airport capacity in response to annual double-digit passenger growth rates. The explosive growth of the Middle East as a hub-and-spoke location is leading to enhanced and ultimately new airport capacity in Dubai and Qatar and the transformation of Abu Dhabi into a major hub.

Still, the general shortage of airport (and airspace) capacity remains one fundamental impediment to the development of a more competitive airline industry. Even though LCCs may favour secondary airports, scarce capacity will continue as a major constraint on market entry while favouring incumbency through the priority given to historic rights in the allocation of scarce runway landing and take-off slots. Consequently, major policy tensions exist between the implementation of more competitive airline industries offering tariff, frequency and service benefits to consumers, and widespread political and public opposition to the construction of new airport capacity. Significantly, airports have also become the 'front line' in the enhanced security measures imposed since 9/11, one effect being to increase the strain on scarce terminal capacity. The LCCs, meanwhile, require much more basic terminal facilities than do the legacy carriers and major airports are now having to construct dedicated low-cost terminals.

The major air transport markets

North America

Deregulation of the US airline industry has consistently been deemed a success by government and industry, while labour and consumer groups have been more sceptical. From 1978 to 2000, there was, in general, a greater quantity of service, more people flying, and lower average fares, although widespread differences occurred across places based largely on level of single carrier domination, size of market demand, and presence or absence of LCCs such as Southwest Airlines (Fuellhart, 2003; Goetz, 2002; Goetz and Sutton, 1997; Vowles, 2000). Very marked concentration was characteristic, the eight largest airlines accounting for 95 per cent of the domestic passenger market, while most hub airports came to be dominated by single carriers controlling more than 70 per cent of the traffic. Today, the principal concern is the worst financial crisis in the history of the US airline industry which lost nearly $35 billion from 2001 to 2005 (Air Transport Association, 2006), and has resulted in substantial industry upheaval as several major carriers sought bankruptcy protection, including United, Delta, Northwest and US Airways (twice since 2002). Meanwhile, TWA was acquired by American in 2001, US Airways merged with America West in 2005, and it is quite possible that additional mergers will reduce the number of 'legacy' carriers. In order to attract more passengers in an uncertain economic and security environment, airlines have been forced to lower fares considerably. Total traffic has consequently rebounded from the post-2001 decline, a record 738 million passengers being enplaned in 2005, but revenues have not been able to outpace costs due to low yields and rising fuel prices.

In addition, there are the concerns as to the future efficacy of the hub-and-spoke strategy and the extent to which the LCC model has permanently altered market conditions so that there is no possibility of the major network carriers being able to reconstitute the market conditions in which high-yield traffic once sustained their revenues. Network carriers have seen their market share shrink from 62 per cent in 2000 to 48 per cent in 2005, while LCC share has grown from 18 per cent to 26 per cent. Although the scale of large hub operations is an effective barrier to market entry, hubs are also undermined by the high labour costs of traditional carriers. As the LCCs grow their networks, more passengers are using them to make connecting journeys (up to one-third in the case of Southwest). Only a few LCCs (such as Frontier and AirTran) have yet found it expedient, however, to establish at least a rudimentary hub as travellers seem prepared to suffer inconvenient, poorly timed connections if the price differential is sufficiently attractive.

Europe

The liberalization of the EU air transport industry was a much more gradual process than in the United States, being introduced through three successive stages of legislation between 1988 and 1997. From April 1997, all EU airlines had open access to virtually all routes within the then 15 member states (plus Iceland, Norway and Switzerland). The only exceptions have been public-service obligation (PSO) routes serving remote regions, cities and islands which are awarded by competitive tender. The enlargement of the Union in 2004 added a further ten states, largely in Central and Eastern Europe, to the Single Aviation Market. European liberalization has had two dominant repercussions.

First, not least because of bilateral constraints, the development of alliances, and the inertia produced both by demand and airport capacity issues, the major European carriers were able to restructure their formerly national networks into trans-European hub-feed systems. The most important are centred on London Heathrow (British Airways – 41.3 per cent of flights in 2005), Frankfurt Main (Lufthansa – 59.8 per cent), Paris Charles de Gaulle (Air France group – 53.6 per cent), Amsterdam Schiphol (KLM – 57.7 per cent) and Madrid Barajas (Iberia group – 49 per cent). The principal EU airports are less dominated than some of their US counterparts (for example, American Airlines has 85 per cent of flights at Dallas/Fort Worth and Northwest has almost 80 per cent at both Minneapolis-St Paul and Detroit), although this trend is being altered by the growing importance of alliances which have particular strengths at particular hubs.

Secondly, and especially since 2001, the European low-cost sector has attracted a bewildering number of new entrants; it is estimated that about 50 LCCs were operating in 2005 (see Dobruszkes, 2006). While not all these will survive, their sheer number is indicative of the problem facing the mainline carriers which, if less damaged by 9/11 and its aftermath than US carriers, are shedding staff and capacity while some are still incurring heavy losses. They are experiencing increasing difficulties in sustaining their networks, especially in terms of the balance between costs and yields and have yet to formulate an effective strategy against the explosive growth of the LCC sector. The Irish national carrier, Aer Lingus, has even transformed itself into a 'superior' LCC (Barrett, 2006), rather like JetBlue in the United States. The two largest LCCs, Ryanair and easyJet, carried 33.4 million and 29.6 million passengers respectively in 2005, placing them both in the world top

20 airlines by this measure. In round terms, LCCs accounted for about 20 per cent of all European air traffic in 2005 (although this rises to about 50 per cent for the British Isles–continental Europe market), while the sector's growth rate is much greater than that of the legacy carriers. Unlike the legacy carriers, the LCCs have opted for linear point-to-point networks although these remain significantly concentrated because of their focus on base airports (Burghouwt, 2005; Burghouwt *et al.*, 2003; Dobruszkes, 2006). Because they often use secondary airports, there has been a marked increase in city-pairs served, often involving quite small provincial towns (Fan, 2006). Both legacy carriers and LCCs, most especially in France, Germany, Spain, Italy and between London, Paris and Brussels, also face intermodal competition from high-speed trains which can compete in terms of time for distances of up to approximately 500 km (Chapter 8).

Asia-Pacific

Asia-Pacific (including Australasia) is the world's fastest-growing air transport market. While the rate of increase is showing signs of slowing down as some of its most dynamic economies – located in the West Pacific Rim – begin to mature, the Chinese and Indian markets are set on a potentially huge expansion as disposable personal income increases through economic development, albeit for limited percentages of their populations. Air transport is of particular importance in this macro-region because of the dominance of its cities in economic development, the rapidly developing importance of tourism and its fragmented geography. While there has been significant liberalization in many countries, most Asian states have as yet stopped short of relinquishing outright control of flag-carriers to the private sector. This reflects the perceived importance of the air transport industry and the desire of governments to pursue selective, pragmatic and deliberate development policies (Bowen and Leinbach, 1995).

The key to understanding the region's air transport infrastructure lies in the same concepts of centrality and intermediacy that have also been seen as diagnostic features of US hubs (O'Connor, 1995). Three southeast Asian airports – Singapore, Bangkok and Hong Kong – act as the dominant nodes. Traditionally, intermediacy was most relevant to east–west traffic (for example, the location of Singapore on the Europe–Australia routes) but the increasing importance of a north–south axis, linking Australia to the West Pacific Rim, particularly favours Hong Kong. Further north, Seoul, Osaka and, above all, Tokyo, are the dominant airports. In addition, Japanese domestic routes are among the densest in the world. The most important international routes link the major cities of the West Pacific Rim to each other and to the dominant east–west axis of global aviation and the largest airline groups ranked by revenue – Japan Airlines, ANA, Qantas, Singapore Airlines and Korean Air – all have hub-and-spoke networks.

As with other air transport regions, there has been a recent upsurge in LCC market entry (Forsyth, 2003; Lawton and Solomko, 2005) although the development of routes across international boundaries remains somewhat constrained by bilaterals (O'Connell and Williams, 2005). It had been thought that the high-density city-pairs, which often necessitate the use of wide-body aircraft, and/or the longer-distance sectors characteristic of the region might militate against the development of LCCs on the North American and European model. As late as 2004, evidence to the contrary was still largely confined to AirAsia in Malaysia and Virgin Blue in Australia. The mainline Asian-Pacific carriers, including Thai International, Singapore Airlines, Air New Zealand and particularly Qantas, have also been more aggressive than those in other regions in establishing their own low-cost subsidiaries and this trend looks set to escalate (Graham and Vowles, 2006). The most dynamic growth of LCCs in Asia seems likely to occur in India where very large aircraft orders by start-up carriers such as Air Deccan and Kingfisher Airlines point to a quantum leap in the provision of air services within the subcontinent, even if some companies eventually fail.

Rest of the world

If the east–west, northern hemisphere, axis that links Asia-Pacific, North America and Europe is the dominant feature of international air transport, the remainder of the world – despite the size of its population – is of relatively minor importance. Australia and New Zealand, of course, are firmly integrated into the Asian-Pacific rim but, despite rapid growth in the Middle East and South America, these regions, together with Africa, remain essentially niche players in global air transport patterns.

The Middle East has a high potential for traffic growth. The Gulf airports used to be important stop-overs on the routes between Europe and Asia but the development of longer-range aircraft allowed these services to be flown non-stop. Although the region then became something of a spoke, and even a little isolated from the principal international air transport flows, the recent development of hub-and-spoke carriers, first Emirates and then Qatar Airways and Etihad Airways, has transformed traffic patterns in the region. All three are estimated to have a 50:50 split between transfer and origin-destination traffic, reflecting the growth of the Middle East as a tourism and business destination but also the dense migrant labour traffic between it and the Indian sub-continent.

Despite a geography of distance and fragmentation that is ideally suited to air transport, Africa accounts for under 3 per cent of world international scheduled air traffic. Intercontinental traffic is largely north–south, reflecting the historical linkages of imperialism. Notably, however, South Africa has been reconfigured in the post-Apartheid era as an African country in terms of its external air transport linkages and now has far more connections across its borders into the rest of the continent than was hitherto the case (Pirie, 2006). International and domestic services within the continent – the Republic of South Africa excepted – are often still poorly developed in absolute and frequency terms. Many airlines remain in state ownership while very few are integrated into global alliances. The principal reason for these trends, of course, is poverty and dependency, reflecting again the central point that propensity to travel by air remains conditional on wealth.

Finally, Central and South America, together with the Caribbean, account for around 7 per cent of world international scheduled passenger traffic. South America shares with Africa an air transport network largely shaped by the dictates of economic and political colonialism. International traffic is dominated by routes to the United States (especially the Miami gateway) and often by the US mega-carriers. European services are channelled through Madrid and, to a lesser extent, Lisbon, the former colonial capitals. Many South American governments opted for abrupt deregulation on the US model and although some carriers, most impressively Lan Chile, have successfully made the transition, other legacy carriers, notably Brazil's Varig, have been decimated by low-cost competition.

Conclusions

The key issue facing air transport – as demonstrated by this brief regional survey – is that the market strategies driven by globalization and liberalization have, at best, unpredictable consequences in terms of sustainability. Airlines as businesses have no rational alternative but to cater to existing demand in ways that are most profitable, while fostering future demand. It can be argued that globalization and liberalization strategies have rationalized the airline industry into a more efficient operation that enhances long-term sustainability. But the alternative interpretation, which argues that globalization and liberalization have resulted in excessive air traffic growth and wasteful competition, is more plausible.

Volatility in demand fuelled by events external to the industry represents no more than short-term declines in a long-term cycle of growth in which the low-cost step change is now the dominant factor. This can function only because the airlines operating this strategy do not pay for the externalities which they create, despite air transport being the most environmentally damaging form of transport per passenger-kilometre. Arguably, the low-cost air transport model is ultimately unsustainable in environmental terms. Like much else in the aviation business, it is a strategy dominated by short-termism and self-interest which has changed the pricing structure of air travel, but has also accelerated the growth rates of a mode that is the fastest-growing cause of transport's contribution to atmospheric emissions. Moreover, in altering the basis of demand for air transport and accentuating the role of mobility, the LCCs have modified the behavioural basis of the demand for air transport. Low fares allow customers to fulfil derived demand in a much wider variety of ways, and more often, while also stimulating latent demand. This is satisfied with relatively small aircraft flying short sectors which could often be served more efficiently by terrestrial modes of transport. Modal substitution is not, however, the ready answer. Givoni (2006) concludes that even if there was high-speed train substitution on all feasible routes, the air transport sector could not meet future demand within current airport capacity provision. Although growth rates may decline as the market matures, it is transparent that the environmental externalities of aviation will grow accordingly. At some point, legislators must address the issue of internalization of external costs and the LCC model can exist in its present form only if this does not occur. Ultimately, its success depends on the assumption that air transport remains heavily subsidized and, unlike other transport modes, is not subject to demand management.

10 | International Maritime Freight Movements

Jean-Paul Rodrigue and Michael Browne

Few transport systems have been more impacted by globalization than freight transport. Paradoxically, in a field dominated by passengers, freight remains fairly unnoticed by the general public, albeit manufacturers and retailers are keenly aware of the benefits derived from efficient distribution. In fact, the profit margin of many retailers and manufacturers is directly dependent on efficient distribution strategies encompassing a wide array of global suppliers. As such, in the last decades, international trade has systematically expanded at a rate faster than economic growth, an outcome of an international division of the production and massive accumulation of new manufacturing activities in developing countries. Maritime transport is at the core of global freight distribution in terms of its unparalleled physical capacity and ability to carry freight over long distances and at low costs. Aside from these well-known characteristics, the maritime industry has changed substantially in recent decades. From an industry that was always international in its character, maritime transport has become a truly global entity with routes that span across hemispheres, forwarding raw materials, parts and finished goods. In fact, it is one of the most globalized industries around:

> A Greek owned vessel, built in Korea, may be chartered to a Danish operator, who employs Philippine seafarers via a Cypriot crewing agent, is registered in Panama, insured in the UK, and transports German made cargo in the name of a Swiss freight forwarder from a Dutch port to Argentina, through terminals that are concessioned to port operators from Hong Kong and Australia. (Kumar and Hoffmann, 2002: 36)

International maritime freight transport is composed of two main segments, the modes which are flexible in their spatial allocation, and the terminals, as locations, which are not. Shipping lines have a level of flexibility in terms of route selection, frequency and levels of service, but port terminals have a fixed capacity that if not used can imply serious financial consequences. Reconciling these two segments remains a challenge, particularly since the volume of maritime freight is steadily growing and since freight distribution is getting more complex as it services many origins, destinations and supply chains. Logistics has done a lot to reconcile the strategies of the maritime actors and in many cases shipping lines have taken matters into their own hands by investing directly in terminal facilities and securing access to hinterlands. Global port operators such as Hutchinson Port Holdings, APM Terminal, Port of Singapore Authority and Dubai Ports International are now managing terminal facilities in almost every single major port around the world.

Ports as locations where maritime and land traffic converges are crucial facilities in the global economy. There are more than 4,500 commercial ports around the world, but only a small share handle a significant amount of traffic (National Geospatial Intelligence Agency (NGIA), 2005). The geography of these ports conditions the global geography of trade and flows since they are locations that cannot be easily bypassed. Ports remain points of convergence and divergence of traffic and their location is constrained by the physical characteristics of their sites. The first physical constraint involves land access and the second concerns maritime access; both must be jointly satisfied as they are crucial for port operations and the efficiency of the maritime/land interface. Thus, poor land and maritime access can impair port operations and port development, although maritime access is the attribute that can be mitigated the least. Activities such as dredging and the construction of facilities such as docks are very expensive, underlining the enduring importance of a good port site, albeit inland access also endures as a factor of importance for maritime freight distribution.

Global production, transport and distribution require the setting of freight management strategies. As such, logistics concern all the activities required for goods to be made available on markets, with purchase, order processing, inventory management and transport among the most relevant. The expansion of production in the global realm induced transport systems to adapt to a new environment in freight distribution where the reliable and timely deliveries can be as important as costs. Logistics has consequently taken an increasingly important role in the global economy, supporting a wide array of commodity chains (Hesse & Rodrigue, 2004, 2006). This is the setting in which maritime transport is increasingly embedded. At the core of this relationship, global commodity chains (GCCs) can be considered as functionally integrated networks of production, trade and service activities that cover all the stages in a supply chain, from the transformation of raw materials, through intermediate manufacturing stages, to the delivery of a finished good to a market (see, for example, Gereffi, 1999). The development of global transport and telecommunication networks, information technologies, the liberalization of trade and multinational corporations are all factors that have substantially impacted GCCs (Dicken, 2003).

In such a new environment, the precepts of international freight transport are being re-defined, both for bulk and for break-bulk cargo. The former comprises homogeneous materials without packaging (ores, coal, grain, raw sugar, cement, crude oil and oil products, and so on) usually for a single consignee and destination, while the latter, often known as general cargo, consists of an almost infinite variety of freight, usually in small consignments for numerous consignees and packaged in a variety of bags, bales, boxes, crates and drums of diverse shape and size. Containers, however, account for the majority of the break-bulk cargo being carried.

Bulk maritime freight

Characteristics

The marine industry is an essential link in international trade, with ocean-going vessels representing the most efficient, and often the only method of transporting large volumes of basic commodities and finished products (Gardiner, 1992). In 2005, approximately 2.9 billion tons of dry bulk cargo was transported by sea, comprising more than one-third of all international seaborne trade. Bulk freight represents the 'traditional' segment of maritime freight distribution with a wide variety of physical characteristics of the cargoes. Each has specific requirements with respect to stowage in the ship, methods of transhipment and inland transport (Table 10.1). Geographically, bulk cargo shows a remarkable stability, particularly in terms of its origins. The extraction and shipment of natural resources, such as minerals and oil, is bound to the geological setting, require massive capital investments and

Table 10.1. Types of maritime cargo. Source: adapted from Hilling & Browne, 1998.

Commodity type	*Examples*	*Maritime Transshipment*	*Inland distribution*
Liquid			
A) Normal pressure and temperature	Crude oil, most oil products, wine slurried coal	Pump/pipe	Pipeline
B) Other pressure and temperature	Liquefied gases (LNG), heavy oils, latex, bitumen, vegetable oils	Pumps, temperature controlled pipelines	Temperature controlled pipelines
Dry Bulk			
A) Flowing	Grain, sugar, powders (alumina, cement)	Pneumatic/suction, conveyor, grabs	Pipes, conveyors, barge, rail wagon, lorry
B) Irregular	Coal, iron ores, nonferrous ores, phosphate rock	Grab, conveyor	Conveyor, barge, rail wagon, lorry
Neo Bulk	Forest products, steel products, baled scrap	Lift-on/lift-off, roll-on/roll-off	Barge, rail wagon, lorry
Wheeled Units	Cars, lorries, rail wagons	Roll-on/roll-off	Rail wagon, lorry
Refrigerated/chilled cargo	Meat, fruit, dairy produce	Lift-on/lift-off	Rail wagon, lorry

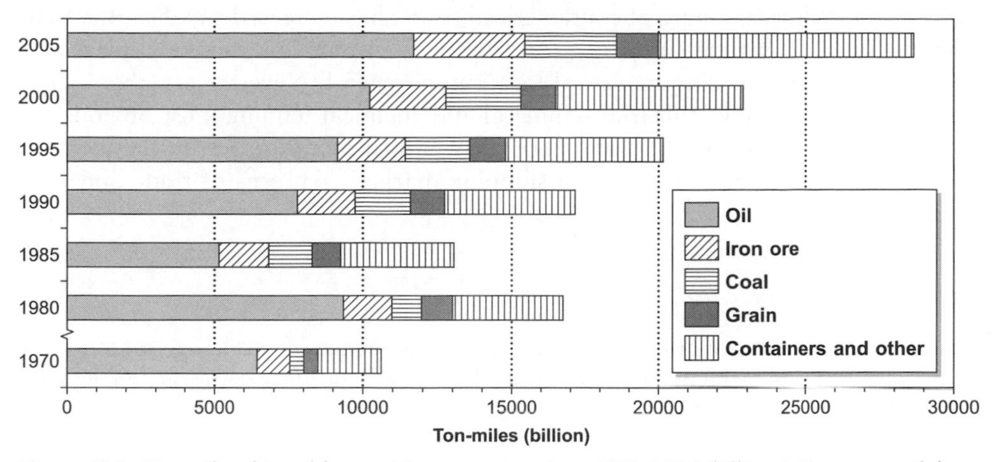

Figure 10.1. Ton-miles shipped by maritime transportation, 1970–2005 (billions). Data sourced from UNCTAD, 2006 and various other dates.

takes place over decades. The maritime traffic associated with these activities is thus highly consistent and varies according to cyclical demand patterns. The same applies for agricultural commodities since agricultural regions are long standing entities with a reasonably consistent output. What is changing more rapidly concerns the destinations of bulk freight as they reflect changes in economic development and the setting of new markets and industrial regions. A remarkable shift thus involved growing demands from the industrializing countries of Pacific Asia and the related changes in the volume of bulk shipping routes.

Dry bulk cargo is shipped in large quantities and can be easily stowed in a single hold with little risk of cargo damage. It is generally categorized as either major bulk or minor bulk. Major bulk cargo constitutes the vast majority of dry bulk cargo by weight, volume and shipping ton-miles generated, and includes iron ore, coal and grain. Minor bulk cargo includes products such as agricultural products, mineral cargoes (including metal concentrates), cement, forest products and steel products.

The cargo type is reflected in the associated port activity. For the higher unit value containerized break-bulk cargo, a port is usually the gateway through which the cargo passes to the hinterland, while for the bulk cargo it acts as a terminal – the cargo is stored and often processed before onward movement. In the case of much bulk cargo, the port site is often an industrial site linked with the transformation and processing of those commodities. While break-bulk flows can be bi-directional (inward and outward), bulk flows are dominantly directional (inward or outward). Bulk cargo is thus imported and processed with the output commonly belonging to a different transport chain that cannot be serviced by the original maritime equipment. Even if the unit and possibly the total value of bulk cargoes may not compare with those of general cargo, the sheer volumes involved give them a special significance in transport systems (Figure 10.1).

Oil, iron ore, grain and coal account for the great majority of ton-miles shipped, about 70 per cent in 2005. Containers and other goods composed the remaining

30 per cent. While the share of containerized traffic has increased significantly, bulk still dominates maritime shipping. It may be argued that raw materials play only a small part in influencing industrial location in general. Nevertheless, there are a number of basic heavy industries – mineral and chemical refining most obviously – where the volume of bulk materials does have a profound impact on the location of processing industries and also on shipping markets, patterns of trade and port activity.

Transport of bulk cargo

Much of the conventional port industry is a consequence of servicing commodities moving in bulk. For the maritime cargo to be moved specific conditions have to be satisfied (Stopford, 1997):

- Transportability. The commodity must have physical characteristics that allow it to be handled and moved in bulk. Liquids such as oil require entirely different, and non-convertible, equipment than solids. Technical improvements have recently permitted the potential to move natural gas in large quantities through the use of liquid natural gas (LNG) carriers, but it still remains an expensive and technically challenging endeavour.
- Costs. The demand and the price for the commodity must be such that the cost of specialized ships and handling equipment is justified. The smallest practicable size of consignment is that of the smallest bulk carrier available. About 1,000 tons remains the minimum threshold for bulk handling.
- Compatibility. The bulk-shipping operation must be adapted to the overall transport system, so it is possible to move commodities along multi-modal transport chains. In many cases, this requires specialized terminals.
- Load size. The individual consignment size must be geared to the stocks that can be held at either end of the transport link. This is related to the actual demand at the consuming end, to the storage space available at each end and to the frequency of shipment.

The availability of storage space is an important determinant of the efficiency and productivity of any port since there is an almost inevitable mismatch between the rate of cargo transhipment and the rate at which it enters and leaves the port on the landward side (Takel, 1981). Storage space acts as the essential buffer to balance the flows on the sea and land sides. This is important for general cargoes, but it becomes critical for large volumes of bulk cargo. The amount of space for storage is a function of the density of the commodity where it must allow for access and handling equipment such as stackers, cranes, conveyors and reclaimers. Additional storage space will be needed where materials are sorted by grade or type (e.g. coals, ores and crude oils) and possibly to accommodate changes consequent upon conditions (e.g. wet and dry ores or coal). A regular flow of bulk raw materials is essential for any industrial process. Storage is vital in reducing the effects of flow variations but storage replenishment for a given tonnage can be either by frequent small shipments or by less frequent large shipments. The choice of vessel size is influenced by the interplay between economies of scale, consignment size and the physical constraints provided by the routes, ports and handling equipment (Stopford, 1997).

Shipping is less limited by size constraints than other modes and is able to capitalize on what has been called the 'cube law': for a doubling of a ship's dimensions the carrying capacity is cubed. In addition, the design, construction and operating costs (crew, fuel) do not increase in proportion with size. A 300,000 ton tanker is able to operate with a crew no larger than that needed for a significantly smaller vessel, although there will be variations depending on national flag regulations, level of automation and company organization (e.g. the amount of emphasis on shipboard maintenance). The same rationale applies for containerships where increasing sizes do not require additional labour and in many cases newer containerships have a smaller crew. There is thus a strong rationale in maritime shipping to achieve economies of scale since they are linked with lower operating costs, particularly for bulk carriers, container ships and tankers.

Yet while consignments of such size may be available from oil and ore producers and acceptable at the processing plant, this would not be the case for all bulk trades. Also, there is no financial advantage in using large vessels if the loading and unloading rate is slow and the vessel is kept unduly long in port. The ultimate constraint on vessel size remains the physical characteristics of the port (channel depths, turning circles, lock gate dimensions, and berth lengths) and the routes along which ships operate, particularly strategic passages. This has led to well known capacity standards such as 'Panamax'[1] and 'Suezmax'.[2] A very large crude carrier (VLCC) of 300,000 deadweight tons can transit the Strait of Malacca but any larger vessel would have to make a much longer voyage by way of the deeper Lombok Strait. These standards have for long shaped global bulk trades and more recently containerised maritime shipping.

Seasonal demand fluctuations influence many of the bulk trades. Steam coal is linked to the energy markets and in general encounters upswings towards the end of the year in anticipation of the forthcoming northern hemisphere winter as power supply companies try to increase their stocks, or during hot summer periods when increased electricity demand is required for air conditioning and refrigeration purposes. Grain production is highly seasonal and driven by the harvest cycle of the northern and southern hemispheres with the largest grain producers being the United States, Canada and the European Union in the northern hemisphere and Argentina and Australia in the southern hemisphere, harvests and crops reach seaborne markets throughout the year. It becomes a matter of fleet reassignment to follow the seasonality.

Petroleum trade

Petroleum transport concerns a tightly integrated distribution system that maintains a continuous flow from the oilfields to its end use, most of it being fuel for transport. There is limited storage taking place outside the maintenance of strategic reserves. The volume of international oil trade has increased as a result of world economic growth and additional demands in energy. Although developed countries such as the United States, Western Europe and Japan account for about 75 per cent of global crude oil imports, the largest growth in demand is mainly attributed to China and India. As of 2004, China became the world's second largest oil importer behind the United States. International oil trade is necessary to compensate the spatial imbalances between supply and demand. Unlike most other countries, which either consume almost their entire production (United States and China) or have privileged partners (Russia and Western Europe), a major portion of the Organization of Petroleum Exporting Countries' (OPEC) oil is traded on international markets.

Each year, about 2.4 billion tons of petroleum are shipped by maritime transport, which is roughly 62 per cent of all the petroleum produced. The remaining 38 per cent uses pipelines, trains or trucks over shorter distances. Most of the petroleum follows a set of maritime routes between producers and consumers. More than 100 million tons of oil is shipped each day by tankers, about half of which is loaded in the Middle East and then shipped to Japan, the United States and Europe. Tankers bound for Japan use the Strait of Malacca while tankers bound for Europe and the United States use either the Suez Canal or the Cape of Good Hope, depending on the tanker's size and its specific destination.

Different tanker sizes are used for different routes, namely because of distance and port access constraints. VLCCs are mainly used from the Middle East over long distances (Western Europe, United States and Pacific Asia). 'Suezmax'[3] tankers are mainly used for long to medium hauls between West Africa and Western Europe and the United States, while 'Aframax'[4] tankers are used for short to medium hauls such as between Latin America and the United States. Transport costs have a significant impact on market selection. For instance, three-quarters of American oil imports come from the Atlantic Basin (including Western Africa) with journeys of under 20 days. Venezuelan oil takes about eight days to reach the United States while Saudi oil takes six weeks. The great majority of Asian oil imports come from the Middle East, a three-week journey. In addition, due to environmental and security considerations, single hulled tankers are gradually phased out to be replaced by double hulled tankers (Rodrigue, 2004).

Coal trade

Coal is an abundant commodity which is mined in more than 50 countries with no world dependence in any one region. Coking coal is used to produce coke to feed blast furnaces in the production of steel. An increase in seaborne transport of coking coal has been primarily driven by an increase in steel production. The increase in import activity has occurred in a number of regions. Currently, Asia and Western Europe are major importers of coking coal. Australia and Indonesia provide a significant amount of coking coal to Asia, while South Africa and the United States are major sources for Western Europe. Steam coal is primarily used for power generation. A number of developing countries have decided to capitalize on the recent dramatic increase in oil and gas prices to build new power plants that utilize coal. This has resulted in significant growth in the steam coal trade. The most dramatic growth has occurred in China and Indonesia, both of which have increased their export capacity in the intra-Asian market (World Coal Institute, 2005).

Coal is traded all over the world, with coal shipped long distances by sea to reach markets. Overall international trade in coal reached 755 million tons in 2004 (compared with 383 million tons in 1994). While this is a significant amount of coal it still only accounts for about 16 per cent of total coal consumed. Transport costs account for a large share of the total delivered price of coal, and as a result international trade in steam coal is effectively divided into two equal-sized regional markets; the Atlantic and the Pacific. Australia is the world's largest coal exporter, exporting over 218 million tons of hard coal in 2004, out of its total production of 285 million tons. Australia is also the largest supplier of coking coal, accounting for 52 per cent of world exports (World Coal Institute, 2005). The USA and Canada are significant exporters and China is emerging as an important supplier. Coking coal is more expensive than steam coal, which means that Australia is able to afford the high freight rates involved in exporting coking coal worldwide.

Grain trade

World grain shipments, which reached 250 million tons in 2004, were almost equally split between wheat and coarse grains such as maize, barley, soybeans, sorghum, oats and rye. Grains include oil seeds extracted from different crops such as soybeans and cottonseeds. In general, wheat is used for human consumption, while coarse grains are used as feed for livestock. Oil seeds are used to manufacture vegetable oil for human consumption or for industrial use, while their protein-rich residue is used as a raw material in animal feed. Total grain production is dominated by the United States. Argentina is the second largest producer, followed by Canada and Australia. In terms of imports, the Asia/Pacific region (excluding Japan) ranks first, followed by Latin America, Africa and the Middle East. The principal vessel classes used in the grain trade are Panamax and 'Handymax'.[5]

The grain market is volatile and highly dependent upon weather patterns and yearly harvest changes. This in turn influences the price of grain and indeed freight rates. The ongoing growth of the global population makes a continuing growth of the maritime grain trade likely, particularly imports from newly industrializing countries, many of which are expected to see a net negative balance in grain production.

Containerized maritime freight

The containerization of maritime transport

The maritime industry has been transformed by more than 50 years of containerization since the first containerized maritime shipment set sail from Port Newark, New Jersey in 1956. It does not come as a surprise that maritime transport was the first mode to pursue containerization since it is the most constrained by loading and unloading operations. Containerization permits the mechanized handling of cargoes of diverse types and dimensions that are placed into boxes of standard size. Thus, non-standard traffic that would have required significant and labour-intensive transhipment activities becomes standardized with time-consuming and costly stevedoring reduced. Instead of taking days to be loaded or unloaded, cargo can now be handled in a much shorter time period as a modern container crane can undertake about two movements per minute. The most common container is 40 feet in length, the equivalent of two 'TEU's.[6] Separate transport systems are becoming integrated by intermodal transport, where each mode tends to be used in the most productive manner. Thus, the line-haul economies of maritime shipping can be combined with the hinterland access provided by rail and trucking. The entire transport sequence is now seen as a whole, rather than as a series of stages, which is changing the role and function of freight forwarders, transport companies, terminal operators and third party logistics providers.[7]

Containerization has been brought about in part by technology and has substantially impacted maritime design with the creation of the containership. While the first containerships were converted cargo vessels, by the late 1960s the containerized market had grown enough to justify the creation of ships entirely designed for such a purpose. Since that time, the construction of containerships has followed incremental improvements in design with economies of scale being the main rationale (Table 10.2); the larger the ship, the cheaper the transport costs per TEU (Cullinane & Khanna, 2000). By the late 1980s, the limitations of the Panama Canal of about 4,000 TEU were surpassed, creating a new class of 'post-panamax' containerships that have a higher capacity but whose draught and transhipment requirements precludes a number of ports (McLellan, 1997). Once this threshold was overcome, the size of container ships entering service quickly increased. In about a decade containership design went from a maximum capacity of 6,700 TEU to 14,500 TEU. Design constraints are now limited by the capacity of port channels to accommodate container ship draughts and as well as the availability of cranes large enough to unload them. In addition, they cause additional pressure on inland transport systems to accommodate the large volume of containers they can tranship. Speed-wise, a threshold of about 25 knots has been reached as excessive energy consumption rules out higher operational speeds.

The access channel depth of many ports is becoming a constraint on the development of larger containerships. Beyond 45 feet (14 metres), the availability of port sites is restrained, imposing limitations on the port calls for post-panamax containerships. One possibility, like in the airline industry (Chapter 9), is the emergence of a limited number of mega-ports linked by high-capacity containerships and serviced by feeder routes. A 'hub-and-spoke' network could imply significant additional costs and delays, however, since it would involve additional transhipments and longer routes. Global containerized trade has surged in recent years from about 88 million TEU in 1990 to 395 million TEU in 2005, and about 70 per cent of the general cargo transported by maritime transport was containerized.

Containerized maritime terminals

From a transhipment perspective, the application of containerization has a very strong rationale since the costs, time and reliability of freight distribution are significantly improved. But due to its physical and logistical characteristics,

Table 10.2. Some major landmarks in containership construction

Year	*Name*	*Capacity (TEU)*	*Yard*	*Length (m)*	*Width (m)*	*Draught (m)*	*Speed (knots)*
1956	Ideal X	58	US	174.2	23.6	?	18.0
1968	Elbe Express	730	B & V	171.0	24.5	7.9	20.0
1981	Frankfurt Express	3,430	HDW	271.0	32.3	11.5	23.0
1991	Hanover Express	4,407	Samsung	281.6	32.3	13.5	23.0
1995	APL China	4,832	HDW	262.0	40.0	12.0	24.6
1996	Regina Maersk	6,700	Odense	302.3	42.8	12.2	24.6
2001	Hamburg Express	7,506	Hyundai	304.0	42.8	14.5	25.0
2003	OOCL Shenzhen	8,063	Samsung	319.0	42.8	14.5	25.2
2005	MSC Pamela	9,200	Samsung	321.0	45.6	15.0	25.0
2006	Emma Maersk	14,500	Odense	393.0	56.4	15.5	24.5

containerization has placed new demands on port terminals. Most major ports have been transformed to become container ports with the construction of new container terminals, often away from the initial port site. Ports unable to adapt to containerization have mostly declined. In other cases, often because of the insurmountable constraints of the existing port site or because of a surging new demand (such as in Pacific Asia), entire new facilities have been built. Several of the world's largest container ports simply did not exist 20 years ago. Ports are facing a growing level of sophistication brought by technical and logistical changes. From simple facilities offering wharves and berths to accommodate ships, terminal facilities have emerged where a complete range of maritime (berthing, mooring, transhipment) and inland services (stacking/warehousing, customs, inland distribution) are provided. Containerized maritime terminals are the facilities in which these range of services are the most developed and extensive, although providing this essential range of services is not without constraints. These are mainly concerned with:

- Space consumption. Container terminals are extensive consumers of space, particularly for the temporary storage of containers waiting to be loaded on ships or picked up for inland distribution. Although containerships need little port time, the growth of containerized trade and their larger capacity have placed pressures on existing terminals to expand laterally. Ports where lateral expansion was possible have thus been able to accommodate the growth of containerized maritime shipping.
- Depth. As the size of containerships increased, so did the port depth requirements to handle them. Post-panamax containerships impose berth depths of at least 14 metres and the new generation of containerships having capacities above 8,000 TEU demands depths in the range of 15–16 metres (at least 50 feet). For several container ports, meeting those requirements has become prohibitive since it would involve the construction of new berth facilities and extensive dredging (often not possible due to environmental restrictions). Doing so also takes place at the risk of attracting enough additional traffic to justify those investments.
- Investment in infrastructure. In the past, port operations involved simple cranes and labour-intensive transhipment activities. The situation has now become very capital intensive while labour requirements have been reduced. Modern containerized operations require a limited amount of labour, but the skill level of this labour is much higher. Each new generation of containerships, in addition to their draught, require larger and more efficient cranes. The level of capital investment becomes very demanding. This takes place in the context of growing hinterland competition where ports are less secure about the stability of their customer base.
- Inland connections. Traditionally, inland access was of lesser importance because most of the warehousing was directly adjacent to port terminals. These facilities stored the freight related to maritime shipping and this freight moved inland as orders were being filled. The situation substantially changed with containerization. Containers are bound directly to their destinations with only temporary stacking at port terminals. The importance has shifted to the capacity of inland transport systems to handle high container throughputs related to port terminals.

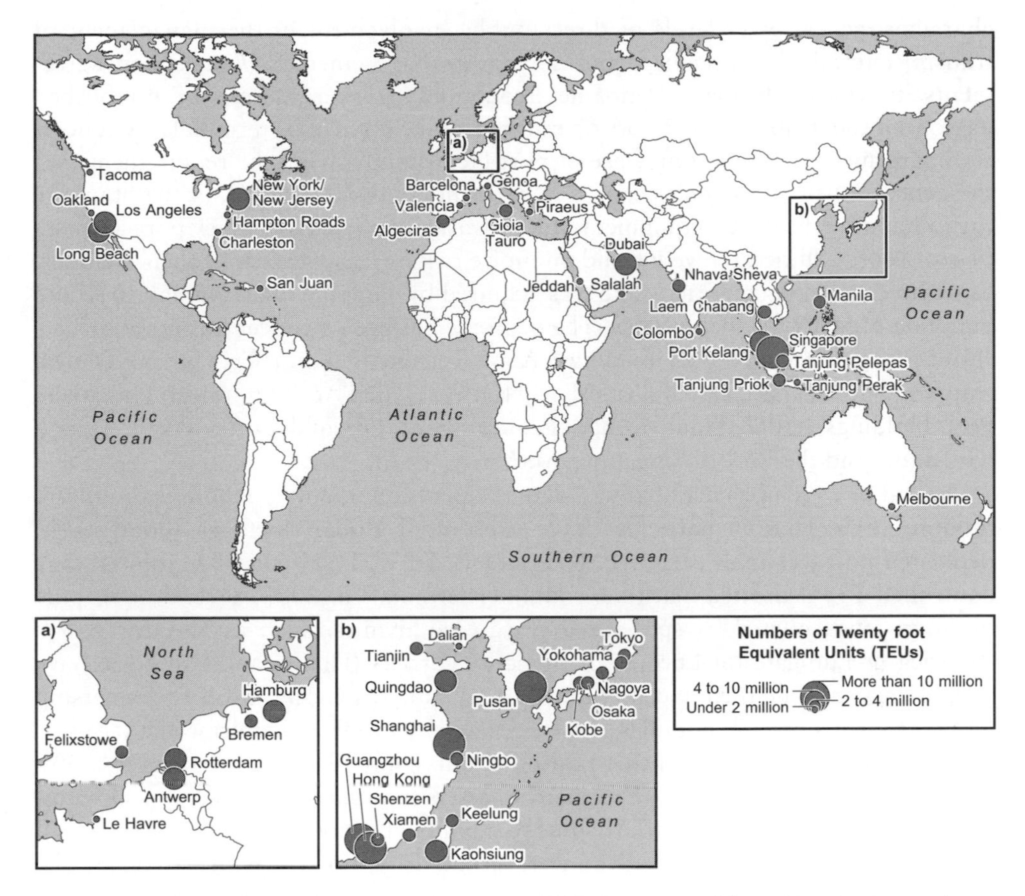

Figure 10.2. The world's 50 largest container ports, 2003. Data sourced from *Containerization International*, various dates.

For large ports, providing inland distribution capacities has become one of the most serious challenges with initiative such as on-dock rail access and transhipments to/from barges.

Figure 10.2 reveals an emerging geography of containerized maritime terminals. Aside from the conventional port clusters in Europe and North America, many corresponding to older ports partly or wholly converted to containerization, new regions and new port clusters have emerged, mainly in Pacific Asia. The most substantial growth took place along the Tokyo–Singapore corridor, where economies have followed the export-oriented model, producing consumption goods highly prone to the use of containerization and bound for the global market.

Global port operators

A significant trend in container port operations has been the increase of the role of private operators (Olivier & Slack, 2006; Slack & Fremont, 2005). In an era characterized by lower levels of direct public involvement in the management of transport terminals, specialized port operators have emerged. Concession agreements in which subsidiary companies (commonly joint ventures) are established have been the major tool for port operators to take control of terminals. A concession agreement is a long-term lease of port facilities involving the requirement that the concessionaire undertakes capital investments to build, expand, or maintain the cargo handling facilities, equipment, and infrastructure. This enables port authorities to secure additional revenue and minimize risk by leasing some of their facilities. Elsewhere, a simple minority stake was acquired by shipping lines, enough to secure handling capacity for their vessels. The outcome has been a concentration of ownership among five major port holdings: APM Terminals (controlled by the Danish shipping line Maersk), Dubai Ports World (DPW, United Arab Emirates), Hutchison Port Holdings (HPH, Hong Kong), Peninsular & Oriental Ports (P&O, United Kingdom) and the Port of Singapore Authority (PSA).

A total of 24 major port holdings were in operation in 2005, handling about 267 million TEUs. Four in particular have substantial global assets of about 30–40 dedicated port terminals each; APM Terminals, DPW, HPH and PSA. Jointly, they controlled 143 dedicated maritime container terminals in 2007. Several other port holdings exist, owned by specialized private companies (such as SSA for North America or Eurogate for Europe) or by ocean carriers (Hanjin and Evergreen have notable assets), but their focus is mostly regional. A concentration of ownership among four major port holdings is taking place, with DPW acquiring the terminal assets of P&O in 2006 to further consolidate its global holdings. However, DPW was required to dispose of P&O's American assets (Baltimore, Miami, New Orleans, New York and Philadelphia) to AIG due to a political controversy; a Middle Eastern company operating major American port terminals was perceived negatively in the post-9/11 setting.

Port holdings are thus the outcome of horizontal integration through expansion and mergers with the process leading to a high level of concentration of the global containerized throughput. For instance, HPH accounted for 26 per cent of world container port capacity with a container throughput of around 51.8 million TEUs in 2005. PSA ranked second, at 19 per cent, handling 41.2 million TEUs for the same year. The main rationale behind the emergence of large port holdings includes:

- Financial assets. Port holdings have the financial means to invest in infrastructures as they have a wide variety of assets and the capacity to borrow large quantities of capital. They can use the profits generated by their efficient terminals to invest and subsidize the development of new ones, thus expanding their asset base and their operating revenues. Most are listed on equity markets, giving the opportunity to access global capital, which realized in the last decade that the freight transport sector was a good source of returns driven by the fundamentals of a growth in international shipments. This financial advantage cannot be matched by port authorities, even those heavily subsidized by public funds.
- Managerial expertise. Port holdings excel in establishing procedures to handle complex tasks such the loading and unloading sequence of containerships and

all the intricacies of port operations. Many have accumulated substantial experience in the management of containerized operations in a wide array of settings. Being private entities, they tend to have better customer service and have much flexibility to meet the needs of their clients. This also includes the use of well-developed information systems networks and the capacity to comply quickly with legal procedures related to customs, clearance and security.

- Gateway access. From a geographical standpoint, most port holdings follow a strategy aimed at establishing privileged positions to access hinterlands. Doing so, they secure a market share and can guarantee a level of port and often inland transport service to their customers. It can also be seen as a port competition strategy where a 'stronghold' is established, limiting the presence of other competitors. Gateway access thus provides a more stable flow of containerized shipments. The acquisition of a new port terminal is often accompanied by the development of related inland logistics activities by companies related to the port holding.
- Leverage. A port holding is able to negotiate with maritime shippers and inland freight transport companies favourable conditions, namely rates, access and level of service. Some are subsidiaries of global maritime shipping lines (such as the A.P. Moller group controlled by the shipper Maersk) while others are directly controlled by them (such as Hanjin or Evergreen) so they can offer a complete logistical solution to international freight transport. The 'footloose' character of maritime shipping lines has long been recognised (Slack, 1994), with a balance of power more in their favour than of the port authorities they negotiate with.
- Traffic capture. Because of their privileged relationships with maritime shipping lines, port holdings are able to capture and maintain traffic for their terminals. The decision to invest is often related to the knowledge that the terminal will handle a relatively secure number of port calls. Consequently, a level of traffic and revenue can be secured more effectively.
- Global perspective. Port holdings have a comprehensive view of the state of the industry and are able to interpret political and price signals to their advantage. They are thus in position to influence the direction of the industry and anticipate developments and opportunities. Under such circumstances they can allocate new investments (or divest) to take advantages of new growth opportunities and new markets.

The main strategy of port holdings consists in the establishment of a network of port terminals capturing the export-orientated traffic of Asia, offshore transhipments opportunities for long-distance trade and improving hinterland access for import markets, particularly at the major continental gateways of Western Europe and North America (Figure 10.3). The issue of port competition has been rendered more complex by this emerging ownership structure. In some ports, the holding controls the entire facility and is thus the port's sole client while other holdings and shipping lines may face a more difficult access in terms of berthing and efficient transhipment. The approach in port competition, notably from other port holdings, thus becomes a strategy of establishing a handhold on a nearby facility that shares a similar hinterland. For instance, A.P. Moller transferred to the port of Tanjung

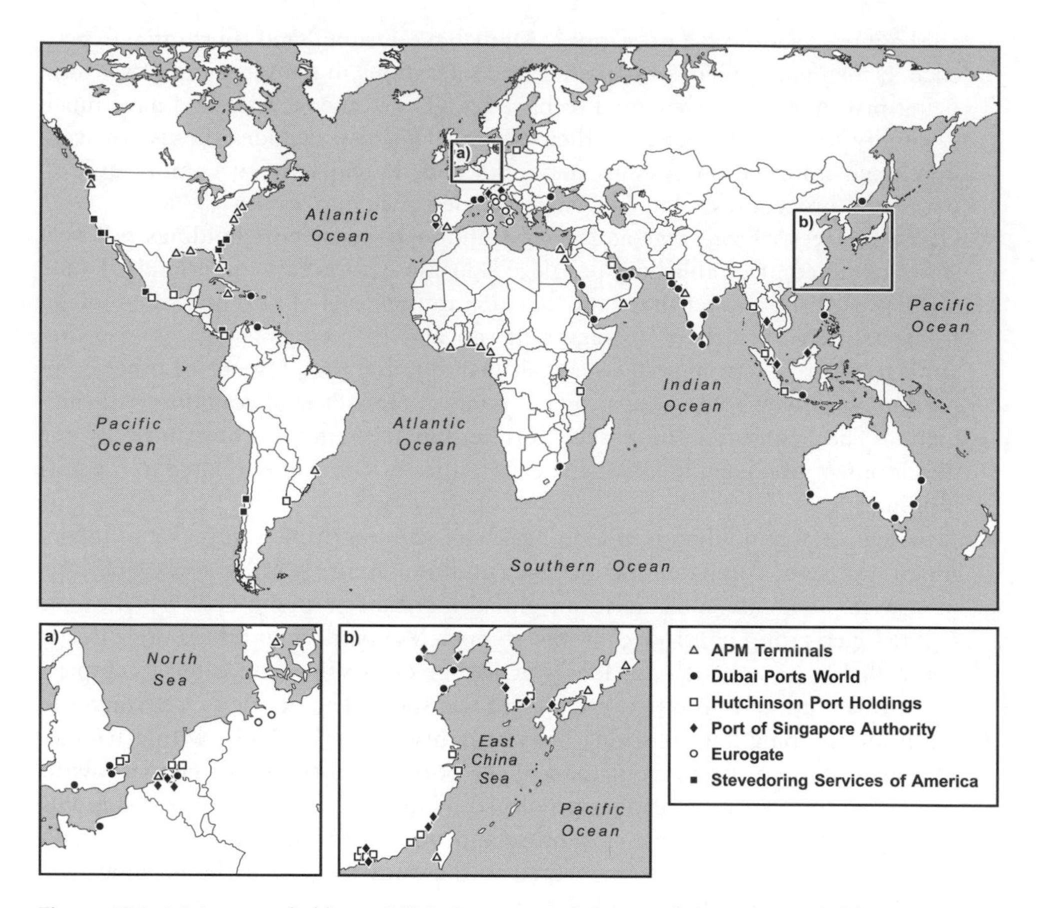

Figure 10.3. Major port holdings, 2006. Data sourced from websites of port holding companies (www.apmholdings.com, www.dpw.co.ae, www.hph.com, ww.psa.com.sg, www.ssamarine.com, www.eurogate.co.uk).

Pelepas near Singapore, because the latter is controlled by PSA. For other ports, particularly large ones, intense competition is the trend as often more than two holdings are present, each owning their own terminals within the port complex (for example, Rotterdam, Antwerp, Laem Chabang, Hong Kong and New York). This strategy is likely to make shipping lines less 'footloose' since they now have vested long-term interests at specific port terminals and selling those interests is much more difficult than negotiating a short-term service contract with a port authority.

Offshore terminals

In a conventional pendulum container service, a maritime range such as the American East Coast could involve several port calls. Pendulum services involve a set of sequential port calls along a maritime range, commonly including a transoceanic service from ports in another range and structured as a continuous loop (Slack,

1998). They are almost exclusively used for container transport with the purpose of servicing a market by balancing the number of port calls and the frequency of services. If the volume is not sufficient, this may impose additional costs for maritime companies that are facing the dilemma between market coverage and operational efficiency. By using an offshore hub terminal in conjunction with short sea shipping services, it is possible to reduce the number of port calls and increase the throughput of the port calls left (Baird, 2006). Offshore terminals can thus become effective competitive tools since the frequency and possibly the timeliness of services can be improved. An outcome has been the growing share of transhipments from around 11 per cent of containerized traffic in 1980 to about 28 per cent in 2005 (Drewry Shipping Consultants, 2006). The world's leading offshore hub is the port of Singapore, where about 91 per cent of its 19.1 million TEU volume was transhipped in 2004. This is mainly attributable to its strategic location at the outlet of the Strait of Malacca, the world's most heavily used shipping route that transits about 30 per cent of the world trade. Other major offshore hubs are Freeport (Bahamas), Salalah (Oman), Tanjung Pelepas (Malaysia) or Gioia Tauro (Italy), Algeciras (Spain), Marsaxlokk (Malta), Taranto (Italy) and Cagliari (Italy). These hubs particularly owe their emergence to a number of factors:

- Location. Offshore hubs have emerged on island locations or on locations without a significant local hinterland to fulfil a role of intermediacy within global maritime networks (Fleming and Hayuth, 1994). They are close to points of convergence of maritime shipping routes where traffic bound to different routes can be transloaded. Offshore hubs tend to be located nearby major bottlenecks in global maritime networks (Strait of Malacca, Mediterranean or the Caribbean) as they take advantage of the convergence effect.
- Depth. Offshore terminals tend to have greater depth since they were built recently in view to accommodate modern containership draughts, placing them at a technical advantage over many older ports. Their selection often involves a long-term consideration of growing containership draughts and the future capacity, in terms of transhipment and warehousing, of the hub to accommodate such growth.
- Land availability. The sites of offshore terminals tend to be less crowded and outside the traditional coastal areas that have seen a large accumulation of economic activities. They often have land for future expansion, which is a positive factor to help securing existing and future traffic.
- Labour costs. Labour costs tend to be lower, since offshore terminals are located at the periphery and they tend to have fewer labour regulations (e.g. unions), particularly at new terminal facilities.
- Hinterland access. Limited inland investments are required since most of the cargo is transhipped from ship to ship with temporary storage on the port facilities. The footprint offshore terminals have on the local or regional transport system is thus limited. In addition, the port operator does not have to wait for local/regional transport agencies to provide better accessibility to the terminal, which is often a source of conflict between the port and the city/region.

- Ownership. Terminals are owned, in whole or in part, by port holdings or carriers which use these facilities efficiently and are free to decide future developments or reconfigurations. Offshore terminals are avoiding a legacy of governance structure controlled by port authorities. They thus tend to be responsive and adaptable to market changes.

In an initial phase offshore terminals solely focus on accommodating transhipment flows and many have a transhipment share exceeding 80 per cent of their container volume. As the transhipment business remains highly volatile, offshore hubs can eventually develop services that add value to the cargo instead of simply moving containers between vessels. This strategy could trigger the creation of logistics zones within or in the vicinity of the port area, in many cases implemented as Free Trade Zone.

Logistics and the maritime/land interface

Maritime logistics

Although maritime companies have always managed their fleet from an operational standpoint, such as scheduling, ports serviced and cargo handling, containerization truly permitted the convergence of maritime shipping and logistics. As such, maritime shipping lines are more related to the requirements of their customers, in terms of price, timing, frequency and level of service. Maritime logistics is thus the convergence of several integration processes including intermodal, economic and organizational in order to add value along the maritime transport chain (Panayides, 2006). Among the many strategies behind freight integration, the provision of 'door to door' services is privileged, which favours a higher application level of maritime logistics.

Thus, new forms of cooperation are emerging in the maritime industry, among shipping lines, but also with actors involved in other modes and other transport services. This cooperation can take many forms, including mergers and acquisitions, joint ventures, contractual agreements and minority stakes. In such a setting, shipping lines each adopt strategies that fit their goals and the management of their assets. Maritime logistics confers a level of flexibility such as how maritime companies allocate their containerships, manage their containers, have access to specific port terminals and inland transport systems and perform a range of value-added activities (Notteboom & Merckx, 2006). Maritime logistics is thus becoming closely integrated with inland logistics.

New services and networks

In an environment of substantial growth in international trade, a small group of very large shipping lines has emerged, along with a concentration of the traffic. While the top 20 carriers controlled 26 per cent of the world's slot capacity in 1980, this figure moved to 42 per cent in 1992 and to 58 per cent in 2003. These carriers are also integrated horizontally, mainly through agreements such as liner conferences, operating agreements (vessel sharing, slot chartering, consortia and strategic alliances) and mergers and acquisitions. Mergers and alliances have been particularly prevalent in the maritime industry. The main rationale in mergers and the formation of alliances is rather simple: increase income and reduce costs. Each member is able to provide existing assets that would be complementary to those in the alliance. Of particular relevance is a geographical complementarity where respective networks and markets are brought together. As private ventures, they aim to establish and maintain profitable routes in a competitive environment. This involves three major decisions about how such a maritime network takes shape (Notteboom, 2004, 2006):

- Frequency of service. Frequency is linked with more timely services since the same port will be called at more often. A weekly call is considered to be the minimum level of service but since a growing share of production is time dependent, there is a pressure from customers to have a higher frequency of service. For a similar total traffic, a trade-off between the frequency and the capacity of service is commonly observed. This trade-off is often mitigated on routes that service significant markets since larger ships can be used with the benefits of economies of scale.
- Fleet and vessel size. Due to the basic maritime economics, large ships, such as post-panamax containerships, offer significant advantages over long distances. Shipping lines will obviously try to use this advantage, keeping smaller ships for feeder services. In addition, a large enough number of ships must be allocated to ensure a good frequency of service. To keep their operations consistent, lines also try to have ships a similar size along their long-distance pendulum routes (see below). This is not an easy undertaking since economies of scale force the introduction of ever-larger ships which cannot be added all at once due to extensive financial requirements and the capacity of shipbuilders to supply them. So each time a bigger ship is introduced on a regular route, the distribution system must adapt to this change in capacity.
- Number of port calls. A route that involves fewer port calls is likely to have lower average transit times in addition to requiring a lower number of ships. Conversely, too few port calls could involve difficulties for the cargo to reach inland destinations remote from the serviced ports. This implies additional delays and potentially the loss of customers. An appropriate selection of port calls along a maritime facade will help ensure access to a vast commercial hinterland.

The emergence of post-panamax containerships has favoured the setting of pendulum services since the maritime land bridge of Panama is no longer accessible to this new class of ships. For instance, pendulum services between Asia and Europe have on average 8–10 containerships assigned and involve 8–12 port calls. Most transatlantic pendulum services have 6–8 containerships and involve 6–8 port calls (Figure 10.4). A pendulum service is fairly flexible in terms of the selection of port calls, particularly on maritime ranges that have nearby and competing ports grouped as regional clusters (e.g. North American East coast, Western Europe). This implies

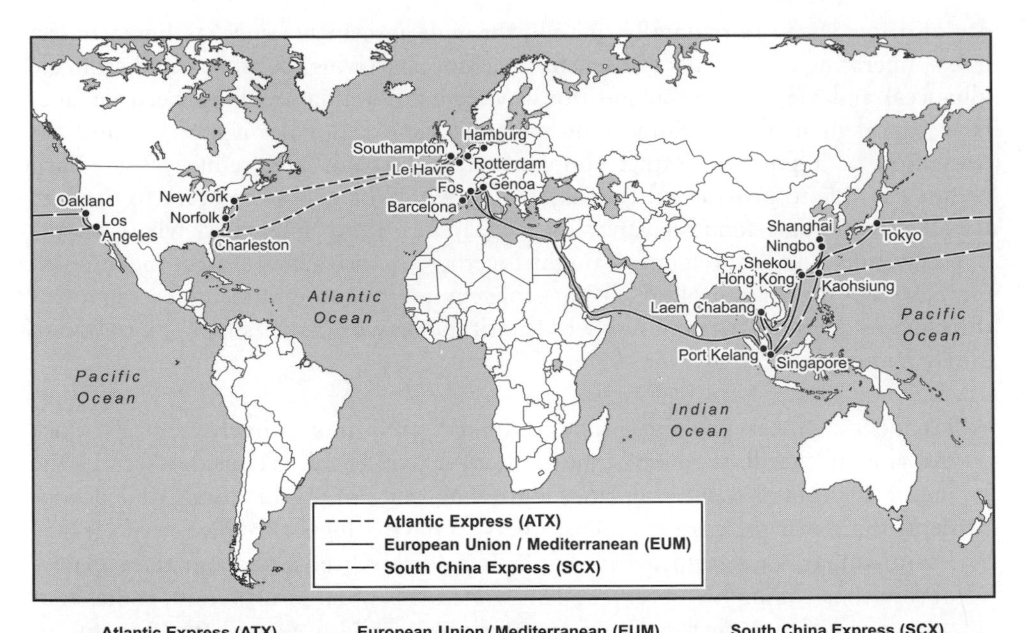

Atlantic Express (ATX)

Origin	Destination	Transit Time (days)
Southampton	New York	8
New York	Norfolk	2
Norfolk	Charleston	2
Charleston	Rotterdam	10
Rotterdam	Hamburg	2
Hamburg	Le Havre	2
Le Havre	Southampton	1
	TOTAL:	27

European Union / Mediterranean (EUM)

Origin	Destination	Transit Time (days)
Port Kelang	Genoa	14
Genoa	Barcelona	2
Barcelona	Fos	2
Fos	Singapore	15
Singapore	Hong Kong	4
Hong Kong	Shanghai	2
Shanghai	Ningbo	2
Ningbo	Shekou	2
Shekou	Hong Kong	1
Hong Kong	Singapore	4
Singapore	Port Kelang	1
	TOTAL:	49

South China Express (SCX)

Origin	Destination	Transit Time (days)
Tokyo	Kaohsiung	4
Kaohsiung	Shekou	1
Shekou	Laem Chabang	4
Laem Chabang	Singapore	3
Singapore	Kaohsiung	3
Kaohsiung	Los Angeles	11
Los Angeles	Oakland	4
Oakland	Tokyo	9
	TOTAL:	39

Figure 10.4. Three major pendulum routes serviced by Orient Overseas Container Line (OOCL), 2006. Data sourced from OOCL website (www.oocl.com)

that a maritime company may opt to bypass one port to the advantage of another if its efficiency is not satisfactory and if its hinterland access is problematic. The shipping network consequently adapts to reflect changes in market conditions. One such example was the abandonment of round-the-world services by Evergreen in 2002, which were instead replaced by three pendulum services that offer more weekly port calls. The frequency of service of two calls per week was judged insufficient to meet the needs of its customers.

The three main markets serviced by pendulum services are Western Europe (Atlantic and Mediterranean Facades), North America (Atlantic and Pacific Facades) and Pacific Asia (Figure 10.4). Each is serviced by a series of port calls where containers are transhipped to offshore hubs or to hinterlands, depending on the function of the port. The itinerary of a pendulum service is thus carefully selected to ensure distribution strategies that reflect the global and regional framework of trade. For instance, in a pendulum service, most of the traffic is bound to the other end of the pendulum. This is particularly the case for the United States where cabotage regulations would forbid a foreign-owned shipping line to carry traffic between American ports. In addition, large high-capacity ships can be used for pendulum services (such as across the Pacific) since they are not required to transit through the Panama Canal and the Suez Canal can accommodate the current generation of very large containerships (up to 16 metres/12,000 TEU).

Because of the capacity limits of the Panama Canal, many shipping companies have changed the configuration of their routes. This became increasingly apparent as a growing share of the global containership fleet was at a size beyond its capacity. The increasing usage of those ships along the Pacific Asia/Suez Canal/Mediterranean route as well as the development of the North American rail landbrige have created a substantial competition to the canal as an intermediate location in global maritime shipping. In addition, estimates indicate that the Panama Canal may reach capacity by 2009–12. A decision was made by the Panama government in 2006 to increase its capacity and accommodate larger containerships. The involves building a new set of locks on both the Atlantic and Pacific sides of the canal with a depth of 60 feet, width of 190 feet and length of 1,400 feet, which would accommodate ships up to 14,000 TEU. The dredging of access channels as well as the widening of several sections of the existing canal will also be required. This will allow Aframax and Suezmax vessels to pass through the canal, thus permitting new opportunities for container services such as the re-emergence of round-the-world services.

Containerized flows and their global imbalances

The production and distribution environment in which maritime transport evolves is also characterized by distortions in international trade patterns caused by globalization and the relocation/development of new manufacturing activities in emerging industrial regions. The Pacific Asian maritime range has been the major recipient of this industrial accumulation, mainly due to the ability to gain from comparative advantages. Container flows are quite representative of global trade imbalances, which have become more acute in recent years (Figure 10.5).

While American containerized imports from Asia increased by 148 per cent between 2000 and 2005, opposite flows have grown to a much lesser extent (30 per cent); there are three times as many containers moving from Asia to the United States than there are in the opposite direction. An imbalance between Asia and Europe is also observed, but to a lesser degree. Thus, production and trade imbalances in the global economy are clearly reflected in imbalances in physical flows and transport rates. On the Pacific, it costs about twice as much per TEU for westbound flows than for eastbound flows, making freight planning a complex task for container shipping companies. In addition, there has been a notable growth in the movement of empty containers.

Empty containers ('empties') account for about 21 per cent of the volume of global port handling (Boile *et al.*, 2006). They pose a logistical challenge to both

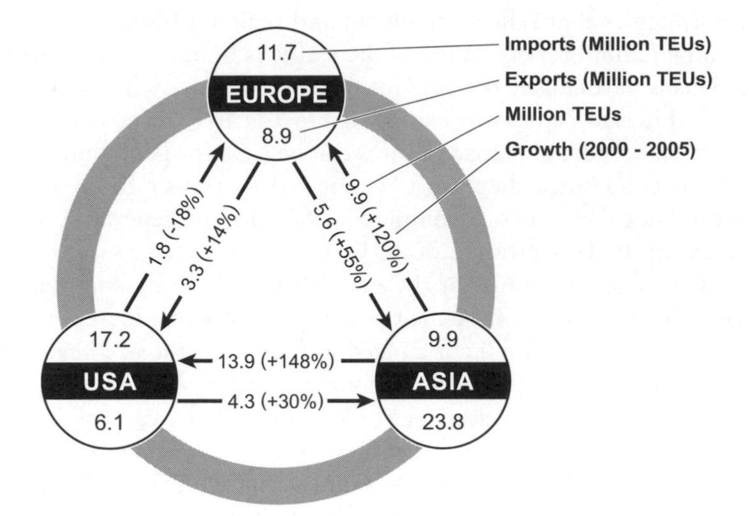

Figure 10.5. Containerized cargo flows along major trade routes, 2005 (in million TEU). Data sourced from UNCTAD, 2006 and various dates.

the maritime and inland segment of freight distribution, an issue being underlined by the fact that at any given time about 2.5 million TEU of containers are being stored empty, waiting to be used. The maritime industry has been hard pressed to address these imbalances, but little can be done since they reflect macroeconomic issues outside their control. Among the strategies being considered is the immediate repositioning of containers that have just been emptied to nearby export-based activities. The use of foldable containers is also a possibility since it reduces the repositioning costs (Konings and Remmelt, 2001). No immediate solution is possible, however, since the core of the issue remains macroeconomic imbalances.

The convergence of inland and maritime logistics

There is a clear trend involving the growing level of integration between maritime transport and inland freight transport systems (Heaver, 2002; Panayides, 2006; Robinson, 2002). Until recently, these systems evolved separately, but the development of intermodal transportation and deregulation provided new opportunities which in turn significantly impacted both maritime and inland logistics. One particular aspect concerns high inland transport costs, since they account for anywhere between 40 per cent and 80 per cent of the total costs of container shipping, depending on the transport chain. Under such circumstances, there is a greater involvement of maritime actors (e.g. port holdings) in inland transport systems. The maritime/land interface thus appears to be increasingly blurred (Notteboom and Rodrigue, 2005). Corridors are becoming the main structure behind inland accessibility and through which port terminals gain access to inland distribution systems. Since transhipment is a fundamental component of intermodal transport, the maritime/land interface relies in the improvement of terminals activities along those corridors.

Strategies are increasingly relying on the control of distribution channels to ensure an unimpeded circulation of containerized freight, which include both maritime and land transport systems. The continuity of the maritime space to ensure a better level of service takes different forms depending on the region. For North America, rail transport has seen the emergence of long-distance corridors, better known as landbridges. The North American landbridge is mainly composed of three longitudinal corridors and is the outcome of growing Trans-Pacific trade and the requirement to ship containerized freight across the continent. For Western Europe, barge systems are complementing trucking with inland waterways accounting for between 30 and 40 per cent of the containers going through major gateways such as Rotterdam and Antwerp. Localized alternatives to improve inland distribution, such as the Alameda corridor,[8] are implemented in addition to trans-continental strategies such as the existing North American landbridge and the planned Northern East–West Freight Corridor spanning across the trans-Siberian to the port of Narvik in Norway with an oceanic leg across the Atlantic.

Conclusion: ports as elements of global logistical chains

The development of bulk and containerized maritime transport has been strongly influenced by technology (Pinder and Slack, 2004). Economies of scale have been achieved through the construction of larger ships and this in turn has affected the optimum shipment size, vessel routings and port selection. While this process is mostly over for bulk transport, it continues unabated for containerized shipping with the introduction of larger ships. Port selection is especially relevant because of the strong link between ports and industrial activity, but particularly between the port and its hinterland. However, technology and vessel design are by no means the only factors at work to influence the patterns of the world maritime shipping; government policy, commercial buying practices and physical constraints such as water depth in ports also play a key role.

Until recently, the individual elements of the transport chain, while functionally related, were operated in a largely disparate way. In the bulk trades, as in maritime transport in general, there is now a realization that the integration of supply chains requires a high level of organizational interdependence. Maritime transport and inland transport must be seen increasingly as functionally integrated. In bulk, as in containerized trades, the reduction of inventory and storage costs by just-in-time shipments and door-to-door services are increasing in significance. Freight transport becomes focused on providing the most extensive services possible within expected cost and reliability parameters. In such a context of flexible maritime networks, such as pendulum services and economies of scale, ports are hard-pressed to act as efficient nodes in global logistical chains. Their responses have involved establishing better hinterland connections and in many cases private terminal operators have stepped in to directly manage the facilities. The global port and maritime landscape is thus adapting to a new environment reflecting economic, technological and political changes.

Notes

1. The largest vessel that can transit the Panama Canal (draught of 12 metres), a bulk carrier of about 65,000 deadweight tons.
2. Draught of 16 metres which can accommodate a loaded tanker of about 200,000 deadweight tons.
3. Between 125,000 and 180,000 deadweight tons.
4. Approximately 80,000 deadweight tons.
5. Vessels of 30–60,000 deadweight tons.
6. Twenty Foot Equivalent Units: the standard unit for measuring container traffic
7. A third-party logistics provider is an asset-based company that offers, often through contractual agreements, logistics and supply chain management services to customers.
8. The Alameda Corridor is a 20-mile rail freight expressway linking the port cluster of Long Beach and Los Angeles to the transcontinental rail terminals near downtown Los Angeles. It was built to provide better rail access to the port cluster, which is the most important in North America in terms of the volume and value of its containerized traffic. The Alameda Corridor consists of a series of bridges, underpasses, overpasses, and street improvements that separate rail freight circulation from local road circulation.

11 | Individual Transport Patterns

Stephen Stradling and Jillian Anable

While previous chapters have looked at transport issues and policy responses across modes and over different spatial scales, these issues and responses are often not fully understood in terms of individuals' transport behaviours and the psychological processes which underpin them. This chapter is concerned with aspects of what Potter and Bailey in Chapter 3 termed second order impacts, 'how societies and economies create and adapt to increasingly transport-intensive lifestyles' from the viewpoint of the individuals who find themselves living those lifestyles. Individuals' travel choices, and the transport experiences that result, are posited as involving the interaction of three factors: obligations ('What journeys do I have to make?'), opportunities ('How can I make those journeys?') and inclinations ('How would I like to make those journeys?') (Stradling, 2003, 2007).

This chapter looks at some of the factors influencing why people travel in the first place, and having made this decision which transport modes they use. It investigates current transport trends and developments from this perspective, presenting data on the amount and type of transport currently consumed – obligations and opportunities – and how these vary with characteristics of person and environment. It examines inclinations towards one private and one public transport mode (car and bus) and how attitudes, values, motivations, experiences and transport patterns vary across separable segments of the travelling public. The potential for the rapid adoption of information and communications technology (ICT) to significantly alter travel decision-making processes and hence travel patterns is also considered. The chapter concludes with an indication of how policy might take account of the 'softer' issues of transport psychology to influence 'smarter' transport choices leading to more sustainable individual travel patterns.

Propulsion, compulsion and consumption

Why do we move around at all? 'Because we can, because we have to, because we like to' is the simplest formulation dividing out the different kinds of motive forces

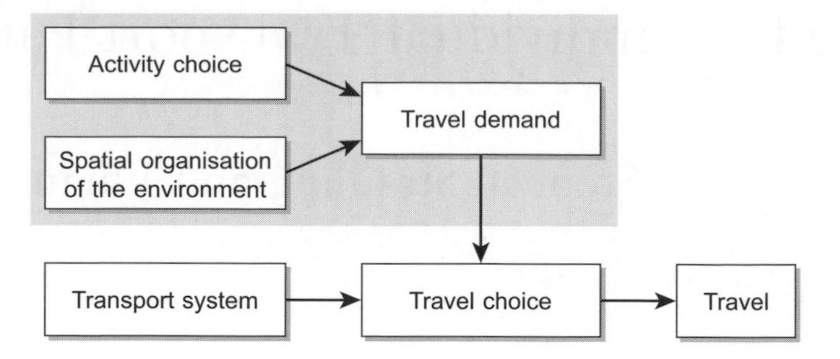

Figure 11.1. Factors influencing travel behaviour. Source: Garling, 1995.

driving travel behaviours and transport choices and thus the creation and maintenance of transport-intensive lifestyles. Transport modes fall into three types: wholly self-propelled modes such as walking, running and swimming; augmented modes which amplify bodily effort, such as rowing, cycling and skiing, or focus natural resources, such as sailing and paragliding; and fuelled modes, whether hay-powered such as horse-drawn carriages and farm wagons or motorized modes such as motorcycle, car, van, lorry, bus, tram, ferry, train and aeroplane which deplete natural resources. Currently, technological effort and expertise is being directed at harvesting more of the earth's natural resources such as biomass and solar energy to source sufficient quantities of renewable fuel to power motorized modes in the future (Chapter 3).

As successive transport innovations have been introduced, the effective radius of people's activity patterns has grown with increases in the speed of transport. Yet there seems to be an average annual travel time budget per person that is relatively constant and has remained so historically and spatially. The UK National Travel Survey indicates that average travel time has held steady at between 350–380 hours per person per year, or around one hour per day, over the past 30 years (Department for Transport (DfT), 2006a). International compilations of travel time data show that this figure of about one hour applies across all cultures and states of development for which data may be discerned (Metz, 2004). Indeed Metz (2004), following Marchetti (1994), argues that the origins of this average travel time of one hour per day may date from the earliest human settlements where the mean area of the territory of long established villages was around 20 km^2, corresponding to a radius of about 2.5 km or one hour's walk from the periphery to the centre and back at 5 kph.

Whether and how people make journeys will be influenced by their anticipation of the various personal costs involved and the resources available to enable and sustain the trip. Garling (2005) characterizes the core determinants of travel behaviour as in Figure 11.1. Thus the temporal ordering of an individual's travel choices in Garling's model is:

1. Activity choice	What shall I do?
2. Destination choice	Where shall I do it?

3. Mode choice — How will I get there?
4. Departure time choice — When should I go?

Activity choice is primary in this formulation. The choice of activities expresses the complex of biological needs and social identities that constitute persons. Travel demand is driven by what goods, services or activities people need or want to consume and where they have to go to do it. The transport system shapes how they might get there and how much time they should allow. Some do and some don't find the travelling rewarding in itself. For example, many motorcycle journeys are from A to A for the enjoyment of riding rather than from A to B to discharge obligations (Broughton & Stradling, 2005); many commuters endure rather than enjoy their journeys, feeling they have no alternative (Anable, 2005); for many embarking on a day trip the choice of destination is a secondary consideration after choosing the activity – 'going out for the day in the car' (Anable, 2002).

Obligations, opportunities and inclinations

Obligations

Aside from these examples of travelling for its own sake (Chapter 12), what activities do people typically travel to undertake? Data gathered from surveys across the western world show a remarkably consistent pattern of increasing car dependence to carry out a relatively stable set of journey purposes. For example, data from the Scottish Household Survey Trip Diary (Scottish Executive, 2004) gives a snapshot of the current quotidian round of daily life for Scottish adults. In total, two-thirds of journeys were for attending place of work (commuting: 24 per cent), re-stocking larder or wardrobe (shopping: 23 per cent), social network maintenance (visiting friends and relatives: 12 per cent) and escorting children to school or leisure activities (8 per cent). Leisure trips have grown considerably in recent decades as a proportion of a typical household's travel portfolio and now account for 40 per cent of the distance travelled in the UK (DfT, 2006a).

Opportunities

Which modes do people use to go about their daily business? In 2003 around 70 per cent of Scottish adults reported that their previous day journeys were by car or van. Business trips were the most car dependent (77 per cent by car) and outings for eating or drinking (24 per cent by car) the least car dependent journey types (Stradling *et al.*, 2005). Table 11.1 gives figures taken from a recent study of almost 1,000 adult residents of Edinburgh who were asked how often they used a range of transport modes. Mode use is categorized here as frequently (once a week or more often), infrequent or never used.

Notwithstanding the exceptionally high bus use in Edinburgh compared with the rest of the UK, these figures show that it is misleading to categorize people as simply and solely 'car users' or 'bus users', as most people use a variety of different modes

Table 11.1. Self-reported frequency of use of 13 transport modes in Edinburgh, Scotland in 2006. Source: Stradling *et al.*, 2007.

	Frequent	*Infrequent*	*Never*
Walk (more than five minutes from house)	94	4	2
Bus	68	29	3
Car as driver	53	15	32
Passenger in car	19	72	10
Bicycle	14	18	68
Taxi	12	84	4
Train	4	84	13
Night bus	2	34	65
Air – domestic	0.7	72	27
Airport bus	0.7	58	41
Air – European	0.4	78	22
Motorbike or moped	0.4	0.7	99
Air – long-haul	0.2	57	43

with varying frequency. Only 2 per cent say they never walk for more than five minutes from their home. Only 3 per cent of this sample said they never use the bus and 4 per cent had never used a taxi. One-third (32 per cent) say they never drive, consistent with the Census figures showing one-third of Scottish households do not have access to a car or van, but only one in 10 say they have never been a passenger in a car. Six in 10 have used the airport bus and only a quarter have not made domestic (27 per cent) or European flights (23 per cent), while 57 per cent have made long-haul flights. While the particular mode use proportions will vary from place to place with the availability of different transport modes (Edinburgh has a good bus system and a local international airport, but no trams (yet), subway, gondolas or jitneys), samples from other urban locations would show similar evidence of transport-intensive lifestyles (Chapter 13).

How many different transport modes do people use? The number of modes that each respondent said they used at all and the number used frequently – once a week or more often – was calculated. While 2 per cent reported using none of these modes at all, at the other end of the spectrum 4 per cent had used 12 of the 13 modes. The mean number of modes ever used was 7.39 and the mean number used once a week or more was 2.33. Of the car drivers in the sample, only 2 per cent could be deemed to be completely car dependent, carrying out all their journeys by car. Overall, the mean number of modes used by frequent car drivers was 8.25, compared to 7.16 for non car drivers – about one mode more. Thus while urban locations like Edinburgh may be characterized as playing host to transport-intensive lifestyles, the amount and variety of transport consumed varies widely across individuals in the same location. While most people, even car drivers, are multi-modal transport users, some people use many modes, others few, to go places.

More generally, transport patterns and choices vary with *person* characteristics such as age, gender and disability, with household characteristics such as income, location and transport availability, with journey purpose, and with attitude/value clusters (Chapter 4). Travel joins up the places where people go to lead their lives and meet their obligations (Stradling 2002a, 2002b; Stradling *et al.*, 2000). Transport patterns and choices also vary with *environment* characteristics or land use effects, the location of trip origins such as homes and trip destinations such as jobs, shops and recreations which will be shaped by the topography of the terrain.

There is, for example, variation with person characteristics such as gender, age and income and with environment characteristics such as settlement size around the observed constancy of one hour per person per day average travel time. In Britain, older people and children spend less time on the move than do people in mid-life; women spend on average rather less time travelling than men; while Londoners allocate more time to travel than do people living elsewhere. Young Londoners in their twenties are particularly mobile, spending on average about 1.4 hours per day travelling (Metz, 2004), though this may be because they have higher activity levels than older adults and are inclined to travel more, or lower incomes than those more established in their careers and they must locate further from the centre of the metropolis and make longer journeys to work and play. Mokhtarian & Salomon (2001) suggested we each have a desired daily travel time budget. But this ideal may be compromised by our current circumstances, with what transport researchers call 'revealed preferences' – the journeys people actually make – not aligned with their 'stated preferences' – the trip types they would like to make.

Transport adequacy, defined by the UK Department for Transport's Mobility and Inclusion Unit (DfT, 2000) as involving affordability, availability, accessibility and acceptability, will facilitate or constrain travel choices. Transport affordability and availability influence the opportunity to travel – how and how much (and thus influence social exclusion: see Chapter 4). Destination accessibility is important for being able to get to places to meet one's obligations. Mode acceptability varies with personal preferences and such inclinations influence mode choice and thus mode share and time spent travelling. Even within highly developed western countries, there are appreciable differences in the travel opportunities afforded by the physical environment and the transport system, and thus on transport-related quality of life measures, between urban and rural populations (Chapters 6 and 7). Living in small accessible towns, accessible rural locations and remote rural locations generates greater distances travelled by both car and bus; those in small towns appear to be best served in terms of convenience of access to local services perhaps because of compact settlement size, while those in rural locations, both remote and accessible, appear the least well served (Stradling *et al.*, 2005).

Inclinations: towards and away from car use

Future historians may well characterize the twentieth century as the century of the private car. During this time, around one billion cars were manufactured, of which over half a billion (500 million) are currently occupying the streets, garages, car parks and grass verges of the planet (the average car spends less than an hour a day in motion: infrastructure thus has to be provided for stationary as well as moving vehicles).

Given the ubiquity and reach of the car it is no surprise that more research has been directed to understanding the attractions of this mode compared to other modes. A number of studies attest to the car as a symbolic object (e.g. Sachs, 1984; Maxwell, 2001) and to the importance of affective motivation rather than instrumental motives such as availability and directness in choosing car over other transport modes (e.g. Abrahamse *et al.*, 2004; Bamberg and Schmidt, 2003; Ellaway *et al.*, 2003; Gatersleben, 2004; Jensen, 1999; Reid *et al.*, 2004; Steg, 2004; Tertoolen *et al.*, 1998) and in influencing driving style (e.g. Lajunen *et al.*, 1998; Stradling *et al.*, 2003). A core component apparent from studies of the attractions of the car is the sense of autonomy, feeling in control (Step Beyond, 2006; Stradling, 2007).

The promise of autonomy implicit in the nomenclature 'automobile' has been shown to resolve into two clearly differentiated dimensions of identity and independence (Stradling *et al.*, 2001) with the young and the poor scoring highest on the identity factor (high loading items ('Driving a car . . . Is a way of projecting a particular image of myself; Gives me the chance to express myself by driving the way I want to; Gives me the feeling of being in control; Gives me a sense of personal safety', and so on), and females over forty scoring highest on the independence factor (high loading items include: 'Driving a car . . . Is a convenient way of travelling; Gives me a feeling of independence; Gives me a spontaneous way of making a journey').

The mobility conferred by the car brings access privileges (Chapter 4). In Scotland, 97 per cent of those in the top household income quintile (top twenty per cent) have access to a car for private use, compared to 32 per cent of those from the bottom quintile (Stradling *et al.*, 2005). Those from households with access to a car travel more often, further, and for longer durations thereby increasing the number and variety of destinations to which they have access. Ratings of convenience of access to local life-support services such as money (bank or building society), food (supermarkets and local shops) and health (GP clinic and hospital outpatients' department) are higher for those with a car; they enjoy more frequent social interactions with their support network of relatives and friends and are thus less likely to suffer social isolation; more visit sports and cultural facilities; they report better health status and fewer of them have disabilities causing difficulties with travelling; they rate themselves higher on indices of civic participation; more of them live in nicer neighbourhoods; and fewer of them had used the local bus service in the past month (Stradling *et al.*, 2005). Of course such advantage does not come without cost. As Featherstone (2004: 2) notes:

> Automobility makes possible the division of the home from the workplace, of business and industrial districts from homes, of retail outlets from city centres. It encourages and demands an intense flexibility as people seek to juggle and schedule their daily set of work, family and leisure journeys . . . on the calculation of the vagaries of traffic flows.

In an era of highly reliable cars we grind towards gridlock as we suffer increasingly from unreliable journey times.

Table 11.2. Ten components of the driving task (1–8 from Panou *et al.*, 2005).

Strategic levels	Activity choice, mode and departure time choice. Discern route alternatives and travel time
Navigation tasks	Find and follow chosen or changed route; identify and use landmarks and other cues
Road tasks	Choose and keep correct position on road
Traffic tasks	Maintain mobility ('making progress') while avoiding collisions
Rule tasks	Obey rules, regulations, signs and signals
Handling tasks	Use in-car controls correctly and appropriately
Secondary tasks	Use in-car equipment such as cruise control, climate control, radio and mobile telephone without distracting from performance on primary tasks
Speed task	Maintain a speed appropriate to the conditions
Mood management task	Maintain driver subjective well-being, avoiding boredom and anxiety
Capability maintenance task	Avoid compromising driver capability with alcohol or other drugs (both illegal and prescription), fatigue or distraction

Driving is a skill-based, socially regulated, expressive activity involving balancing driver capability and task difficulty to avoid loss of control together with real time negotiation with co-present transient others with whom the driver is sharing the public highway to avoid intersecting trajectories, while maintaining or enhancing the driver's self-image (Stradling, 2005). There are task demands at the strategic, tactical and control level which will consume calories, concentration and concern. Panou *et al.* (2005) characterize eight operational levels to the driving task; Table 11.2 shows these and adds two more.

Driving today is not, however, an unalloyed pleasure. Dudleston *et al.* (2005) found that while 84 per cent of a large sample of Scottish car drivers agreed that 'I like travelling in a car', 45 per cent agreed that 'I find travelling by car can be stressful sometimes', 43 per cent that 'I am trying to use the car less' and 57 per cent that 'I would like to reduce my car use but there are no practical alternatives'. Yet 29 per cent of a separate sample of Scottish car drivers (Stradling, 2006) disagreed that 'Driving to me is just a way of getting from A to B' and young, male novice drivers, who often drive for pleasure to get rid of their frustrations or to seek thrills, drive faster, more aggressively and competitively and with smaller safety margins than others. They also have higher road traffic accident involvement (Chliaoutakis *et al.*, 1999, 2002; Fuller *et al.*, 2006), making a disproportionate contribution to the 1.2 million motor vehicle accident deaths suffered annually across the planet.

Research on responses to questions about attitudes to car use and the environment has identified four driver segments in the UK: Die-Hard Drivers (DHDs), Car Complacents (CCs), Malcontented Motorists (MMs) and Aspiring Environmentalists (AEs). The segments differ in the extent to which they exhibit attachment to the car, are willing to consider alternative modes, are already multi-modal, feel willing and able to reduce their car use, are aware of transport issues, acknowledge the transport contribution to environmental problems and say they are prepared to bear additional cost for continuing car use (Anable, 2002, 2005; Dudleston *et al.*, 2005; Stradling, 2007):

- Die-Hard Drivers like driving and would use the bus only if they had to. Few believe that higher motoring taxes should be introduced for the sake of the environment and many support more road building to reduce congestion.
- Car Complacents are less attached to their cars, but currently see no reason to change. They generally do not consider using transport modes other than the car and faced with a journey to make will commonly just reach for the car keys.
- Malcontented Motorists find that current conditions on the road, such as congestion and the behaviour of other drivers, make driving stressful, would like to reduce their car use, but cannot see how. They say that being able to reduce their car use would make them feel good, but they feel there are no practical alternatives for the journeys they have to make.
- Aspiring Environmentalists are actively trying to reduce their car use, already use many other modes and are driven by an awareness of environmental issues and a sense of responsibility for their contribution to planetary degradation.

Table 11.3 shows examples of differences between the four car driver groups.

Most motorists, especially DHDs and even the AEs who are keen to cut car use, like travelling in a car – cars are comfortable, convenient, convey autonomy as well as mobility and promise the benefits of speed, which is why cutting car use is a challenge. But many drivers – except the DHDs – find car use can be stressful and thus, potentially, to be avoided. Most of the CCs – and more than in the other segments – don't consider mode options but just get in the car. While equivalent proportions of MMs and AEs are trying to use the car less, hardly any of the MMs think it will be easy, unlike the AEs. MMs see themselves as willing but unable, they have the inclination to cut car use but lack the opportunity. More of the DHDs and CCs would like more roads built to ease congestion; over twice as many DHDs as MMs and AEs support unrestricted car use and the 'right to mobility' (Chapter 5) and also think global warming threats are exaggerated, while more AEs think car users should pay higher taxes, and more say they are prepared to pay them if the revenue is hypothecated to public transport improvements. In short, these different kinds of drivers have different attitudes, values, motivations and transport experiences when behind the wheel of their car.

Inclinations: towards and away from buses

Segmentation studies have also divided non-car users into three types: Car Sceptics (CSs), Reluctant Riders (RRs) and Car Aspirers (CAs), each comprising around 8 per cent of Scottish adults (Dudleston *et al.*, 2005):

Table 11.3. Support for car use and environmental attitude statements by Die-Hard Drivers (DHD), Car Complacents (CC), Malcontented Motorists (MM) and Aspiring Environmentalists (AE). Source: Dudleston *et al.*, 2005.

	DHD	*CC*	*MM*	*AE*
N	155	142	213	159
% of 669 Scottish car drivers	23.2	21.2	31.8	23.8
% Male (all car drivers in the sample: 51.6%)	55	47	55	48
% Strongly agree + Agree				
I like travelling in a car	98	89	77	77
I find car driving can be stressful sometimes	9	55	54	62
When I am getting ready to go out, I usually don't think about how I'm going to travel, I just get in the car	68	87	49	34
I am trying to use the car less	7	11	74	70
It would be easy for me to reduce my car use	21	5	4	70
It is important to build more roads to reduce congestion	58	71	33	33
People should be allowed to use their cars as much as they like, even if it causes damage to the environment	37	23	14	16
Environmental threats such as global warming and deforestation have been overexaggerated	35	22	14	18
For the sake of the environment, car users should pay higher taxes	7	11	18	24
I would be willing to pay higher taxes on car use if I knew the revenue would be used to support public transport	15	22	30	33

All attitude items p = .000 on chi-square comparisons.

- Car Sceptics are travel aware, environmentally aware, managing – typically voluntarily – without a car, more likely to use bicycles and to support constraints on unfettered car use.
- Reluctant Riders tend to be older and less well off, involuntarily dependent on public transport and, where possible, travelling as passengers in others' cars.
- Car Aspirers, more of whom are unemployed, from social class DE, and environmentally unaware, need better access to destinations than their current high bus use provides and for this and other reasons aspire to car ownership.

Local buses are the most used form of public transport in Britain with 4.6 billion passenger journeys in 2004/05, representing around two-thirds of all public transport journeys (DfT, 2006a). That said, bus use has been falling for many years for a variety of reasons – not least deregulation (Chapter 5) – and the car remains the dominant transport mode (DfT, 2006a). One barrier to increased bus patronage is held to be the image of bus services: 'a transport mode that has become associated with young people . . . elderly people . . . and people on low incomes . . . i.e. a mode of last resort' (Bus Partnership Forum 2003: 9). A recent study in Edinburgh (Stradling *et al.*, 2007), however, found image to be the factor of least concern to bus users: in descending order of endorsement the factors that generated dislike or discouragement of bus use were: feeling unsafe (e.g. 'Drunk people put me off travelling by bus at night'); preference for walking or cycling (e.g. 'I prefer to walk'); problems with service provision (e.g. 'No direct route'); intrusive arousal (e.g. 'The buses are too crowded'); cost (e.g. 'The fares are too expensive'); preference for car use (e.g. 'I feel more in control when I drive'); disability and discomfort (e.g. 'There are not enough hand rails inside the bus'); and finally self-image (e.g. 'Travelling by bus does not create the right impression').

These factors all show social and affective concerns with the quality of the urban bus travel experience as well as more instrumental reasons for service dissatisfaction. Drunks and groups of youths on the bus were perceived as threatening and the effect was amplified during the hours of darkness. The uncertainty of waiting for the bus, especially at night, was also a source of anxiety to some bus users. Such factors may, in the limiting case, induce avoidance behaviour: 'I refuse to travel to Leith or West Edinburgh by bus at night. I don't feel safe and other passengers can be very intimidating' (Female aged 28); 'I would like to travel to and from Leith in the evening, but I don't because the direct buses are infrequent and I am fearful of the bus stops and of drunks' (Female aged 56).

One respondent in this study indeed disliked the core premise of pubic transport – that she had to travel with the general public – although for some public transport was an opportunity to engage in positive interpersonal interactions with fellow passengers (social exchange) whether with friends, acquaintances or co-present strangers. Engwicht (1993: 19), in characterizing cities as inventions to maximize exchange opportunities and minimize travel, regarded 'streets as a dual space for both movement and exchange' with 'plenty of opportunities for spontaneous exchanges on the walk to the public transport stop, and while riding with others'. For the bus, however, as with other forms of public transportation, there is permanent tension between the exchange and movement roles. On a bus the rules of social exchange including the etiquettes of co-presence apply when 'having to travel with the general public' and endure enforced proximity whilst respecting private space, whereas for the bus as occupier of a movement space destination choice is constrained by routes, system time scheduling and timetabling determine duration and frequency, and fare collection and verification of travel entitlement govern place and pace of entry.

Information, communication and travel substitution

Might people be less inclined to move around as technology alters the way we communicate, shop, work, play and conduct our everyday affairs? ICTs – especially the mass adoption of computer networks, the Internet, residential broadband access, wireless technology, mobile computers, Personal Digital Assistants and mobile phones – increasingly provide digital services and virtual resources which diminish the importance of spatial location, allow activities to be undertaken without the need to travel and alter the activities that can be undertaken en route.

Table 11.4 shows the considerable expansion in the ownership of some of these technologies in the UK in recent years. Since 2004/05, household internet

Table 11.4. Percentage of households in the UK owning selected technologies. Source: ONS, 2005a.

	1998/99	*2004/05*
Home computer	33	62
Internet connection	9	53
Mobile phone	27	79

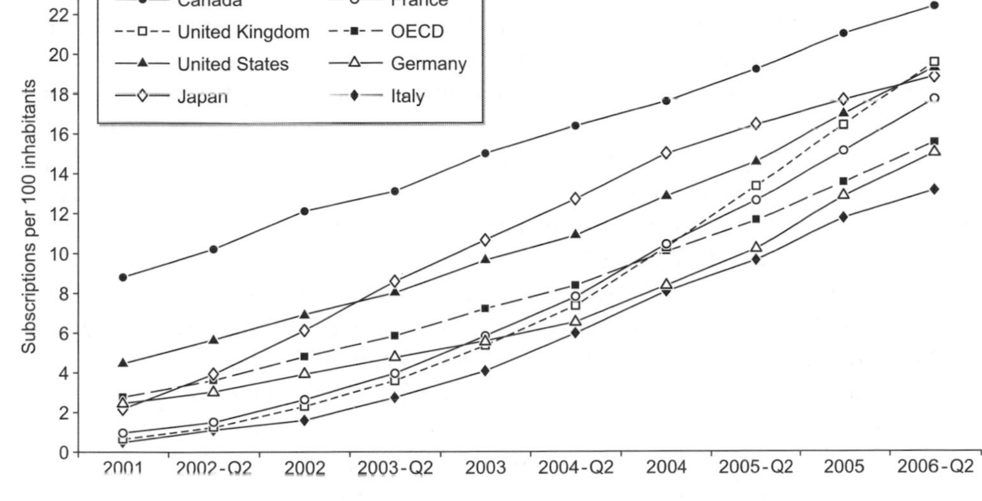

Figure 11.2. Broadband penetration in G7 countries since 2001. Source: OECD, 2006b.

connections in the UK have increased to 57 per cent, growing at a rate of about 5 per cent per year. More than two thirds of these households now have a broadband connection (Office for National Statistics (ONS), 2006). In Organisation for Economic Cooperation and Development (OECD) countries, the number of broadband subscribers increased by 33 per cent in just one year between 2005 and 2006 with an average of 15.5 subscribers per 100 inhabitants in 2006 (OECD, 2006b). Within the OECD, Northern European countries have the highest penetration rates (Figure 11.2). These technologies are also becoming cheaper, flexible and more mobile to allow network connectivity while on the move. For instance, as of April 2006 an estimated 29 per cent (UK) of mobile users had accessed the Internet from their mobile phones (Mobile Herald, 2006). Consequently, it has been suggested that,

> if the age of rail made a profound impact on 19th Century development patterns, and the age of the motor car was inextricably interwoven with social evolution in the 20th Century then 'The Information Age' will be the innovation which history will deem to have been the driver of society in the earliest 21st Century. (Derek Halden Consultancy (DHC), 2006, citing Guest, 1999)

Of course, the potential for telecommunications technology to affect travel behaviour is not new; the invention of the telephone in the 1870s was widely expected to eliminate at least some travel and the energy crisis of the 1970s reinforced the idea of telecommuting and teleconferencing as ways of reducing energy consumed by travelling (Mokhtarian, 2000). But despite these developments and the recent ICT explosion, there has been no perceptible decrease in travel, at least at the macro level, as per capita car ownership and distances continue to rise. Indeed, the relationship between telecommunications and travel behaviour is complex. On the one hand, if journeys to work are being replaced by teleconferencing, business journeys by tele- or videoconferencing, shopping by e-commerce and activities such as social networking by various forms of digital communication, there is the potential for travel to be substituted and therefore reduced. In reality, however, there is a complex matrix of second order effects which in many cases reduce or even negate the traffic reduction effects to be expected from the simple substitution of real journeys by virtual journeys (Cairns *et al.*, 2004). This is because travellers modify their behaviour and their inclinations to take into account the opportunities available to them. If an opportunity is available which does not involve travel then people may travel less. If more opportunities become available and people can do more things then they may travel more (DHC, 2006).

Some of the types of questions that need to be answered include: to what extent will new trips be made by the teleworker or by other family members during the course of the day that would otherwise not have been made? To what extent does teleshopping replace or supplement store shopping? What is the net effect of home-shopping delivery vehicles on overall travel demand and energy use? How, proportionately, will transport substitution affect different traffic modes (that is, will regular public transport users become occasional car users)? Will ICT affect longer-term locational decisions so that people will tend to live further from their places of work and offices will decentralize (see Helminen and Ristimäki, 2007)? To what extent will latent demand be realized by other road users taking advantage of freed-up road space and of cars not being used for commuting (DHC, 2006)? Is the aforementioned travel time budget, that has been so consistent over the decades, altering as people are willing to budget more time for the daily commute as a result of mobile ICT and the ability to undertake work, networking and recreational activities en route?

To understand the net impact of all these factors involves taking into consideration four different major kinds of impact (Figure 11.3):

- Substitution: telecommunications replace travel.
- Complementarity – enhancement: telecommunications directly stimulates travel by providing opportunities for people and businesses to achieve more, make more contacts and participate in more activities.
- Complementarity – efficiency: information through intelligent transport systems optimizes the network and improves travel by making the network more efficient (e.g., by avoiding travelling at congested times of day).
- Complementarity – indirect impacts: impacts on land use and economic development which in turn affect travel, lifestyle changes with reductions in work travel being replaced by increases in leisure travel.

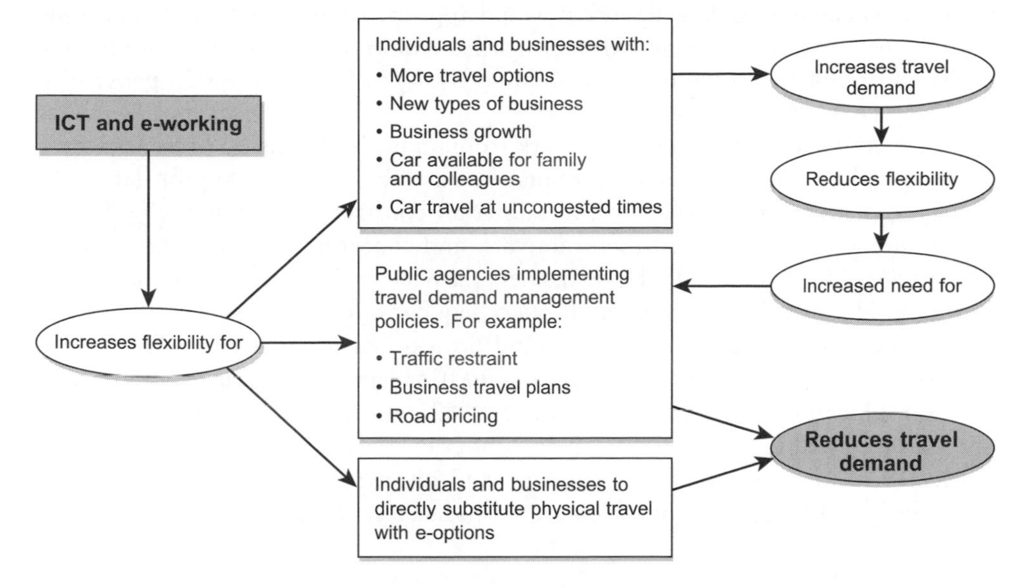

Figure 11.3. ICT and e-work influences on travel demand. Source: DHC, 2006.

Overall, the evidence is mixed on the central question of substitution versus complementarity. Mokhtarian (2000) synthesizes the research evidence by explaining that the substitution effect is more likely to be short term and direct, whereas the complementarity effect is likely to be long term and occur outside the scope of the behaviour being studied. The evidence would suggest that when the broader and longer-term effects are studied, on balance ICT generates more communication, including new travel. That said, supportive policy measures such as grants, charges and taxes (including road pricing), institutional support for workplace travel plans and flexible working patterns could 'lock-in' the potential travel reduction benefits.

Most research in this area has concentrated on teleworking (see DHC, 2006, for a review). The concept of teleworking emerged about 20 years ago and referred to the practice of working from home using telecommunications links to replace commuting (telecommuting). Since then, the use of mobile ICT and electronic networking has expanded the concept to include people working in other remote locations such as neighbourhood centres, cafés, hotel rooms, on trains and even in cars (ONS, 2005b; Laurier, 2004). In the UK, the proportion of the working population that is teleworking – defined as those who work mainly in their own home or in different places using home as a base who use both a telephone and a computer to carry out their work – is growing rapidly. Since 1997, the number of teleworkers in the UK has increased by on average 13 per cent a year, giving an overall increase between 1997 and 2005 of 158 per cent (ONS, 2005b). In 2005, 2.4 million (8 per cent of the workforce) could be classed as teleworkers, an additional 1.0 million work at home occasionally and a further 0.7 million work at home without relying on ICT to do so. By combining these figures, it would appear that around 4.1 million people (or 13.7 per cent of the UK workforce) worked from home some or all of the time in the UK in 2005. This compares with an equivalent 2004 figure of 20 per cent in the USA (DHC, 2006).

Not everyone is able or willing to work from home. The decision to telecommute is influenced by both external constraints such as job suitability, appropriate technology and management approval, and internal constraints such as the desire for social interaction, the ability to self-motivate, and concerns about visibility and job security (Mokhtarian, 2000). These barriers may mean that there is a natural limit to the uptake of this behaviour and its consequent effect on travel patterns. The highest incidence occurs in managerial and professional occupations (DTI, 2002), those working in very small and large organizations and in large urban and accessible rural locations (DHC, 2006). Also, some 62 per cent of teleworkers in the UK are self-employed (ONS, 2005b). Analysis of case study research from around the world reveals a wide range of advantages and disadvantages to the individual of this way of working, some of which are reproduced in Table 11.5.

Table 11.5. Advantages and disadvantages to the individual of teleworking. Source: adapted from DHC 2006.

	Advantages	*Disadvantages*
Worker Efficiency and Satisfaction	▪ Improves worker efficiency: ▪ 'More time at home = balance of home/work priorities = less stress = better performance.' (James, 2003) ▪ Distraction free workspace aids concentration. ▪ Increases the number of productive working hours (commuting one hour each day equates to six. full work-weeks a year) ▪ Can lead to improvements in diet, exercise and other health related variables. ▪ Gives employees greater choice of where to live.	▪ Reduced social interaction can lead to social & professional isolation. (Mesner, 2002) ▪ Tainting of the home environment with work related anxiety. ▪ Mental health e.g. increased levels of depression in full time teleworkers. ▪ Increase burden of household duties for the teleworker, i.e. taking on a chauffeuring role relinquished by others in the house. ▪ Potential to over-work, over-eat and under-exercise. Desk-potato or 'fridge factor' syndrome. ▪ Blurring of distinctions between work and leisure which is deleterious to both (Cairns *et al.*, 2004) ▪ Work/life balance may be affected with reported increases in work addiction.

Table 11.5. *Continued*

	Advantages	*Disadvantages*
Equity	■ Gives jobs access to those in geographically remote areas. ■ Increases employability for disabled and mobility-restricted workers. ■ Can help decrease social exclusion by diffusing services in the community. Virtual mobility provides accessibility opportunities both substituting for physical mobility and enabling access where previously there was an accessibility deficit. (Kenyon *et al.*, 2002) ■ Recognises changing family needs, particularly single parent/dual career households. Permits more convenient childcare arrangements. ■ The flexibility afforded to teleworkers is more conducive to both partners in a relationship being able to develop careers and accommodate childcare and domestic matters. (Lyons *et al.*, 2000).	■ Equality issues with colleagues in posts which are not tele-amenable. ■ Issues for people with unsatisfactory home lives (including domestic violence or oppressive caring duties) for whom the workplace has hitherto been a haven. ■ Perceived discrimination against people without sufficient working space at home or uncomfortable home environments. ■ Traditional forms of home working have often led to exploitation in the form of very low pay and the avoidance of normal workers' rights.

Despite broadly similar findings worldwide, there are some interesting locational differences. For example, avoiding high property costs and traffic congestion are more important in the UK and the Netherlands than elsewhere (James, 2003). In Britain, teleworkers are apparently likely to work longer hours than others. Programmes to promote the uptake of working from home in countries like Canada and Australia are often driven by an economic need to make the most of geographically dispersed populations (DHC, 2006). Studies from Asia or southern Europe on teleworking are scarce and there are indications that there is a cultural bias against this way of working. Indeed, DHC (2006, citing Huws *et al.*, 1999) speculate that teleworking mainly finds favour in Anglo-Saxon cultures where a culture of individuality and work ethic means that working in isolation is possible and may even be desirable.

Overall, the promotion of ICT may well be regarded as a good thing, even if increases in travel demand exceed the reduction and efficiency benefits. Flexible working practices and improvements in communication technology can enhance accessibility, choice and employment opportunities for individuals and offer new business opportunities and productivity enhancements in the commercial sector. Nevertheless, whilst e-working may well have a variety of roles to play in meeting functional and even emotional requirements of activities, they may fall short of meeting other social or psychological requirements (DHC, 2006). Indeed, the barriers to the uptake of behaviours such as teleworking as well as the resulting travel reductions from ICT are likely to be psychological, social or institutional rather than technical (Mokhtarian, 2000). As has been demonstrated in this chapter, this is because the generalized cost of travel – the monetary and time 'cost' of the journey – only partly explains travel choices. Instead, other factors such as the quality of the location-based experience, the desire for social interaction and the ability to fulfil multiple activities at the real location such as combining social and recreational activities with a business trip may result in the travel and location based alternative being chosen above its virtual counterpart (Mokhtarian, 2000).

Conclusions: smarter choices in changing travel behaviour

What determines why people travel and the modes they choose to use? The interaction of people's activity needs, the spatial organization of the environment and the characteristics of the transport system as set out by Garling (2005) are experienced by the individual in the form of obligations, inclinations and opportunities (Figure 11.1). In addition, three dimensions to the travel experience itself can be considered: the utility of arriving at a destination; the utility of the activities that can be conducted whilst travelling and the utility of the travel itself (Mokhtarian, 2000). Person variables (e.g. attitude, values, motivation, age, gender, income, health, physical ability, etc.), interpersonal variables (e.g. social networks) and physical and social environments interact together and influence behaviour. Environments condition the range of emitted behaviours by promoting and sometimes demanding certain actions and by discouraging or prohibiting others. Persons have needs and preferences, some which they hold in common (most car drivers enjoy travelling by car: Table 11.3), some of which differ (few people have ever ridden a motorcycle: Table 11.1) but all of which are based, the evidence suggests, on affect – feelings – at least as much as rational computation of instrumental utility.

A major current concern in the context of global warming and 'peak oil' is that we are not going to be able to sustain our transport-intensive lifestyles and will have to change (Chapter 3). Can traffic-reduction policy draw on the different inclinations towards and away from car use? We have seen that with regard to cutting car use Die-hard Drivers are the least amenable and Aspiring Environmentalists the most amenable. Consistently across a variety of types of survey, including empirically measured changes in actual behaviour, there would appear to be a core figure of around 20 per cent of car drivers in the UK who either express a willingness to change or have already changed, and a further 20–40 per cent who have some inclination to change their travel behaviour (Anable, 2005; Cairns *et al.*, 2004; Dudleston *et al.*, 2005; Goodwin, 1995; Stradling, 2006; Transport for London (TfL), 2006).

Combining this evidence with the emerging consensus that technological breakthroughs are not going to provide the 'silver bullet' for the mitigation of climate change and energy security threats caused by the transport sector (Hickman and Banister, 2006; Pridmore *et al.*, 2003), the indication for policy makers is that travel behaviour change is best galvanized by the local intensification of 'soft' policy measures directed at individuals' travel patterns. These would address concerns at the community, workplace, school gate and household level by engaging the public in a way that taps into their inclination to reduce car dependency for the full range of instrumental, symbolic, affective and moral reasons.

Demand management strategies can focus on regulation, policy and pricing, land use management policies and practices, minor infrastructure, awareness and travel behaviour (Pinkard 2005) and a whole variety of bespoke transport policy initiatives at the local level which rely on encouraging and stimulating the right conditions for voluntary travel behaviour change – known as soft factors – have been shown to assist the travelling public in making smarter choices (Cairns *et al.*, 2004). Nevertheless, traffic reductions and improvements achieved through soft measures have to be 'locked in' with hard measures (Cairns *et al.*, 2004) such as denying access to cars through pedestrianization, bus lanes or cycle lanes, or by charging for access by car through tolls, parking or congestion charges. We have discussed here the rapid expansion of ICT and the mixed evidence on the degree to which it is reducing overall travel demand. At the moment, however, if new technology does reduce travel it will be a fortuitous accident rather than a conscious outcome driven by policy. In contrast to the single acts of infrastructure building that have previously characterized transport thinking, the penetration of ICT requires continuous monitoring and adjustment (DHC, 2006).

The segmentation of travellers outlined in this chapter suggests that combinations of hard and soft policy measures to change car use will be most effective when targeted at the differing attitudes, beliefs and values of the segments (Dudleston *et al.*, 2005). A better understanding of the travel experiences which inform people's readiness or reluctance to change, in order to enhance the journey quality of the more sustainable modes of transport, will facilitate change in individual travel patterns and choices.

12 | Transport, Tourism and Leisure

Derek Hall

Tourism and recreation represent a vitally important set of spatially expressed economic, social, cultural and environmental change processes that are both reliant on transport and generate considerable transport demand in their own right. The relationship between transport and tourism thus embraces a considerable range of issues pertinent to transport geography. As such, this chapter complements and provides links with a number of the other chapters in this volume.

In this chapter, 'tourism' is interpreted as being both an activity undertaken away from home involving an overnight stay (the formally recognised definition), and as a recreational pursuit that can be undertaken within a day trip. The latter is often referred to as excursionism, and although usually dominated by domestic activity (that is, undertaken within the participants' own country), some cross-border excursionism takes place, particularly, not surprisingly, by those living close to a permeable international boundary (such as a mainland EU border). Because most excursionism does not cross international boundaries it tends to be poorly enumerated in official statistics, as is domestic tourism generally. But whereas attempts to enumerate domestic tourism may be derived from accommodation statistics, excursionism, by its very nature, is of course excluded from such data.

The relatively poor statistical enumeration of domestic tourism, together with the fact that international tourism explicitly impacts on national balance of payments positions, tends to result in most (economic) analytical emphasis being placed on international rather than domestic tourism, even if in many countries the latter is numerically much greater. In reality, of course, on any given transport mode, and at any transport-related tourism attraction, one may find international and domestic tourists, overnight stayers and excursionists together, indistinguishable in their pursuit of similar recreational activities.

This chapter takes both a conceptual and a practical approach to understanding the tourism phenomenon from the perspective of transport geography. It briefly examines the central role of transport in tourism, addresses the gap between the tourism and transport literatures in relation to such key concepts as sustainability, and considers some of the practical implications of this. The increasing importance of transport forms acting as a focus of visitor interest in their own right is also examined from a geographical perspective. Exemplification is based on a range of environmental contexts.

This brief review necessarily focuses on a limited number of issues. Unfortunately, destination marketing, service quality, travel welfare, gender issues, walking and cycling, and the diffusional role of tourism transport – in terms of culture contact, the spreading of disease, and employment generation – all express important spatial dimensions that are largely neglected in this short appraisal.

The global tourism context

Freedom to travel is considered a basic human right and is enshrined in international agreements as such. Over a number of decades, and in some case centuries, as many nations' levels of economic development have increased and periods of paid leisure time have been established, the growth of domestic tourism and recreation has become supplemented by international travel (further supplemented by development of the almost oxymoronic notion of business tourism). While the combination of (increasing) disposable income and free time has been the obvious driving force fuelling a continued, inexorable growth of tourism, ease of mobility (car ownership, fast, mass global air travel and relatively diminishing costs) and access to information (notably latterly through the Internet) have facilitated such growth.

The World Tourism Organization (WTO, 2005, 2006) recorded globally an 83.4 per cent increase in international tourism arrivals between 1990 and 2005, from 441 to 808 million (Table 12.1). Tourism receipts increased by 119 per cent between 1990 and 2004, from US$280 billion to US$623 billion. In 2004 43 per cent of international arrivals were by air, recording a numerical growth of 15 per cent over 2003. The WTO (1997) has projected international arrivals to reach over 1.56 billion by 2020. Of these, 377 million (24 per cent) will be long-haul travellers, representing a long-haul growth of 5.4 per cent per annum. The actual and potential scale of global tourism activity is clearly enormous, and the integral role of tourism transport as an agent of environmental, economic and social change within such processes would appear critical.

Policy for tourism-related transport is important at a number of levels and intersects many key transport issues. Examples may include: inter-modal co-ordination and integration; traffic management, pedestrianization and the management of urban spaces; retail centre planning and access; supporting the economic and cultural survival of mainland and island peripheries; social exclusion from tourism and leisure through poor access to appropriate transport; relationships between tourism marketing and new route development; the competitive nature of landward access to airports; the economic and physical impacts of cruise liner visits, particularly in smaller, peripheral communities; managing transport as a heritage attraction; individual access to and activities within remote and fragile rural locations (for example,

Table 12.1. Recent international tourism growth, by region. Source: WTO, 2006.

	International tourist arrivals			
	Year (arrivals in millions)		*Annual percentage increase*	
Region	*2004*	*2005*	*2003/04*	*2004/05*
Europe	424.5	441.6	4.3	4.0
Northern	49.7	52.9	8.4	6.6
Western	139.0	142.7	2.2	2.6
Central/Eastern	86.3	88.0	10.0	2.0
Southern/Mediterranean	149.5	158.0	1.9	5.7
Asia & the Pacific	145.4	156.7	27.2	7.8
North-East	79.4	87.6	28.6	10.8
South-East	48.3	50.6	30.1	4.8
South	7.6	8.0	18.5	5.4
Oceania	10.1	10.5	12.1	4.0
Americas	125.9	133.6	11.2	6.1
North	85.8	90.1	10.9	4.9
South	16.2	18.1	17.2	11.6
Central	5.7	6.5	17.2	14.1
Caribbean	18.1	18.9	5.9	4.3
Africa	33.4	36.8	8.4	10.0
North	12.8	14.3	15.1	12.2
Sub-Saharan	20.7	22.4	4.7	8.6
Middle East	36.3	39.7	19.9	9.5
WORLD TOTAL	766	808	10.0	5.6

using off-road SUVs and snowmobiles); and individual and group access to marine resources (for example, see Davenport & Davenport, 2006).

Such a list may represent an opportunity for a holistic policy approach, but can also exemplify the fragmentation of a sector which is characterized by a range of activities undertaken in diverse locations with both single and multiple purposes which may or may not be inter-related. Certainly in the UK, for example, there may often appear to be little policy coherence at national level, while local authorities, government and non-government agencies (NGOs) attempt to patch together strategies, simultaneously being compelled to compete with each other for scarce resources.

Sustainability is a key strategy concept in both transport and tourism policies, and acts as an area of practical implementation binding the two functions and policy arenas together such as recreation-related strategies within local transport plans. But its attainment continues to be constrained by several factors. These include: our relatively poor knowledge and understanding of the complex patterns of tourism-related activity; the almost universal separation of responsibilities for transport and tourism governance between different (and often competing) central and local government agencies; and the fragmented nature of the tourism industry into myriad private, public and voluntary bodies involved in implementation.

Tourism transport roles

In relation to the supply side of tourism transport, perhaps four general spatially expressed roles can be identified: (i) linking the source market with the host destination; (ii) providing mobility and access within a destination area/region/country; (iii) providing mobility and access within an actual tourism attraction; and (iv) facilitating travel along a recreational route, where both the transport form and nature of the route may combine or act singly to provide the tourism experience. A fifth role is represented by transport forms located in a particular place, and which, although static or providing limited movement, act as the focus of recreational and heritage interest, such as in museums and at rallies.

For the first role the transport modes and infrastructure used for connecting source markets and host destinations may not be wholly or significantly employed or designed for the purposes of tourists. Limited research has been undertaken on how route choice and travel to a destination affects perceptions of it (see, for example, Blomgren & Sørensen, 1998; Butler, 1997; Elby & Molnar, 2001; Jacobsen & Haukeland, 2002; Murray & Graham, 1997). In relation to the third role noted above, site-specific transport requirements for particular tourism attractions may be so specific or unique to that particular location that the ability to conceptually extrapolate from particular empirical experience may be severely limited (for example, monorail and shuttle transport at Disneyland in Anaheim, California).

Facilitating travel along an explicitly recognised recreational route may involve transport acting as a tourist attraction in its own right, either complementing or even dominating the travel experience. Examples here vary considerably from a steam train or 'heritage' coach service, such as in the Scottish Highlands, to a Caribbean cruise liner (e.g. Robbins, 2003). Yet often, transport for tourism – whether bicycle, open-top bus or lake cruise boat – has been viewed merely as a means to an end, conceptualized as a consumer service rather than as a producer service (Debbage & Daniels, 1998), modelled as a cost rather than a benefit.

Examination of the second role, regional or national level provision of mobility and access, is confronted with a paradox. On the one hand, the enormous growth of both domestic and international tourism in the past half-century should have provided us with at least substantial empirical evidence of the spatial patterns and transport demands of tourists' activities in terms of what they do and where they go once they arrive at a particular destination (Prideaux, 2000). Yet this appears largely not to be the case, despite the historic concern over the local impacts of especially mass tourist activity. Most evidence relates to fragments of aggregated information derived from different attractions and places of accommodation, rather than coherent overviews of whole visits. Research based on diary records, for example, is limited, in terms of both methodology available and the usefulness of data rendered. A scattered case study literature has provided some insight into pedestrian flows in urban areas (e.g. Chadefaud, 1981), including the activities of cruise tourists onshore (e.g. Jaakson, 2004), the importance of pedestrianization in central areas for tourism and recreational shopping (e.g. Hass-Klau, 1993), opportunities for physical activity (e.g. Badland & Schofield, 2005), traffic management and congestion in national parks (e.g. Daigle & Zimmerman, 2004; Holding & Kreutner, 1998; Steiner & Bristow, 2000), modal use (e.g. Downward *et al.*, 2000) and time-budget studies (Debbage, 1991). But little is known about the details of combinations or sequence of visits in order to provide a geographical understanding of the interrelationships of different places and activities. Most time-budget studies, for example, have tended to emphasize how much of resort visitors' time is spent in their accommodation.

Transport at the tourism destination

Within tourism destination areas, we can conceptualize transport demand in terms of three ideal categories of transport users, all potentially competing with each other for transport and transport space in terms of public transport access and road space, including parking, cycling and pedestrian space (Hall, 1999, 2004b):

- The host community: residents of the destination area who are not directly involved in the tourism industry but who find themselves competing with both tourists and tourism industry employees. Indeed, they may be socially excluded from tourism and recreational activity, perhaps because of poor access to appropriate transport, ironically, as the result of being physically displaced by the growth of those very tourism-related activities and infrastructure. The demonstration effects of tourists' mobility may exacerbate in the host community a sense of relative deprivation and potential antagonism;
- Employees of the tourism industry, who may be drawn from the host community or may be incomers from other parts of the region or country, or even migrant workers from another country, and who may have particular transport needs because of the demands of tourism's unconventional working hours; and
- Tourists themselves: who may or may not use specialized, differentiated transport. Their direct competition with the other two groups for transport and transport space will vary considerably according to their scale, types and location of activities.

There may often exist a complex range of relationships between, and relative identities, of 'host' and 'guest' (e.g. Kohn, 1997). None the less, this threefold categorization does offer a framework for examining relative access to transport and transport space, and thus for exploring the social sustainability of transport access and use in tourism destination areas (see also Chapters 4 and 6). However, such a framework requires detailed knowledge of the local and wider environmental contexts within which such competition for, and use of, transport for tourism takes place.

Only relatively recently has there begun to be shown some synergy between the transport (planning) and tourism debates on such issues of common concern as the impacts of transport externality costs borne by host populations and environments in tourism destination areas (e.g. Black & Nijkamp, 2002; Lumsdon, 2000; Lumsdon & Page, 2004). This is in no small measure the result of increasing alarm over climate change and the role of tourism transport – and both directly and indirectly – in relation to this (e.g. Becken, 2005a, 2005b; Palmer-Tous *et al.*, 2007; Patterson *et al.*, 2006).

Transport, tourism and 'sustainability'

Identifying tourism transport

One obvious constraint on the development of a literature on the sustainability dimensions of tourism transport is our ability to actually identify tourism transport as a discrete functional entity for analytical and policy purposes (Halsall, 1992; Hall, 1999, 2004b, 2005; Page, 1998, 2005). This dilemma reflects the obvious logistical problems that:

- Tourism transport embraces a diverse range of modes (Figure 12.1), functions, spatial expressions and relationships, and ownership patterns;
- Transport forms may be employed for tourism purposes exclusively (charter aircraft, tour coaches, cruise liners, 'heritage' transport), partly (scheduled air services, long-distance and local destination train services, ferries, express buses, taxis, hire cars, cycles, motorcycles), occasionally (private cars, local public transport in a seasonal tourism destination), or rarely (private and public commuter transport); and
- The tourism function of transport may be expressed explicitly (charter aircraft, touring coaches, cruise ships), or may be anonymous (hire cars, private cars, cycles, motor-cycles) (e.g. see Lumsdon, 2006).

Recognition of these analytical and conceptual issues helps to set the context within which to identify the way in which transport may amplify and articulate

LAND	AIR	WATER
Bicycle/walking	Scheduled/'budget' flights	Ocean/river cruises**
Car/taxi	Charter/inclusive flights	Scheduled passenger vessels
Bus/motor coach*	Personal air transport (private jet)	Ferries
Tram*	Scenic flights**	Canal barges/cruisers**
Caravan/mobile home	Hot air balloon**	Ocean-going yachts**
Train*	Gliding**	Small recreational craft**
	Hang- and para-gliding**	
	Parachuting/free-falling**	

* may have high intrinsic heritage role
** may have high intrinsic experiential role

Figure 12.1. Potential tourism transport modes. Adapted and developed from Collier, 1994.

sustainability questions in international tourism. The conceptualization of such relationships will be modified by the cultural context (Hall, 2004b). Notable in this respect is the simplistic dichotomy between:

- Less developed societies where tourism transport may be seen as an instrument and/or symbol of differentiation and inequality between tourist and resident (e.g. Simon, 1996), and
- Economically more developed societies where the very lack of differentiation between transport of host and guest, commuter and recreationalist, may pose practical administrative and planning problems for addressing ways of tackling the external costs of different transport functions (e.g. Oberholzer-Gee & Weck-Hannemann, 2002).

For example, the effectiveness of implementing congestion charging policies would be enhanced by knowledge of the nature and extent of discretionary leisure travel, which could then be priced away from the busiest periods and areas. Yet, in debates on road tolls, congestion pricing, carbon taxes and other mechanisms seeking the repayment and/or reduction of road transport externalities, the differentiation of tourists and non-tourists competing for road space is rarely addressed. This appears to be largely because of the potential difficulties of implementing any discriminatory policy (e.g. see Feitelson, 2002). Similarly, the competition between tourists and hosts for public transport use has only occasionally attracted critical analysis (e.g. Robbins, 1997).

Transport in tourism

Given that sustainability is internationally recognised as a supposedly central tenet of policy both for transport (e.g. European Commission, 2003a; Goldman & Gorham, 2006; Haynes *et al.*, 2005; Organisation for Economic Cooperation and Development (OECD), 2002a, 2002b; Tight *et al.*, 2005) and tourism (e.g. Butler, 1999; Choi & Sirakaya, 2006; Ko, 2005; United Nations Economic and Social Commission for Asia and the Pacific (UNESCAP), 2001), attention to tourism transport could provide a coherent and holistic framework for sustainability policy. But in practice, although the sustainability of tourism activity has long been a much-discussed and contested concept, much of the debate has tended to focus on particular destination sites, regions and cultures, rather than on the manner in which tourists actually travel to their final destinations. Implicit and rarely explicitly expressed, certainly never by the tourism industry, is the environmentally beneficial effect that less travel would bring.

According to the WTO (2005), 45 per cent of all international tourist arrivals in 2004 were by road. Motor vehicles and their externalities thus represent an important element of the impact of tourism-related activities. While the automotive industries may argue that technological solutions can overcome current environmental fears (see Chapter 3), Steg and Gifford (2005: 59–60) provide two caveats. They suggest that the mitigating effects of new technologies tend to be overshadowed by the continuing growth of car use, and, once purchased, drivers may be tempted to use energy-efficient cars more often because of their supposed environmental friendliness and lower running costs (the rebound effect or Jevons principle). Social measures to reduce overall car use, such as car sharing and car clubs/pools, appear less realistic for tourism purposes, although vehicles used for recreational journeys are more likely to have a higher occupancy rate than those employed for commuting and other regular point-to-point trips.

Transport problems and identified measures tend to be framed by policy makers as technical-economic problems or challenges. This is perhaps one reason why this area of policy does not relate well to tourism practice. Behavioural responses tend to be presented in terms of obstacles and deviations from an 'ideal' pattern that is usually defined in a predominantly rational and uniformly informed world instead of in a world of irrationality and distortion. Further, when policies are developed to mitigate negative impacts of transport in one domain, they may involve different institutional actors from those taking decisions in another, and conflicting policies may result.

More comprehensive sustainability policies easily become entangled in the dominating views that exist in trade policy circles, reflecting the complicated policy mix that tourism represents – embracing the overlapping and often-competing policy arenas of transport, trade, regional development and human welfare and behaviour. One way to engage in more constructive discourse with other policy circles is to create better datasets that enable overviews of the contribution of transport to the tourism sector and to the national economy such as through the creation of transport satellite accounts (Link, 2005).

Sustainable transport?

Although there is no wide agreement on a definition of 'sustainable transport', the term can be understood as referring to a transport system which in itself is structurally viable in an economic, environmental, and social sense, and which does not impede the achievement of overall sustainability of society (Donaghy *et al.*, 2005; Richardson, 2005). This last phrase appears to be something often glaringly absent in discussions of 'sustainable tourism', but is pivotal in any critique of the sustainability of tourism, particularly that requiring long-haul travel. This is not least because international and domestic air travel is highly energy-intensive and its speed enables and encourages people to travel long distances (e.g. Olsthoorn, 2001; Schipper *et al.*, 2001) (Chapter 9).

Despite the claimed increasing energy efficiency of new passenger aircraft, the impact of air travel can only be reduced by diminishing passenger numbers, and thus frequency of flights. Given the current global dominance of 'free-market' frameworks, it might appear that this can only be achieved by making the price of tickets reflect the true environmental costs of air travel (Himanen *et al.*, 2005). The rapid growth and use of low cost carriers (LCCs) (e.g. Mintel, 2005a, 2005b) suggests that just the reverse has been taking place over the past decade (Box 12.1). Requiring airlines to pay tax on aviation fuel would be a necessary first step in positively acknowledging the sector's external costs. (Perceived) global terrorism and its implications can exert an impressive short-term restraining effect.

Box 12.1. Low Cost Carriers and tourism. Sources: Alderighi *et al.*, 2004; Bieger and Wittmer, 2006; Decker, 2004; Dobruszkes, 2005, 2006; Druva-Druvaskalne *et al.*, 2006; Endzina & Luneva, 2004; EU Business, 2005; Hall, 2004c; Hall & Müller, 2004; Klein & Loebbecke, 2003; Naish, 2004; Richards, 2006; Roberts & Harrison, 2006; Ryanair, 2006a, 2006b.

Integration and European spatial restructuring has been aided by Low Cost Carriers (LCCs) in relation to a range of traditional and relatively recent European mobilities, rendering earlier models of civil aviation roles and practice – particularly in relation to leisure-related mobilities and income elasticity – almost redundant. The market for air transport users has been substantially deepened, both domestically and internationally, reflecting relatively low fares (partly resulting from 'stripped down' operating costs and by web-based yield management fare pricing mechanisms, use of regional airports, employment of 'supermarket'-style marketing and promotion, and the exploitation of a rapidly achieved high level of personal access to information technology.

Thus 95 per cent of easyJet's tickets are said to be purchased online, 75 per cent of the low-cost market in Europe has represented new customers, and by the first quarter of 2005 LCCs carried 19 per cent of European air passengers compared to five per cent in 2000. For 2006, Ryanair was claiming to be carrying 42 million passengers annually, with plans to double its size in the next five years following an agreement with the government of Morocco to establish up to 20 routes and carry close to one million passengers per annum on flights to Morocco. Domestically, the company was arguing that in 2007 it would deliver 9 m passengers through Dublin airport generating a tourist spend of €2.5 bn.

But deeper market penetration has largely involved providing access to the lower end of the market, with potential image problems for some destinations, particularly those seeking a positive new or rejuvenated image projection. For example, several firms specialize in 'stag' and 'hen' weekend packages to Latvia, taking advantage of LCC operation from Germany and the UK since the country's 2004 EU accession. This is at a time when Latvia is still trying to overcome a half-century's legacy of being part of the Soviet Union. Riga, the capital, has been targeted as a 'hot spot' that offers amazing nightlife, fantastic cheap local beer – and the 'most stunning-looking women' on the Continent. The *Riga Visitor's Guide*, freely available at Riga airport and in hotels, advertises Riga as an entertainment city and contains a polythene-wrapped insert section devoted to nightlife, offering clear opportunities for sex tourism.

While some of these processes may not be sustainable, they may act as the first-stage basis for diversification and adding value to tourism products.

Indeed, at least one report has suggested that what low-cost air travellers are saving in fares they are spending to upgrade the quality of their accommodation. The growth of LCCs may also be seen to have facilitated regular travel to second homes, easier international student exchange mobility, and labour mobility for the tourism industries.

Transport in tourism 'sustainability'

Reviewing the literature on environmental impacts of leisure-related activities (e.g. Becken *et al.*, 2002; Ceron and Dubois, 2003; Gössling, 2000, 2002a, 2002b; Peeters, 2003), Gössling *et al.* (2005) drew three major conclusions:

- Whether employing measures of energy consumption, greenhouse gas emissions or area-equivalents as basis for calculations, a substantial share of tourism is revealed as unsustainable;
- The use of fossil fuels and related emissions of greenhouse gases is, from a global viewpoint, the most pressing tourism-related environmental problem; and
- Transport contributes more than proportionally to the overall environmental impact of leisure-tourism: this is between 60 and 95 per cent at the journey level, including local transport, accommodation, and activities (see also Dubois & Ceron, 2006; Peeters & Schouten, 2006).

Under these circumstances, Gössling *et al.* re-emphasize the need to reduce greenhouse gas emissions, particularly in the transport sector. Yet case study analyses on the impacts of the tourist industry from an ecological efficiency (or eco-efficiency: Cramer, 2000) perspective, found great variations, depending upon source and destination countries, tourist cultures and environments (Gössling *et al.*, 2005). For example, to generate one unit of financial value in the Seychelles, concurrent emissions of carbon dioxide-equivalent are seven times larger than the world average, while in France, some types of tourism have an eco-efficiency ratio less than one-tenth of the world average. Employing ecological footprint analysis (EFA) (Holden & Høyer, 2005; Hunter and Shaw, 2007), Gössling *et al.* (2002) found that for European tourists flying to the Seychelles, more than 97 per cent of their energy footprint was the result of air travel. This suggested that efforts to make tourist destinations more sustainable, such as through energy saving devices or with the use of renewable energy sources, could only contribute marginal savings for long-haul destinations.

Travel distance and mode of transport are the most important factors influencing the eco-efficiency of tourist behaviour, which can be positively influenced by an extended length of stay and higher expenditures per day. Short travel distances are a pre-condition for sustainability: 'Any strategy towards sustainable tourism must thus seek to reduce transport distances, and, vice versa, any tourism based air traffic needs *per se* to be seen as unsustainable' (Gössling *et al.*, 2002: 208). Indeed, up to

75 per cent of EU citizens are said never to travel beyond EU borders (European Commission, 2003b).

Despite a recent and greatly welcome special issue of the *Journal of Sustainable Tourism* (e.g. Becken, 2006), it continues to be notable how the sustainable tourism and sustainable transport literatures have remained largely separate. This is partly because the tourism industry, in its promotion of the (claimed, potential) 'sustainability' of tourism, focuses on destination activity and largely chooses to ignore the means by which tourists actually travel to the destination from their home (e.g. see Gössling *et al.*, 2002; Høyer, 2001). Thus, with a few notable exceptions, 'tourism', within the industry's interpretation and definition of its 'sustainability', appears to be a process and activity that applies almost exclusively to destinations, deliberately and simplistically abstracted from the wider spatial context of travel and its complex web of mobilities. This approach appears to dishonestly fail to identify the wider sustainability implications of tourism activity, and notably the externalities of motor car and air travel, and thus defeats the notion and purpose of sustainability being a holistic concept by focusing on just one spatially expressed dimension of tourism activity (Hall and Brown, 2006).

There is an urgent short-term need for a menu of policies, such as shifting short-haul air travel to rail, raising fuel efficiency, and fiscally regulating motor vehicle and air travel. Total external costs of air transport are currently estimated at around €55 per 1,000 passenger/km. Turning these external costs into a congestion charge for the skies by adding them to ticket prices would result in €5.5 per passenger/100 km. In the US such a scheme has been rejected on the grounds of exerting 'unfair pressure on a struggling sector' (Skuse, 2005: 45), but the EU has reaffirmed the need for airline climate taxes as a necessary part of a climate policy drive to reduce emissions.

Yet there currently appears to be almost a conspiracy of procrastination and denial that urgent action is needed to reduce the global environmental impacts of transport and travel. The WTO conference held on tourism and climate change in Tunisia in April 2003 was largely concerned with the impacts of climate change on tourism, rather than the converse (Nicholls, 2004). Despite an industry (public relations) promotion of 'sustainable tourism', such mixed signals may unwittingly encourage tourists to ignore the environmental consequences (and other externalities) of their travel behaviour (notably in relation to the rapid growth of low-cost airlines in Europe and elsewhere). For example, Becken (2004) found in Australia and New Zealand that while 'tourism experts' interviewed saw a changing climate as a potential threat to tourism, they could not necessarily see tourism's fossil fuel consumption and carbon dioxide emissions as contributing to climate change. Not surprisingly, half of all tourists surveyed questioned any link between tourism and climate change.

Geopolitical issues of access to travel and recreation resources will become an important focus of tension and potential conflict in the face of rapidly increasing global pressures. Although at a global level there remain obvious constraints on certain types of activities and restrictions on access in certain areas, generally there is relative freedom of movement and access around the globe, subject only to the ability to pay. But such freedoms are available largely because they are restricted to a privileged minority of the world's inhabitants. And although current carbon rationing schemes for mitigating environmental impacts seek an equitable outcome, what will be the result when, later this century, living standards have risen sufficiently in the currently very populous less developed countries of Asia, Latin America and Africa for their citizens, justifiably, to want to join those of us who have long enjoyed the privileges of travel and become fellow global tourists? (See also Chapters 3 and 13.)

Transport as tourism experience

Until relatively recently, transport for tourism was viewed as a means to an end, rather than a self-contained experience in its own right. As noted earlier, it was conceptualized as a consumer- rather than a producer-service. The intrinsic value of the spatial in recreational travel has not been well researched (Sørensen, 1997). Yet, there appear to be an increasing number of contexts in which transport can provide a tourism end in itself. Not only is transport a key element of the tourist experience (Pearce, 1992), but the transport experience can actually comprise the tourism experience (Hall, 2005; Lamb and Davidson, 1996). Thus not only is transport important as an essential underpinning for tourism, and helps to place tourism within a wider societal and environmental context, but it can be a focus of interest both because of its functions and also because of its intrinsic attributes. A tourism-related experiential continuum for selected transport modes is expressed in Figure 12.2.

The transport experience can be the primary if not exclusive tourism experience, embracing qualities of heritage, nostalgia, education, uniqueness, added-value and entertainment. The commodification of transport and of travel experiences has generated new niche products, such as hot air ballooning or snowmobiling. There are perhaps three categories to be explored here: unique transport experiences,

Utility —— **TRANSPORT ROLE** ——→ *Tourism Experience*

Local car/taxi Urban bus, tram & metro	Intercity rail Flights Local ferry Longer road, road-rail & road-sea journeys Local cycling/walking	Scenic road routes/trails Motor coach tours Urban sightseeing bus routes River cruises Walking/cycling trails Transport museums & rallies	Walking and cycling holidays Heritage railways Other trains with staged experiences Ocean cruises Speciality activities: e.g. ballooning, hang-gliding, whitewater rafting

Low —— **INTRINSIC EXPERIENTIAL VALUE** ——→ *High*

Figure 12.2. Tourism-related experiential continuum for selected transport modes. Adapted and developed from Lumsdon and Page, 2004.

added-value experience within transport services, and the intrinsic attraction of transport itself.

First, certain forms of transport offer a unique experience based on the nature of that transport and the location it is set in, thereby contributing to the spirit of place. Diverse examples include Venetian gondolas plying the Grand Canal, San Francisco cable cars climbing Nob Hill, or Poland's Ostróda-Elbląg Canal, providing cruises which involve the unique experience of the vessel being hauled overland on skipways to overcome an elevation difference between waterways (Furgala-Selezniow *et al.*, 2003; Ostróda-Elbląg Navigation, 2003). Such transport modes may be intended to be, or have become, a predominantly recreational attraction and experience rather than a practical means of getting from one place to another: being in or on the particular transport mode in its geographical setting provides the primary experience.

Secondly, added-value experience within transport services can embrace one-off prestige experiences such as travel in one direction on the opulently restored Venice Simplon-Orient-Express (Orient Express, 2006), or an erstwhile day excursion in Concorde to Lapland for Santa Park at Christmas. The latter perhaps reflects how commodification can be achieved by the marketing of place or icon myths, with images of transport as treasured spaces and/or experiences as an important ingredient. Such commodification fuels the desire to experience transport and results in demand for visitor experiences rather than merely the products and services that support them. The concept of the experience economy, as articulated by Pine and Gilmore (1998, 1999), is based upon an understanding that visitors need experiences with which they and the provider can engage and thereby both derive added value.

Thirdly, transport acts as a focus for significant recreational sub-groups whose interest and source of travelling is actual transport forms. Such compelling interest may focus on nostalgia, heritage, engineering and/or industrial archaeology. Attendances at transport museums (e.g. Divall and Scott, 2001), rallies and excursions, and membership of transport enthusiasts' organizations, emphasize that those with an intrinsic interest in transport represent a significant leisure sub-culture and substantial niche market. In the UK, for example, some 200 bus and/or tram rallies and special events are held annually (Anon., 2006a, 2006b). A voluminous literature exists aimed at all levels of transport enthusiasts and specialists, and is becoming increasingly international in scope, not least through the burgeoning of enthusiasts' websites (see, for example, Cape Railways Enthusiasts' Association (CREA), 2004; European Train Enthusiasts – Eastern New England (ETE-ENE), 2004; Indian Railways Fan Club (IRFC), 2005), organized visits programmes and merchandising (e.g. Interessengemeinschaft Eisenbahn (IGE), 2006). In the UK and elsewhere the growth of recreational interest and activity focused on inland waterways has encouraged the recreational development and promotion of such transport engineering 'icons' as the Anderton Boat Lift in north-west England (British Waterways, 2006) and the Falkirk Wheel in central Scotland (British Waterways Scotland, 2006). In such cases, gastronomic, retailing and educational opportunities are emphasized to provide a vital infrastructural complement to the main attraction.

The transport enthusiast leisure sub-culture tended to be perceived as representing a male-dominated specialist, esoteric, often far from fashionable, interest. But as tourism development in the 1970s and 1980s drew increasingly on heritage and nostalgia (Dann, 1994), so existing transport attractions responded, albeit somewhat slowly, with innovations to appeal to broader market segments. New transport and heritage attractions introduced products to encourage further segmentation. This has been noticeable in destinations responding to new (not necessarily indigenous) market demands, as in Central and Eastern Europe (e.g. MÁV Nostalgia, nd, 2000; Rosiak & Szarski, 1995; Ronedo, 1998).

Repositioned transport attractions such as 'heritage' steam railways (e.g. Höhmann, 2002; also Prideaux, 1999), have been incorporated into local and regional networks and place promotion collaboration. They have developed family-oriented interpretation and marketing, such as 'Thomas the Tank Engine' and 'Santa Special' days and weekends coupled to specialist shopping opportunities (Tillman, 2002). Such 'preservationism' has been almost pejoratively viewed as part of a collective nostalgia for the (usually masculinized) skills, symbols and certainties of our recent industrial past (Strangleman, 1999; Urry, 1995). But Halsall (2001) has contended that the social and cultural identity which has become attached to heritage railways is stronger than mere nostalgia. Thus in the UK, for example, some 140 restored railways (Butcher, 2006; Douglas, 2001) now promote themselves with a strong emphasis on heritage, place and interactive experience involving both entertainment and education (Divall, 2002) (Figure 12.3).

The attraction of transport (infrastructure) and its associated locations, (re)generated by popular culture, can enforce the concept of a fictional heritage place (Couldry, 1998). For heritage railways, providing a 'front stage' set for filming has the multiple value of generating income from the film company for use of the location, providing out of season employment for rolling stock, and generating more visitors through identification in and association with particular films. Perhaps the UK model for this was the Keighley and Worth Valley Railway, situated in 'Brontë Country' (West Yorkshire), which featured heavily in the sentimental 1970 film *The Railway Children*. The Kent and East Sussex Railway claims to have provided the location for the shooting of some 40 film and television productions (Tenterden Railway Company, 2006).

Two important implications arise from these trends. First, transport attractions need to be continually improving the quality and range of visitor experience and evolving their promotion and marketing in order to be suitably positioned to appeal to continually fragmenting and fusing market segments. Secondly, transport attractions need to complement, and be promoted alongside, other attractions in their region as part of 'place promotion' (Kotler *et al.*, 1993), and as a means of becoming embedded within perceived attraction clusters and circuits in order to better attract family-based markets. They also need to maintain close collaborative links both nationally and internationally with other transport attractions in order to reach the now globalized 'niche' markets. A major challenge for policy makers and resource managers is to render both attractions and their markets sustainable.

Figure 12.3. Location of UK restored railways. Modified from Butcher, 2006.

Summary and conclusions

Transport as both a tourism facilitator and tourism object performs a number of roles that continue to evolve as markets segment or fuse and as new technologies are adopted and adapted. As noted in the introduction, this chapter has been able only to offer a brief discussion of some of these roles and the issues arising from them pertinent to transport geography. The chapter has taken both a conceptual and practical approach to understanding the tourism phenomenon from the perspective of transport geography. It has briefly examined the central role of transport in tourism, and has addressed the gap between the tourism and transport literatures in relation to such key concepts as sustainability, considering some of the practical implications of this. The increasing importance of transport forms acting as a focus of visitor interest in their own right was also examined from a geographical perspective.

The ability to travel, to be able to reflect on one's own life and culture in relation to the first-hand experience of others', is a major and significant welfare consequence of our privilege to be able to travel across the globe. Perceptions of travel and tourism as a positive force have viewed it, ethnocentrically, largely from that position of privilege and dominance. Many would consider access to a motor car and to air travel to be two of the most liberating aspects of their lives, not least for recreational purposes. Yet, with our growing understanding of the forces behind climate change, the negative welfare externalities of these activities impact not merely on tourism destinations, but on all global environments in an inextricably interlinked and irrevocable way. As a result, in order to be able to respond to major questions concerning the continuity of such largely unfettered global travel, informed, holistic geographical appraisals are essential.

Part 3 | Future Transport Geographies

13 | Transport Directions to the Future

Glenn Lyons and Becky Loo

The broad set of endeavours that represent the transport geographies explored in this book share some enduring principles and key issues of concern that together set a defining architecture for the sub-discipline. For example, people need and desire the mobility required to meet other people such as their friends and families, and to access the goods, services and other opportunities distributed across space that shape their daily lives. Although (remarkably) the amount of time taken up by travel over the years is consistent – evolving transport technology has allowed us to travel further within the same time budgets (Chapters 8 and 11) – there are other significant considerations that have emerged to influence the relationships between time, place and human interaction. Amongst these are the exponential advances in information and communications technologies (ICTs) which translate place-to-place interaction into person-to-person interaction independent of distance, or the heightened imperative of lessening the environmental impacts of overall mobility, which is leading change in the legislative and policy drivers that influence the behaviour of developers, firms and individuals (Chapter 3). This chapter poses and offers answers to five questions as a means of both reflecting upon some of the themes explored in the book and providing some contemplation and exploration of directions our transport systems might take in future.

What are current transport trends?

By way of setting our discussion in context, it is worth restating current transport trends in the world's main transport markets (Chapters 6–10). Table 13.1 shows how the overall amount of passenger travel and how this travel across modes has changed in the recent past within the generally high-income (or rapidly developing) economies of the United States and the European Union. As at 2004, there were 469 cars in Europe per 1,000 people compared with 770 per 1,000 in the USA.

Table 13.1. Changes in passenger travel in the EU25 and in the USA between 1995 and 2004 (after European Commission, 2006c).

	Billion passenger km 1995 (mode share – %)	*Billion passenger km 2004 (mode share – %)*	*% change from 1995 to 2004*
Passenger car			
EU25	3787 (77)	4458 (77)	+18
US	5702 (87)	7165 (86)	+26
Bus (and coach)			
EU25	474 (10)	502 (9)	+6
US	219 (3)	226 (3)	+3
Rail			
EU25	324 (7)	352 (6)	+9
US	17 (0)	22 (0)	+29
Air			
EU25	324 (7)	482 (8)	+49
US	650 (10)	896 (11)	+38
All four modes			
EU25	4909	5794	+18
US	6588	8309	+26

Note: Mode share is the share of passenger km across the four modes considered.

Correspondingly, it can be seen that the car is by far the most dominant motorized mode in terms of distance travelled and this has remained the case for some substantial time. In absolute terms, use of the car also accounts for the substantial growth in overall travel over the ten years. Similar trends are evident in freight transport, with road haulage accounting for the largest and a consistently increasing share of tonne-kilometres (Figure 13.1). While some surface modes (such as short-sea shipping) other than the car and lorry have seen some increase in use over the last ten years, what also stands out is the huge growth in air travel – nearly 50 per cent in Europe and nearly 40 per cent in the USA. The scaling up of mobility levels in terms of how far we travel is continuing.

In other regions, however, the picture is somewhat different. In the more economically diverse Asia-Pacific region, for example, there is a less clear correlation between increasing incomes and increasing motorization, perhaps (at least in part) explained by the extreme density of many of the region's cities (Newman & Kenworthy, 1999) (Figure 13.2). There are signs that this might be changing, since the rate of growth in motor vehicle density is increasing faster in the Asia-Pacific than elsewhere, leading to very intense congestion problems in particular locations at

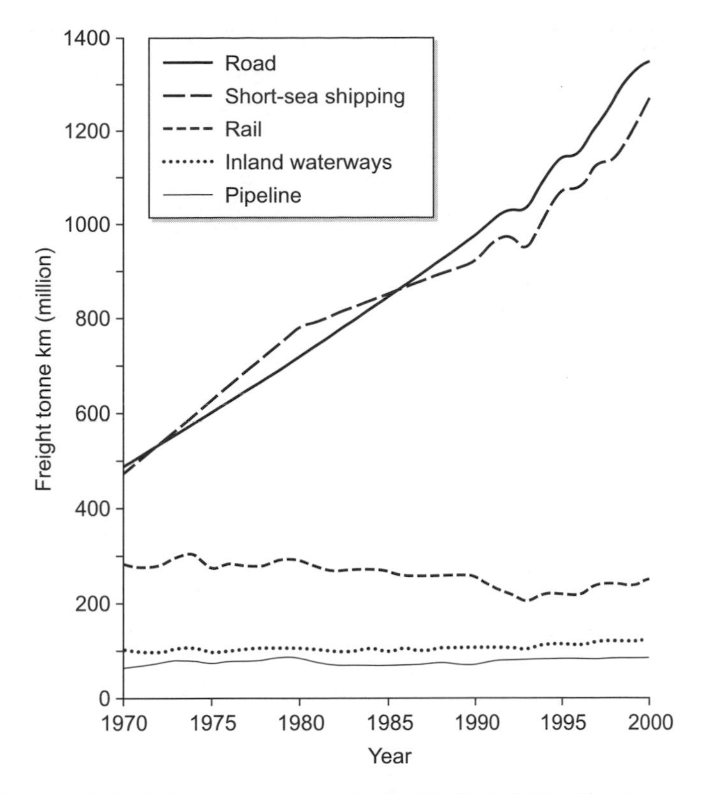

Figure 13.1. Modal split for freight transportation in the EU, 1970–2001. After European Commission (2006c).

particular peak times (Høyer, 2006) (Table 13.2). Air travel, expressed as air passenger intensity (that is, air passengers per 1,000 people), remains low for some high-income economies like Korea, although it is far higher in 'Western' countries such as New Zealand (indeed it is significantly higher than the UK). The movement of freight is becoming an increasing challenge as many Asia-Pacific economies continue to develop as major exporters of industrial goods to the rest of the world. Key issues include how to develop an integrated intermodal freight transport system and optimize supply chains (Loo & Liu, 2005).

Why do we travel, in the ways that we do?

This book has touched a number of times explicitly and implicitly upon this question – which has particular resonance in the light of the trends outlined above – across its chapters (see especially Chapter 11). Put simply, personal travel is generally considered to be a derived demand: derived from the need or desire to access and participate in activities at destinations. Historically, travel has been a necessity to overcome distance and to connect people to goods, services, opportunities and other

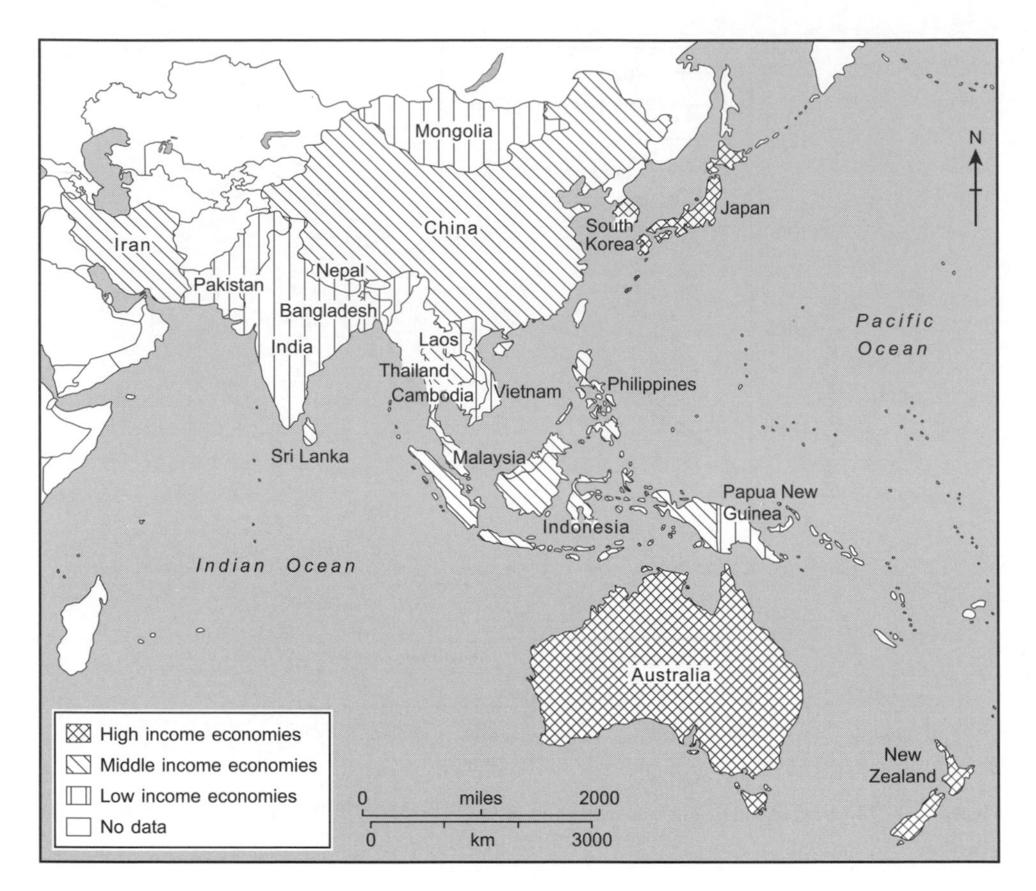

Figure 13.2. Principal countries in the Asia-Pacific region.

Table 13.2. Average annual percentage growth rates of major motorization indicators, 1993–2003.

	Motor vehicles per 1,000 people	*Motor vehicles per km of road*	*Passenger cars per 1,000 people*	*Road traffic (million vehicle-km)*	*Two-wheelers per 1,000 people*
Asia-Pacific region	4.20	4.06	6.12	10.58	3.96
Others	2.71	2.44	3.05	2.39	–0.86
Total	2.95	2.67	3.05	10.59	0.31

people. Distance and speed coupled with an individual's or firm's time, financial and energy constraints have strongly defined the level of access afforded to an individual for a given spatial configuration of people, goods, services and opportunities.

There appears to be an established societal understanding that the key 'currency' of travel is time (Schafer, 1998). National statistics for Great Britain reveal that for the last 30 years the average amount of time spent travelling per person per day has remained remarkably constant at around one hour (Department for Transport, Local Government and the Regions (DTLR), 2001), although it is important to note that such averages do not necessarily reflect a constant distribution of time spent travelling across the population. Theoretical explanations of this possible constancy include: biological programming (evolution has left us biologically programmed to spend a fixed amount of time on travel); utility maximization (an optimum point is reached that reconciles increased travel time to access a larger supply of activities with reduction in time to undertake such activities caused by increased travel times); and social routine (everyday life is full of settled routines of which travel becomes a part and takes its share of the allocation of time between all parts of the routine) (Höjer and Mattsson, 2000).

Earlier in the book there have been considerations of the use of travel time itself (Chapters 1 and 8), a key factor underpinning the rise of the new mobilities paradigm and work on the behaviour of people inside vehicles (see, for example, Brown & Laurier, 2005; Urry, 2004b). Jain and Lyons (2008) have examined the notion that travel time is a 'gift' both to others – in terms of one's commitment to overcome distance to be co-present – but also to oneself in that it provides 'transition time' and 'time out'. The first of these refers to the need to physically experience crossing space in time to achieve the sense of distance and difference; and the temporal opportunity to translate, adjust or prepare for a different social setting and social identity at the destination. Time out is seen as travel's provision of an impermeable 'protected' space away from the world around; time for selfish indulgence with a reprieve from the different roles assumed in life.

If time (and/or financial means and/or energy) are constraining factors on our overall level of travel, then speed extends our spatial 'reach' which in principle affords us a greater range of opportunity and quality of access for a given (travel time) budget. Herein has lain what has become an increasingly evident relationship between the speed of movement and the spatial configuration of society. The desire for greater 'reach' in terms of being able to access more places has quite literally driven the emergence of motorized travel modes. Initially, these afford individuals more choice and greater opportunity for access creating the prospect for economic advantage or enhancement to well-being.

As choices are made, however, they begin to define patterns of behaviour and social practice that rely increasingly upon the ability to overcome distance with speed. Land use developments fall into this process such that it becomes increasingly necessary for individuals and firms to consume motorized mobility to participate in the economy and in society. Thus in the case of the motor car there is a peculiar coexistence of both the independence afforded by it and the dependence created by it (which is made more complicated still by rising congestion, and its impacts on the utility of the private car). In an age when social inclusion is an important political goal in many societies, it is tempting to suggest that greater mobility allows an individual to be more included in society. Yet some commentators argue that greater (motorized) mobility, perversely, is playing a key part in creating a more exclusive society, especially when goods, services and people are located such that access to a car is essential to reach them (Kenyon, 2003).

The defeat of distance has not stopped with the train or car. In many parts of the developed world, (shorthaul) aviation has been the fastest-growing form of transport in recent years. Proponents of further growth in air travel advocate the public desire for its consumption and its economic importance, with much of the literature on innovation and economic growth pointing to the comparative advantage of those locations with the best international accessibility by air (see, for example, Cooke & Morgan, 1998; Knowles, 2006a; Kresl & Singh, 1999; Simmie, 2001). But the debate concerning aviation must be more comprehensive and far sighted. One can draw strong parallels between car dependence and what one could now term 'plane dependence'. The availability of air travel creates new opportunities such as the ability for (more) holidays abroad and more international business meetings. Arguably, these activities remain discretionary, although such opportunities can be less discretionary once they have been pursued and turned into established activity – for example, second homes abroad, personal relationships that transcend national boundaries and island economies reliant on international tourism.

Who is responsible for current transport trends?

The short answer to this question is that they are shaped both by individuals and firms making choices and performing behaviours and by those who define legislative and motivational levers that govern or influence behaviour (Chapter 5).

It could be assumed that the transport system exists to *serve* society. Indeed, it may be argued that policy makers have for many years espoused this assumption. The policy era known as 'predict and provide' endured for many years, during which the logic was that as society's demand for (motorized) travel increased so too should the capacity of the transport system be increased to accommodate that demand and thus avoid the congestion that would otherwise result. As time has passed it has become increasingly clear that the base assumption is only a partial reflection of reality (Goodwin, 1999; Docherty, 2003). In practice society *shapes* transport but so in turn does transport shape society and space (Knowles, 2006a; Lyons, 2004). There are unintended consequences of the assumption that transport serves society: in providing more transport system capacity it had not been intended that this would, in itself, be responsible for generating additional traffic, even though it was made easier to transcend distance through faster movement (Standing Advisory Committee on Trunk Road Assessment (SACTRA), 1999). One arrives at a conclusion that transport must *support* society rather than merely serve it – in other words, developments in transport must be encouraged in such a way that their wider consequences for society are understood and accounted for.

This revised viewpoint is becoming more evident in transport policy thinking, at least in those countries where the political system recognises the concept of market failure and the need for some sort of state intervention in transport (Glaister, 2001; Docherty *et al.*, 2004). In 1998 a Transport White Paper in the UK was published which recognised a need to influence as well as accommodate travel behaviour (Department of the Environment, Transport and the Regions (DETR), 1998). One of the lines of reasoning within this paper was that to tackle congestion would call for 'everyone doing their bit'. This reasoning is sound to a point and indeed helpfully – at least for the government – both shares out ownership of transport problems facing society whilst also respecting a belief in individuals' right to choice. The reasoning is weakened, however, if the enactment of stated policy is reliant more upon individuals taking greater responsibility for their actions than upon policy measures which can significantly influence the choices of individuals.

In the absence of sufficiently stringent policy measures, the choice making of individuals falls victim to social dilemmas. For instance, Goodwin (1997b) pointed to the irony that if everyone took the 'slower' mode of transport to work (the bus), they might all get to work more quickly than by taking (as they do) the 'faster' mode (the car). As individuals we can appreciate the logic of a shift in collective behaviour arising from 'everyone doing their bit'. But the dilemma arises because as individuals we find it difficult to believe that a change in our actions alone can make any difference and to believe that other individuals would likewise change their own behaviour to help realise such difference.

The problem of 'everyone doing their bit' is further compounded by habitual behaviour. Individuals settle into routines of behaviour in which making choices is not seemingly governed by utility maximization but by familiarity and satisficing behaviour – for instance, we find a way of getting to work which is 'good enough' and struggle subsequently to find the time and motivation to give conscious consideration to whether or not this behaviour is or remains optimal. Our travel choices are also not made in a vacuum but in the context of a wider set of choices governing our lifestyle and well-being both as individuals and as units or networks of individuals. This is likely to explain, for instance, why there has been a substantial increase in the amount of long-distance commuting in recent years – the proportion of people in Britain with one-way commutes of 35 miles or more has nearly doubled in the last 20 years (Lyons & Chatterjee, 2007).

This said, our travel choices and the wider lifestyle choices that shape them do change over time, and therefore can be changed if the sociopolitical will exists. As people move through different life stages and encounter key life events, habitual behaviours are weakened and behaviour change at the individual level can be quite substantial (Dargay & Hanly, 2003). For instance, Stanbridge and Lyons (2006) have investigated travel behaviour change associated with residential relocation. Their survey results revealed that over one-quarter of people changed their main mode of travel to work following a relocation. An important observation is that while it can appear that it is human nature to resist change, people are in general very adaptable when it is thrust upon them. The September 2000 fuel crisis in the UK bore this out (Lyons & Chatterjee, 2002), as has the imposition of the London Congestion Charge, which has confronted the social dilemma and brought about a 30 per cent reduction in congestion (Transport for London, 2006; see Chapter 3).

Responsibility for shaping transport geographies, in light of the above, must rest significantly with government – although there are parallels between the choices of individuals and politicians. Just as individuals are inclined to exhibit selfish behaviour so too do politicians pay careful heed to the electorate and voting behaviour to ensure their political survival (Goodwin, 1997b). Just as individuals make choices about transport in the context of wider lifestyle and social considerations so too do politicians face choices about transport that must be taken in the context of a wider set of (sometimes conflicting) economic, social and environmental considerations. Highlighting this matter, 2006 bore witness to the publication of significant UK government reports on the economics of climate change (HM Treasury, 2006), transport's role in sustaining productivity and economic competitiveness (Eddington, 2006a) and land use planning (Barker, 2006).

How is the information age redefining accessibility?

The period marking the end of the twentieth century and the start of the new millennium is notable for the acceleration in the developments of ICTs and their penetration into the functioning of society. Important new choices have become available to individuals in terms of their access to people, goods, services and opportunities and indeed may, in turn, provide new options to policy makers: the Internet allows us to transcend distance in near zero time in order to reach things.

To date governments' ever-present concern for the maintenance of economic prosperity has paid heed to an apparent link between economic growth and traffic growth with a presumption that to restrict the latter may also restrict the former. An important objective is therefore to examine ways in which the two can be 'decoupled'. One partial means of doing so in principle is through land use planning, whereby people are physically closer to those things they wish to access, thus reducing the need for (as much) motorized travel. Acknowledging people's complex lives and the choices available to them, it seems unlikely that in practice land use planning alone can achieve a decoupling. Yet the advent of 'virtual mobility' (Kenyon *et al.*, 2002) challenges the presumption that accessibility is necessarily correlated with mobility. Indeed, it can be suggested that, in fact, economic growth is and always was coupled with growth in *access*. Accordingly as accessibility becomes less dependent upon mobility so in turn economic growth and traffic growth can arguably be decoupled.

We saw in Chapter 11 that the information age might impact on travel habits in four ways (Mokhtarian, 1990): substitution (telecommunications decreases travel); enhancement (telecommunications directly stimulates travel); operational efficiency (e.g. telecommunications improves travel by making the transportation system more efficient); and indirect, long-term impacts (e.g., telecommunications may ultimately affect land use, which will affect travel). Kenyon *et al.* (2003) have additionally suggested that telecommunications can also *supplement* travel – providing an increase in the level of access for an individual without the need for further travel.

This suggests that the internet is able to play a part in promoting greater social inclusion without placing further pressures on the transport system.

In passing, an important caveat is required – not everyone is embracing or has yet embraced the information age fully, either because of choice or because of lack of the available means. Thus while the penetration of home (broadband) Internet access has been rapid and will continue to increase, it is unlikely ever or for a number of years to be inclusive directly of all in society, even in the developed world. The relationships between distance and access will accordingly remain varied across and between populations.

One can also point to a *fusion* between telecommunications and travel – namely that it is now possible thanks to mobile communications to simultaneously travel and telecommunicate (see Lyons & Urry, 2005).[1] This is further complicating the ways in which transport geographies are being shaped: as place-to-place communication increasingly becomes person-to-person, it is increasingly difficult to judge what a person is doing by where they are. Social networking is supported increasingly by telecommunication as well as co-presence (Larsen *et al.*, 2006) especially amongst younger generations (Hashimoto, 2005). Indeed, the popularity of email and text messaging suggests that, for many, asynchronous communication, with its implicit lack of co-presence, is a popular choice. One can consider how the workplace communications culture appears to be transforming – an employee may be as likely to be noted by their manager for the time their first email is sent as for their arrival at the office in the morning, the so-called culture of 'presentee-ism'.

Yet in policy terms, while overtures are made to the possibility that ICTs might help address some of the pressing challenges of increasing physical mobility, virtual mobility has not yet been seriously embraced as part of an integrated approach to transport.[2] One of the reasons appears to be a significant difficulty in being able to secure clear empirical evidence upon which to base policy measures – research is struggling to keep pace with the developments in ICTs and their (changing) impacts on society.

What future possibilities lie ahead and how might we prepare for them?

Two observations are important when seeking to answer this question. First, the future is not predetermined – it is ours to shape as active agents of public policy and societal change. This acknowledges that the task of looking to the future is not merely one of prediction, but a realization that an infinite number of possible future pathways lie ahead, only one of which will be taken depending upon the collective impact of choices by governments, business and individuals. The second observation is to make reference to a quote attributed to the science fiction writer, William Gibson – 'the future is already here. It's just unevenly distributed.' His point was to suggest that there are small pockets of the future existing in the present.

Taking the second observation, one can consider a specific example. There exist already today, over the Internet, virtual worlds in which real people create their own 'avatars' or representations of themselves and interact through the graphical

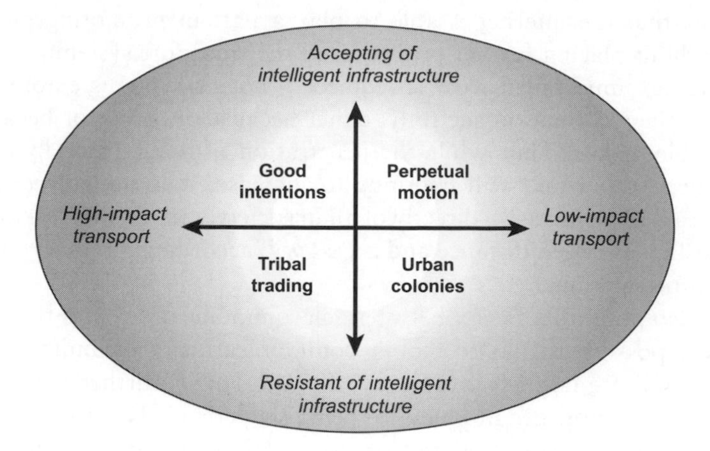

Figure 13.3. The four Intelligent Infrastructure Systems scenarios described in Table 13.3 expressed alongside 'axes of uncertainty' on a two-dimensional grid. After OST, 2006: 42.

representation offered by computing. There are press stories of people being addicted to 'living' in these worlds and of others who are achieving financial success through interaction and transaction in such worlds (see, for example, http://secondlife.com which bills itself as 'an online society within a 3D world, where users can explore, build, socialize, and participate in their own economy'). While these worlds are growing in popularity, for most of us they remain closer to the realms of science fiction – they are intriguing but unfamiliar and seem at odds with the social practices we are routinely engaged in. Yet just as email was a crude communications mechanism amongst academics in its early incarnations, who is to say that virtual worlds will not find their place in mainstream society?[3] This is not to suggest that they would replace the 'real' world as we know it, but they could coexist in ways that redefine the nature of our individual and collective reliance upon access as defined by space and time.

Returning to the first observation, there are specific methods at our disposal to try and examine future possibilities in order to inform present-day decision making. In particular scenario planning is a tool which explicitly acknowledges uncertainty and the myriad of possible future pathways (Curry *et al.*, 2006). The UK Government Office of Science and Innovation (formerly the Office of Science and Technology (OST)) has a Programme called Foresight which employs this method within its projects. One such project which reported in January 2006 (OST, 2006) was *Intelligent Infrastructure Systems*. This used scenario planning coupled with input from some 300 (social) scientists and other experts to examine the future of transport and the role of technology to a time horizon of 2056. By considering varying extents of people's acceptance of 'intelligent infrastructure' and extents of transport's environmental impact, four scenarios were developed and given the following names: 'perpetual motion', 'urban colonies', 'tribal trading' and 'good intentions' (Figure 13.3). Table 13.3 summarizes the four scenarios or 'sociologies of the future'.

In such an exercise it is important to maintain a distinction between visions and scenarios. A scenario is a possible future whereas a vision is a future being striven

Table 13.3. Summaries of the four Intelligent Infrastructure Systems scenarios. Source: OST, 2006.

Good Intentions	**Perpetual Motion**
The need to reduce carbon emissions constrains personal mobility. Traffic volumes have fallen and mass transportation is used more widely. Businesses have adopted energy-efficient practices: they use wireless identification and tracking systems to optimize logistics and distribution. Some rural areas pool community carbon credits for local transport provision, but many are struggling. Airlines continue to exploit loopholes in the carbon enforcement framework.	Society is driven by constant information, consumption and competition. In this world, instant communication and continuing globalization has fuelled growth: demand for travel remains strong. New, cleaner, fuel technologies are increasingly popular. Road use is causing less environmental damage, although the volume and speed of traffic remains high. Aviation still relies on carbon fuels – it remains expensive and is increasingly replaced by 'telepresencing' for business, and rapid trains for travel.
Tribal Trading	**Urban Colonies**
The world has been through a sharp and savage energy shock. The global economic system is severely damaged and infrastructure is falling into disrepair. Long-distance travel is a luxury that few can afford and for most people, the world has shrunk to their own community. Cities have declined and local food production and services have increased. There are still some cars, but local transport is typically by bike and by horse. There are local conflicts over resources: lawlessness and mistrust are high.	Investment in technology primarily focuses on minimizing environmental impact. Good environmental practice is at the heart of the UK's economic and social policies: sustainable buildings, distributed power generation and new urban planning policies have created compact, dense cities. Transport is permitted only if green and clean – car use is energy-expensive and restricted. Public transport – electric and low energy – is efficient and widely used.

for. A policy response to scenario planning may either be to use it as a stimulus for establishing a vision or to make policy decisions which are judged to be robust against the diversity of future possibility as illustrated by the scenarios.

Conclusion

In looking at the present and evolving future, this chapter by its nature is not deeply grounded in proven theory and empirical evidence. While it has revisited or illustrated a number of important trends and principles, its purpose has been to highlight the dynamic nature of the world in which we live and to point towards the changing array of choice that faces society. Motorization has overwhelmingly defined the transport geographies of the twentieth century. Whether or not this can continue to be the case in the longer term remains to be seen. Acknowledgement of the effects of climate change may necessitate more stringent policy responses to influence (if not dictate) the behaviours of individuals in terms of their carbon consumption in future. The concern over both the economics of climate change and the economics of congestion is giving growing credence to the likelihood, if not inevitability, of more widespread road pricing and more sophisticated taxation regimes for aviation. Heightened concern over transport security will present governments with an additional problem to tackle. In less developed areas, such as much of the Asia-Pacific region, dependency on fossil fuels and the economic impacts of peak oil production could play out differently (Greene *et al.*, 2003). A combination of older vehicle fleets and local variations in transport mode – such as the ubiquity of the powered two-wheeler – when combined with less available capital to invest in the greening of surface transport could produce profound implications for the hitherto rapid economic growth of many Asia-Pacific economies.

The opportunities presented by the information age and technological advance may be embraced by government and/or individuals as a means of supplementing or replacing the availability of motorized mobility in order to transcend distance to participate in society. Our greatest challenge, it would seem, is the pace of change, both in terms of the imperative of climate change and problems of congestion alongside the rapid transformation of society by a continually evolving information age.

Notes

1 Editors' note: indeed, substantial parts of this book were edited on the move by putting in-transit travel time on rail and air journeys to productive use. Communicating in real time via the Internet – often between different continents (at the risk of undermining our sustainable transport credentials!) – was an extremely important part of this process.

2 Some emphasis was given in a recent UK White Paper (Department for Transport (DfT), 2004b) to the possibilities for 'smarter choices' such as home shopping and teleworking to reduce motorized mobility.

3 Consider also the parallels between the famous 'communicator' of 1960s *Star Trek* and today's mobile phones.

14 | Revitalized Transport Geographies

John Preston and Kevin O'Connor

To the uninitiated, transport geography appeared to have had its heyday in the slipstream of the quantitative revolution some 30 years ago. In the UK, this was reflected by a series of textbooks on the subject in the 1970s and early 1980s (Hay, 1973; Robinson, 1976; Robinson & Bamford, 1978; White & Senior, 1983) and regular annual reports on the sub-discipline, written by Alan Hay, in the (then) new journal *Progress in Human Geography* (1977 through to 1981). More sporadic pieces, written by Peter Rimmer, appeared in 1985, 1986 and 1988. Influential North American texts of the time included Eliot-Hurst's (1973) *Transportation Geography: Comments and Readings* and Taaffe and Gauthier's (1973) *Geography of Transportation* which also focused on the quantitative-based ideals of spatial science.

After a period during which its profile receded somewhat, transport geography has in recent years enjoyed a noticeable revival. Arguably, this process started first in North America, with the Transportation Geography Speciality Group (TGSG) of the Association of American Geographers (which was founded in 1980) producing the *Geography of Urban Transportation* (Hanson, 1986),[1] followed by, for example, the second edition of *Geography of Transportation* (1996), Black's (2003) *Transportation: a Geographical Analysis* and Transport Geography on the web.[2] In the UK, the revival is more closely associated with the 1990s, with textbooks by Hoyle and Knowles (1992, 1998) and Tolley and Turton (1995), several research-based edited volumes (Gibb, 1994; Hall, 1993; Hoyle, 1996; Tolley, 1990; Whitelegg, 1992) and the launch of the *Journal of Transport Geography* (*JTG*) in 1993 (Knowles, 1993).

The new millennium has heralded further optimistic developments. These include the global reader edited by Tolley and Turton (2001), the influential collection of policy analysis *A New Deal for Transport?* (Docherty & Shaw, 2003) published in the RGS-IBG book series, the ongoing debate on research needs in the *JTG* (see, for example, Horner & Casas, 2006) and a 'state-of-the-nation' paper on transport geography in the USA by Goetz *et al.* (2004). The Transport Research Group of the German Geographical Society has been increasingly active (Gather *et al.*, 2001; Nuhn & Hesse, 2006) and in France transport geography appears to be prospering under the umbrella of the Comité National Français de Géographie (see, for example, Bavoux *et al.*, 2005). Finally, the recent accreditation of the *JTG* by Thomson Scientific for inclusion in its Social Science Citation Index is further evidence of revival, which this book seeks to underpin.

The aim of this concluding chapter is to outline our views on how this renaissance in transport geography might be sustained, although it is certainly not our intention to be prescriptive.[3] The crux of our argument is that transport geography's revival will be most effectively promoted through a genuine commitment to interdisciplinarity, expanding the content of its enquiry whilst at the same time adopting complementary advances in methodology. With regard to content, it seems that there are two important fields of analysis: 'first round effects' which will continue to be driven by changing forms and patterns of mobility, and 'second round effects' which will shape and be shaped by these new mobilities. We use globalization, the dominance of extended metropolitan regions and the reconfiguration of urban areas as examples. A need for more inclusive theoretical engagement and pluralistic methodologies arises if the more variegated transport geographies that are likely to emerge from these processes are to be understood. Holistic, genuinely interdisciplinary approaches will be required to emphasize the importance of underlying socio-economic processes as well as spatial and temporal patterns; sustainability, in its various guises, may provide an appropriate, but by no means perfect, integrating concept.

First round effects: mobility transitions

Successive chapters of this book have charted the huge increase in the mobility of people and goods across the world as markets have widened and transport technology improved (see especially Chapters 6–10 and Chapters 12 and 13). But this is not to say that everyone and everywhere has experienced massive increases in mobility: the young, the elderly and those on low incomes, particularly in developing countries, may have missed out. For the developed world, the UK – lying between the continental European and North American poles – is perhaps typical of trends in advanced countries. In 1952, the average person in the UK travelled around 4,300 km. By 2002, this had increased almost threefold to 12,700 km. Equally, it is important to recognise that these growth trends are not new: Pooley *et al.* (2005) examined the changes in everyday mobility in Britain from as far back as 1890 by examining life histories, undertaking in-depth interviews and analysing secondary data. Similarly, Grubler and Nakicenovic (1991 – cited by Haggett, 2001) analysed data for France going back to 1800 and found that total personal mobility had grown a thousand-fold whilst population had only increased two-fold. A 500-fold increase in per capita mobility over the last 200 years is consistent with exponential growth of 3.2 per cent per annum. A three-fold increase in mobility over the last 50 years is consistent with an exponential growth of 2.2 per cent per annum.

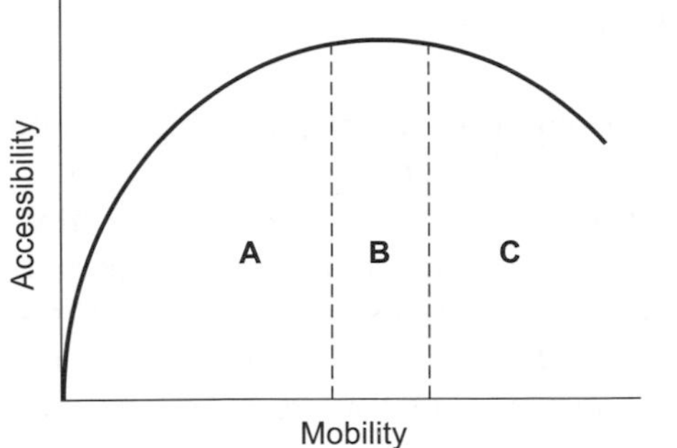

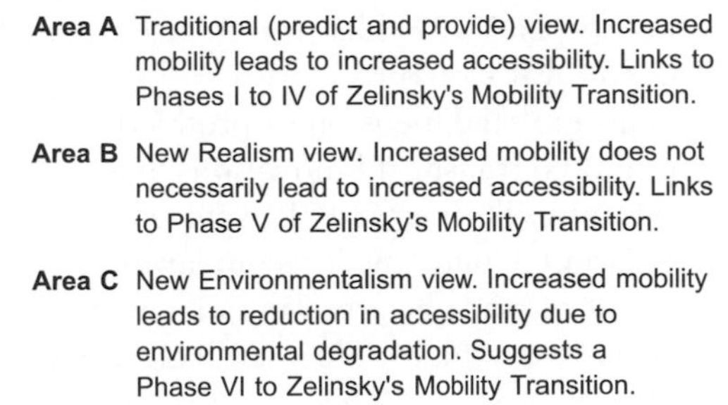

Figure 14.1. Relationships between mobility and accessibility.

This is suggestive of a substantial reduction in mobility growth rates over time. This is confirmed by evidence of reducing rates of mobility growth, at least in advanced economies, from the literature on transport intensity (the ratio of mobility to GDP – see Tapio, 2005).

As has also been noted in the book, there is a close interaction between mobility (the ease of moving) and accessibility (the ease of reaching socio-economic opportunities). This relationship might, however, be expected to exhibit diminishing returns and even a tipping point where increased mobility is associated with reduced accessibility, due to, for example, congestion, resource depletion and/or excessive centralization of facilities. An important research issue is where and when (if ever) this tipping point is reached. Proponents of the 'new realism' contend that this point has now been reached (Goodwin *et al.*, 1991; see Chapter 1), whilst radical environmentalists argue that it has long been passed (Hillman, 2004). These views are illustrated by Figure 14.1, which draws parallels with the concept of the mobility transition (Zelinsky, 1971) and in particular the projected phase V in which non-economic travel becomes dominant and mobility becomes counter-productive (at least in economic, if not in social, terms). A key role for transport geographers may therefore be in providing mappings of the differentiated patterns of mobility and accessibility transitions and their physical consequences in terms of changes in journey times and costs, congestion, accidents and pollution.

Second round effects: spatial (and socio-economic) transformations

Recent 'first round' effects in mobility have been closely associated with the 'second round' effects of functional and spatial fragmentation of social and economic activity, from the local to the global scales. This fragmentation has evolved rapidly, especially thanks to innovations in information and communication technologies (ICT), to the extent that it represents a global shift in the international organization of production (Dicken, 2003). Outcomes of this process focus on the increasing separation of production and assembly from specialist research and development and management functions, reflecting in large part the so-called relational assets that Storper (1997) has stressed are embedded in local and regional social, political and cultural structures.

This global dimension of production and consumption, enabled by maritime and intermodal transport innovations, has re-shaped world trade and therefore transport geography. Storper (2000) went on to demonstrate that trade is no longer simply the rational outcome of comparative advantage: rather, there is a considerable amount of trade between nations with similar products and similar cost structures, especially for high value services (O'Connor & Daniels, 2001). Innovation and creativity are increasingly important in generating economic success, with the concentration of knowledge in particular places where the most creative people want to live identified as a critically important explanatory factor of the reorganisation of production (Florida, 2005).

To explore these ideas, it is necessary to understand the spatial impact of supply chain restructuring at the global scale, something that Thierstein and Schnell (2002: 71–2) believe is '. . . very rarely evaluated', yet an area of enquiry in which transport geography might be expected to play a key role. For example, in their review of the electronics industry, Barnes *et al.* (2000) point to the reduction in the complexity of the supply chain, with a small number of firms manufacturing many of the products of name-brand suppliers. This in turn leads to a concentration of production in space, so that the global system reshapes around fewer, larger centres; currently in the USA, more than two-thirds of all shipments travel 600 miles or fewer (Colography Group, 2004). Some of these changes reflect the increased cost of fuel, whilst others are associated with issues of the declining reliability of long distance transport flows. The end result, however, is that logistics structures are more embedded within the wider pattern of transport flows and spaces than ever before (Hesse, 2004).

First and foremost, recognising this place embeddedness requires a new way of looking at cities akin to the idea of the extended metropolitan region (McGee, 1994), global city region (Scott, 2001) or 'one hundred mile city' suggested by Sudjic (1992). Gordon *et al.* (1998) link these new organizational structures to the changing geography of agglomeration economies, which are now more widely spread across the metropolitan region rather than being concentrated in the core (Figure 14.2). Giuliano and Small's (1999) study of clusters of jobs across local centres in Los Angeles lends weight to this analysis, providing evidence for the positive impact of transport in creating higher levels of labour market mobility and increased choice and flexibility for employers as a result. Lang's (2003) research on office activity shows that shares of new office building construction are greater in the suburbs than the central city in all of the USA's metropolitan areas other than New York and Chicago, in part due to the decisions of some international firms to (re)locate to suburban areas, especially those near to airports (Muller, 1997; Button & Stough, 1998). Where the logistical connections between firms are organized over even longer distances they can reinforce the growth of the 'megalopolis' such as those identified in Figure 14.3 for the USA, which include Silicon Valley and Route 128 (Saxenian, 1994; Lang & Dhavale, 2005).

Despite these trends, and the increasing importance of ICT as an alternative to physical mobility (Graham & Marvin, 2001), the role of the Central Business

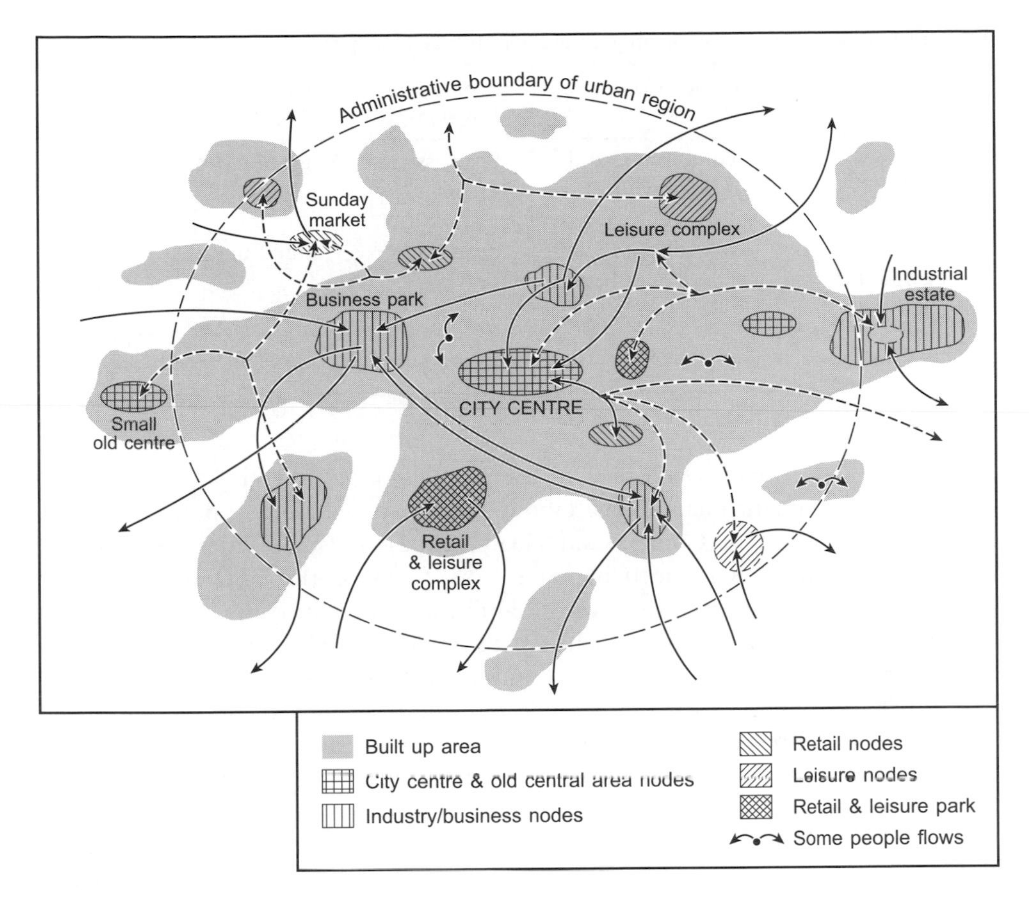

Figure 14.2. A model of an extended metropolitan region. After Healey, 2001.

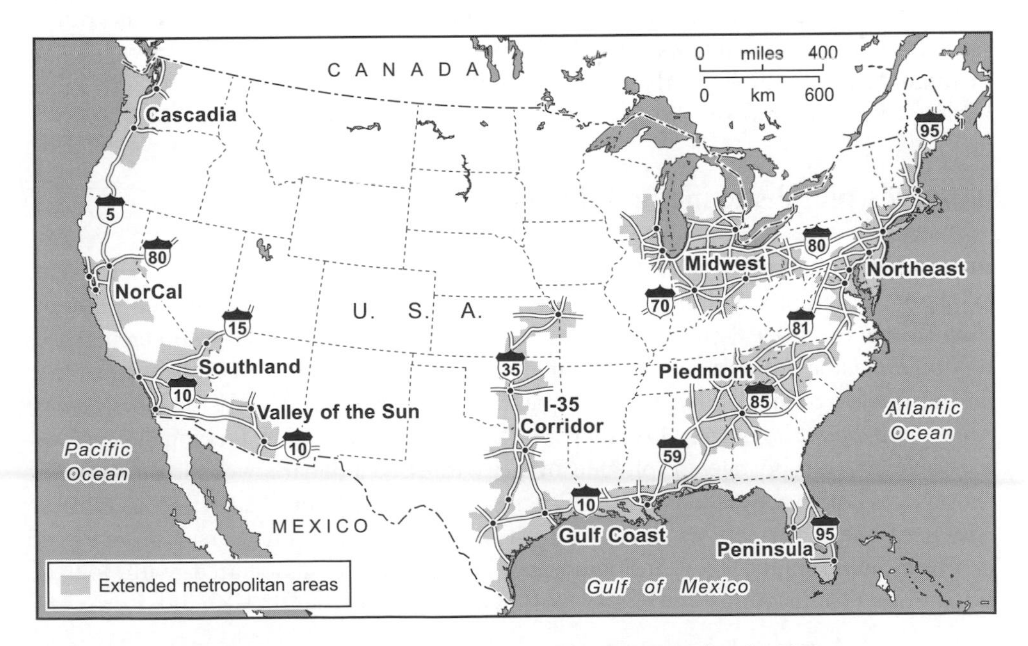

Figure 14.3. Extended metropolitan areas of the USA. After Lang & Dhavale, 2005.

Table 14.1. Production and consumption connections in extended metropolitan regions.

Spatial Links	*Goods/Components*	*People and Information*	*Labour Casual and Fulltime*	*Consumers*
Global National	To markets, and links with other firms	Business Travel	Migration	Tourism
Extended Metropolitan Region	Warehouse to shop Linkages between firms	Business Travel	Commuting	To Major Centres
Local				To Local Centres

District (CBD) is being reinforced in many cities, especially those for which it remains the dominant focus of the labour market, or where corporate activity is especially important (O'Connor and Healy, 2002). In some cities, such as the example of Portland, Oregon (Chapter 6), this focus on the CBD is a deliberate policy attempt to achieve urban reconcentration and reduce the need to travel. Empirical research also notes how clusters of services in these inner urban environments benefit from face-to-face contact and quick exchange of knowledge, and the dense activity connections made possible by pedestrian movement and motorized courier services (Storper & Venables, 2004). The end result of these two processes is '. . . increased intra-metropolitan specialization depending on producer services' reliance on face-to-face contacts for conducting their business' (Boiteux-Oraim & Guillian, 2004: 573), with an increasing differentiation of economic functions across the metropolitan space (Searle, 1998). The level of mobility required for the movement of goods, information and labour into, across and out of extended metropolitan regions therefore suggests an increasingly complex set of connections, predictability of time and flexibility of delivery, which has been labelled city logistics (Taniguchi *et al.*, 2001). Table 14.1 identifies the key transport movements for a variety of activities at several scales within the extended metropolitan region.

Towards pluralistic methodologies

In seeking to make sense of these first and second order effects it is likely that transport geography will increasingly rely upon a range of theoretical and methodological perspectives. Transport geography, like the spatial sciences more generally, has on occasion stood accused of being somewhat a-theoretical (Johnston, 2003), an impression confirmed by, for example, the recent *Handbook of Transport Geography and Spatial Systems* (Hensher *et al.*, 2004). Rather than lacking theory per se, however, transport geography has in fact witnessed a debate over the most useful theoretical approaches consistent with many other geographical sub-disciplines. Much of the American tradition, with its roots in (quantitative) spatial science, is often underpinned by ontologies and epistemologies associated with logical positivism (Taaffe and Gauthier, 1994). On the other hand, UK and European

Table 14.2. Some Developments in Analytical Methods in Transport Geography.

	Traditional Methods	*Newer Methods*
Quantitative	Hydrodynamic traffic models Equilibrium assignment Logit models Gravity/Spatial Interaction models (Micro) Simulation models Conjoint analysis/Contingent valuation Real world trials/case studies Graph Theory Systems Analysis	Mixed logit Computable general equilibrium models Geographically Weighted Regression. GIS-T Agent-Based Models Complexity theory
Qualitative	In-depth interviews Focus Groups	Discourse Analysis Time-space diaries Digital observation Multi-media representation Ethnographic methods Deliberative research
Hybrid	Gaming Delphi Studies	Q Methodology Meta analysis/Benefit Transfers Personal GPS Theory of Planned Behaviour

transport geographers, especially those interested in social processes, have long recognised that there are alternatives to the logical positivist approach (Root, 2003; Whitelegg, 1981). None the less, the sub-discipline has been relatively unaffected by the impact of Marxist and other (post-)structuralist viewpoints, despite the clear interplay between structures and agents in the transport sector, the importance of institutional and social constraints and the resonance of Marx's famed phrase 'the annihilation of space by time' (see, for example, Leyshon, 1995). Similarly, more humanistic approaches, exemplified by the pioneering work of Jacobs (1961) on American cities, have been somewhat neglected, despite the importance of individual agency and challenges to the conceptual bases of neo-classical economics and modern planning.

But times are changing. More critical and humanistic approaches have culminated in the 'new mobilities paradigm' (Urry, 2004b), which highlights the wider relationships between transport and society, the importance of the movement of ideas, images and information and the intrinsic sociocultural benefits of the movement of people and goods. An important recent application of humanistic approaches has been in studies of the impact of information technology on travel and the re-working of time and space, not least by the emergence of virtual mobility (Hanson, 1998; Kenyon *et al.*, 2002). Moves to a more multi-method approach, with a mix of the quantitative and the qualitative, have permitted the emergence of more reflexive and self-critical transport geographies: Table 14.2 contrasts key 'old' and 'new' methodologies.

Sustainability as the key to interdisciplinarity?

A key theme of this book has been how future transport geographies (in the developed world at least) might evolve in response to the era of hypermobility characterising the late twentieth century (Adams, 1999). The current situation is one of increasing congestion, environmental degradation and deteriorating quality of life, and there is now a rapidly developing consensus that action is required to arrest these trends, even if the exact form of that action remains unclear. The search for better sustainability might therefore provide an appropriate (normative) conceptual foundation for forward-looking transport policy, although it can be criticized as an overused and unworkable concept (Simon, 1989) and it is hardly novel, dating back at least to the 1980 *World Conservation Strategy*. Symptomatic of this is the fact that there has been an interminably long literature on defining sustainability from which few workable definitions have emerged.[4] One of the better attempts comes from Richardson (1999: 27), who defined a sustainable transport system as 'one in which fuel consumption, vehicle emissions, safety, congestion, and social and economic access are of such levels that they can be sustained into the indefinite future without causing great or irreparable harm to future generations'.

As an academic device, sustainability has a number of potential advantages. First, there is a substantial literature on sustainable transport that is global in its coverage and inclusive of transport geography (see, for example, Banister *et al.*, 2000; Black, 1996; Black and Nijkamp, 2002; Newman and Kenworthy, 1999). Secondly, there is considerable institutional support for the concept of sustainable transport, particularly from supra-national bodies such as the World Bank (1996) and the European Conference of Ministers of Transport (ECMT) (2002). Thirdly, as illustrated by Figure 14.4, sustainability, by highlighting the inter-relationships between society, economy and the environment, provides a useful interdisciplinary core for transport geography (Docherty, 2003; Price, 1999).

Understandings of the links between transport, wider socio-economic processes and the environment are incomplete, despite recent attempts by transport geographers and others to improve the representation of these interactions (Chapters 2–4). One such effort has been in the area of transport and the economy, the subject of a recent major governmental report in the UK (Eddington, 2006a) and in which transport geography is reaffirming links with the broader traditions of (the new) economic geography. Two broad approaches can be identified here: the first focuses on transport and economic development, and is typified by Banister and Berechman (2000). The emphasis is mainly on how, why and where economic activity develops, and new growth theories (Romer, 1986) and the empirical work on the impact of transport infrastructure investment stimulated by Aschauer (1989) are drawn upon. The second approach focuses on the links between transport and land-use, with a focus on transport's impact on the location of economic activity that can be traced via Alonso, Lowry and Herbert and Stevens back as far as Weber and Von Thünen (Wilson, 2000). An important contribution by the new economic geography has been to bring these approaches together, with computable general equilibrium models the key methodological tool (Fujita *et al.*, 1999). The parameterized diagrams that are produced give some important insights, which can be improved both

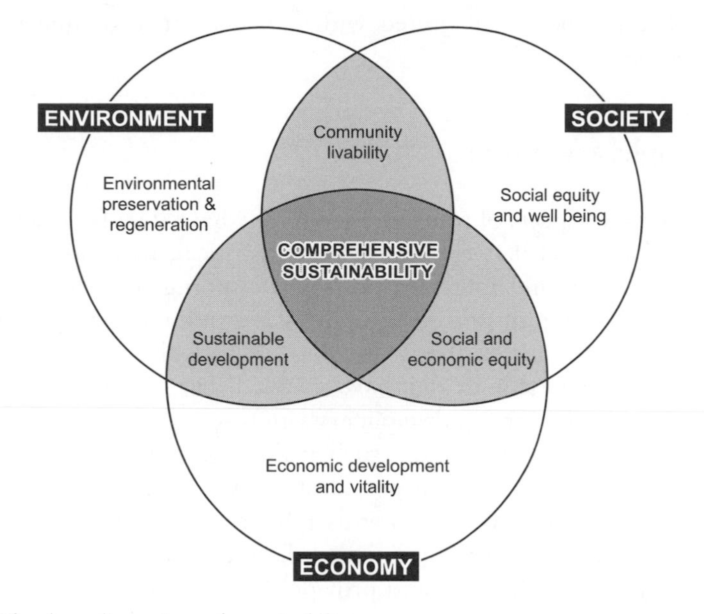

Figure 14.4. The three dimensions of sustainability.

quantitatively, through developing the evidence base, and qualitatively, through in-depth understandings of the key transmission mechanisms that can only be provided through methods such as close dialogue interviews.

Transport geographers and others are also beginning to get to grips with the interaction between transport and society. This is not a new phenomenon – geographers have been interested in issues of social justice at least since David Harvey's seminal work on the topic (Harvey, 1973). What is new is the language (with the emphasis in the UK on social exclusion/inclusion and in the USA on environmental justice) and the intensity of policy interest, epitomized by the work of the UK's Social Exclusion Unit (2003). A flavour of the research response by transport geographers and others can be gleaned from the recent books by Hine and Mitchell (2003), Lucas (2004) and Rajé (2004).

But progress towards a research future focused around sustainability is by no means guaranteed. The new mobilities paradigm, and its analysis of the positive benefits of mobility, reflects the growing emphasis placed by the European Union on sustainable mobility (Banister *et al.*, 2000). Although in the majority of cases, mobility is not highly valued in and of itself, the intrinsic utility of mobility may be increasing, as people use their time more flexibly and as technology allows them to achieve more on the move. Yet as long as transport is highly dependent on non-renewable fossil fuels, sustainable mobility is even more of an oxymoron than sustainable development. Sustainable accessibility has been put forward as a more promising and consistent policy goal (see, for example, Le Clercq and Bertolini, 2003). In operationalizing sustainable accessibility, the emphasis to date has been placed on environmental issues, but to be consistent with other innovations, this will need to become better integrated with a wider range of underlying socio-economic processes.

An interdisciplinary future

The interdisciplinary aspects of transport geography have long been acknowledged. For example, Williams (1981) emphasized the historical, morphological, environmental, economic, social and political aspects of transport geography. More recently, it could be argued that transport geography is repositioning itself in a genuinely interdisciplinary framework so that it can best unpack the complex, multi-scalar shifts in mobility and accessibility that have emerged. The growing focus on sustainability, the need to apply various disciplinary lenses to fully understand its implications and the emergence of innovations such as the new mobilities paradigm, means that these interdisciplinary links will only deepen over time.

Figure 14.5 conceptualizes these interdisciplinary links. There have been some fluctuations in the fortunes of the individual disciplines that make up this framework. With the demise of 'predict and provide' (Chapter 1), at least for the roads sector (Owens, 1995), civil and structural engineering have become less influential – although given the increased policy focus on congestion, traffic engineering, along with related disciplines in mathematics and computer science, remains in the ascendant. The influence of traditional strategic land use planning may too be on the wane, not least given the increased awareness of planning failure in the past (see, for example, Flyvbjerg *et al.*, 2003). By contrast, there seems to be an increasing role for political science (see, for example, Dudley and Richardson, 2000; Vigar, 2002). Other disciplines currently in the ascendancy include sociology (of central importance to the new mobilities paradigm), anthropology (particularly given renewed interest in ethnographic approaches – see, for example, Brown and

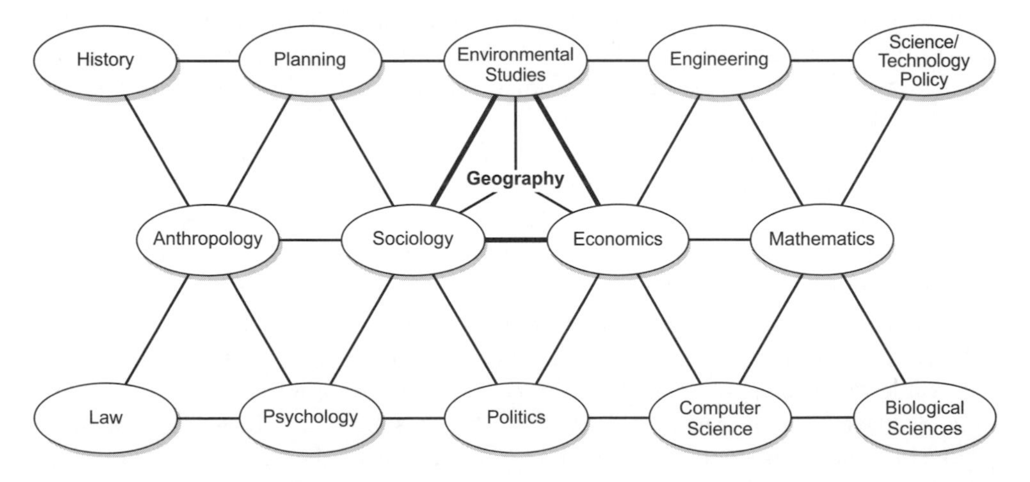

Figure 14.5. An interdisciplinary framework for transport geography. Other disciplines can be added to this mesh and the framework morphed to suit.

Laurier's (2005) work on cars as habitable space), biological sciences (not least for the development of agent-based and similar models) and science and technology policy analysis. Psychology deserves a special mention given, for example, the recent popularity of applications of Azjen's Theory of Planned Behaviour (Jopson, 2004; see also Chapter 11).

Geography can be a key interlocutor between these disciplines, providing essential linkages between the physical and the human and providing the necessary emphasis on both spatial and temporal scales. Taken together, these developments suggest a future research agenda focusing on globalization – or, more precisely, multi-scalar transformations – and information, communications and technological change (Janelle, 1997; Janelle & Beuthe, 1997; Hodge & Koski, 1997). This in turn requires more careful consideration of the links between transport and the wider economy (Preston, 2001), societal change (Giuliano & Gillespie, 1997) and the environment (Royal Commission on Environmental Pollution (RCEP), 1995). Underpinning all of this is the likelihood that many of the recent certainties of modern transport geography will be replaced by the multiple uncertainties of post-modern geographies of mobilities, flows and spaces. The economic, social, environmental and political implications of these uncertainties will provide inspiration to transport geographers specifically, and human geographers more generally, in terms of both what we research and our approaches to investigation and analysis.

Notes

1 Now in its third edition (2004) and co-edited by Giuliano, following a second edition in 1995.
2 See www.people.hofstra.edu/geotrans. Now published as: Rodrigue, J.-P., Comtois, C. and Slack, B. (2006) *The Geography of Transport Systems*. London: Routledge.
3 The chapter is informed by a series of conference presentations (and the responses to them) exploring the future of the sub-discipline (Preston, 2003b, 2004; Preston & Rajé, 2005).
4 See, for example, www.sustainableliving.org/appen-a.htm.

References

Abrahmse, W., Steg, L., Gifford, R. and Vlek, C. (2004) *Psychological factors influencing car use for commuting*. Paper presented at the 3rd International Conference on Traffic & Transport Psychology, Nottingham.

Actions on the Integration of Rural Transport Services (2004) *Rural Transport Handbook*. www.rural-transport.net. Accessed 14 December 2006.

Adam, B. (2001) The value of time in transport. In: Giorgi, L. and Pohoryles, R. (eds), *Transport Policy and Research: What Future?* Aldershot: Ashgate, pp. 130–43.

Adams, J. (1999) *The Social Implications of Hypermobility*. Paris: OECD.

AEA Technology (2004) *Evaluation of the Multi-modal Study: Final Report*. AEA Technology and Institute of Transport Studies, University of Leeds for Department of Transport.

Air Transport Association (ATA) (2006) *Annual Revenue and Earnings, US Airlines – All Services*. Washington, DC: ATA.

Airbus (2002) *Global Market Forecast, 2001–2020*. Blagnac: Airbus S.A.S.

Airey, A. (1985) The role of feeder roads in promoting rural change in eastern Sierra Leone. *Tijdschrift voor Economische en Sociale Geografie* 76, 192–201.

Airline Business (2006a) The world airline rankings 2005. August, 57–86.

Airline Business (2006b) Airline alliance survey 2006. September, 45–87.

Airline Business (2006c) Airports 2005. June, 49–60.

Åkerman, J. (2005) Sustainable air transport – on track in 2050. *Transportation Research Part D* 10, 111–26.

Alamdari, F. and Fagan, S. (2005) Impact of the adherence to the original low-cost model on the profitability of low-cost airlines. *Transport Reviews* 25, 377–92.

Albrechts, L. and Coppens, T. (2003) Megacorridors: striking a balance between the space of flows and the space of places. *Journal of Transport Geography* 11, 215–24.

Alderighi, M., Cento, A., Nijkamp, P. and Rietveld, P. (2004) *The Entry of Low-cost Airlines*. Discussion paper TI 2004-074/3. Rotterdam: Tinbergen Institute. www.tinbergen.nl.

Allen, J., Massey, D. and Cochrane, A. (1998) *Rethinking the Region*. London: Routledge.

Anable, J. (2002) Picnics, pets and pleasant places: the distinguishing characteristics of leisure travel demand. In Black, W. and Nijkamp, P. (eds), *Social Change and Sustainable Transport*. Bloomington: Indiana University Press, pp. 181–90.

Anable, J. (2005) Complacent car addicts or aspiring environmentalists? Identifying travel behaviour segments using attitude theory. *Transport Policy* 12, 65–78.

Anon. (2006a) Bus rally calendar and museum guide. Part one. *Buses* 612, March supplement, i–xvi.

Anon. (2006b) Bus rally calendar and museum guide. Part two. *Buses* 613, April supplement, i–xvi.

Aschauer, D. (1989) Is public expenditure productive? *Journal of Monetary Economics* 23, 177–200.

Atkins, S. (1989) Women, travel and personal security. In Grieco, M., Pickup, L. and Whipp, R. (eds), *Gender, Transport and Employment: the Impact of Travel Constraints.* Aldershot: Gower, pp. 169–89.

Audit Commission (1999) *A Life's Work: Local Authorities, Economic Development and Economic Regeneration.* London: HMSO.

Badland, H. and Schofield, G. (2005) Transport, urban design, and physical activity: an evidence-based update. *Transportation Research Part D: Transport and Environment* 10, 177–96.

Bae, C. (2001) *Cross-border impacts of growth management programs: Portland, Oregon, and Clark County.* Paper presented at 17th Pacific Regional Science Conference, Washington, DC.

Baggott, S., Brown, L., Milne, R., Murrells, T., Passant, N., Thistlethwaite, G. and Watterson, J. (2005) *UK Greenhouse Gas Inventory, 1990 to 2003: Annual Report for Submission Under the Framework Convention on Climate Change.* Harwell: AEA Technology.

Bailey, I. and Rupp, S. (2005) Geography and climate policy: a comparative assessment of 'new' environmental policy instruments in the UK and Germany. *Geoforum* 36, 387–401.

Baird, A. (2006) Optimizing the container transhipment hub location in Northern Europe. *Journal of Transport Geography* 14, 195–214.

Bamberg, S. and Schmidt, P. (2003) Incentives, morality or habit? Predicting students' car use for university routes with the models of Ajzen, Schwartz and Triandis. *Environment and Behavior* 35, 264–85.

Banister, D. (2000) Sustainable urban development and transport – a Eurovision for 2020. *Transport Reviews* 20, 113–30.

Banister, D. (2005) *Unsustainable Transport: City Transport in the New Century.* London: Routledge.

Banister, D. and Berechman, J. (2000) *Transport Investment and Economic Development.* London: UCL Press.

Banister, D. and Berechman, J (2001) Transport investment and the promotion of economic growth. *Journal of Transport Geography* 9, 209–18.

Banister, D. and Stead, D. (2004) Impact of information and communications technology on transport. *Transport Reviews* 24, 611–32.

Banister, D., Stead, D., Steen, P., Akerman, J., Dreborg, K., Nijkamp, P. and Schleicher-Tappeser, R. (2000) *European Transport Policy and Sustainable Mobility.* London: Spon.

Barker, K. (2006) *Barker Review of Land Use Planning: Final Report – Recommendations.* London: HMSO.

Barkham, P. (2007) New tickets to ride – and they'll cost you. *The Guardian*, 2 January.

Barnes, E., Dai, J., Deng, S., Down, D., Goh, M., Hoong, C. and Moosa, S. (2000) *Electronics Manufacturing Service Industry.* The Logistics Institute – Asia Pacific. White Paper Series. National University of Singapore. www.tliap.nus.edu.sg/tliap/Research_WhitePapers.aspx Accessed 2 February 2007.

Barnes, P. and Barnes, I. (1999) *Environmental Policy in the European Union*. Cheltenham: Edward Elgar.
Barrett, S. (2006) Commercialising a national airline – the Aer Lingus case study. *Journal of Air Transport Management* 12, 159–67.
Baumol, W. (1982) Contestable markets: an uprising in the theory of industrial structure. *American Economic Review* 72, 1–15.
Baumol, W. and Oates, W. (1988) *The Theory of Environmental Policy*, 2nd edition. Cambridge: Cambridge University Press.
Bavoux, J., Beaucire, F., Chapelon, L. and Zembri, P. (2005) *Géographie des Transports*. Paris: Armand Colin.
Baxter, J. and Eyles, J. (1997) Evaluating qualitative research in social geography: establishing 'rigour' in qualitative analysis. *Transactions of the Institute of British Geographers* 22, 505–25.
Becken, S. (2004) How tourists and tourism experts perceive climate change and carbon-offsetting schemes. *Journal of Sustainable Tourism* 12, 332–45.
Becken, S. (2005a) Harmonising climate change adaptation and mitigation: the case of tourist resorts in Fiji. *Global Environmental Change Part A* 15, 381–93.
Becken, S. (2005b) Towards sustainable tourism transport: an analysis of coach tourism in New Zealand. *Tourism Geographies* 7, 23–42.
Becken, S. (2006) Editorial: tourism and transport: the sustainability dilemma. *Journal of Sustainable Tourism* 14, 113–15.
Becken, S., Simmons, D. and Frampton, C. (2002) Segmenting tourists by their travel pattern for insights into achieving energy efficiency. *Journal of Travel Research* 42, 48–52.
Begg, I. (2001) *Urban Competitiveness: Policies for Dynamic Cities*. Bristol: Policy Press.
Behnen, T. (2004) Germany's changing airport infrastructure: the prospects for newcomer airports attempting market entry. *Journal of Transport Geography* 12, 277–86.
Bell, P. and Cloke, P. (eds) (1990) *Deregulation and Transport: Market Forces in the Modern World*. London: David Fulton.
Beuret, K. (1994) Taxis: the neglected mode in public transport planning. In *Provision for Accessible Transport Services*. Proceedings of Seminar F held at the PTRC European Transport Forum, London, Unpaginated.
Beuret, K. (1995) Call a cab – chance would be a fine thing: the importance of taxi transport for elderly and disabled people and the problems of provision. *Proceedings of 7th International Conference Mobility and Transport for Elderly and Disabled People*, 16–19 October, Reading, UK, 317–22.
Bhalla, A. and Lapeyre, F. (1997) Social exclusion: towards an analytical and operational framework. *Development and Change* 28, 413–33.
Bieger, T. and Wittmer, A. (2006) Air transport and tourism – perspectives and challenges for destinations, airlines and governments. *Journal of Air Transport Management* 12, 40–6.
Bird J. (1971) *Seaports and Seaport Terminals*. London: Hutchinson.
Bird, J. (1982) Transport decision makers speak: the seaport development in the European Communities Research Project – Part II. *Maritime Policy and Management* 9, 3–102.
Bird, J., Lochhead, E. and Willingale, M. (1983) Methods of investigating decisions involving spatial effects including content analysis of interviews. *Transactions of the Institute of British Geographers* 8, 143–57.
Bishop, S. and Grayling, T. (2003) *The Sky's the Limit – Policies for Sustainable Aviation*. London: IPPR.
Bjerrgaard, R., Bangemann, M. and Papoutis, C. (1996) *Auto Oil Programme Document, COM(96) 248 Final, Sheet 23*. October, Brussels: European Commission, DGXII.

Black, W. (1996) Sustainable transportation: a US perspective. *Journal of Transport Geography* 4, 151–9.
Black, W. (1998) Sustainability of Transport. In Hoyle, B. and Knowles, R. (eds), *Modern Transport Geography*, 2nd edition. Chichester: Wiley, pp. 337–51.
Black, W. (2001) An unpopular essay on transportation. *Journal of Transport Geography* 9, 1–11.
Black, W. (2003) *Transportation: a Geographical Analysis*. New York: The Guilford Press.
Black, W. and Nijkamp, P. (2002) *Social Change and Sustainable Transport*. Indiana: Indiana University Press.
Black, W. and Nijkamp, P. (eds) (2002) *Social Change and Sustainable Transport*. Bloomington: Indiana University Press.
Blomgren, K. and Sørensen, A. (1998) Peripherality – factor or feature? Reflections on peripherality in tourism research. *Progress in Tourism and Hospitality Research* 4, 319–36.
Boardman, B. (1998) *Rural Transport and Equity*. London: Rural Development Commission.
Boeing (2003) *Boeing: Current Market Outlook 2003*. Seattle: Boeing Commercial Airplane Group.
Boile, M., Theofanis, S., Golias. M. and Mittal, N. (2006) *Empty Marine Container Management: Addressing Locally a Global Problem*. Proceedings of the 85th Annual Meeting of the Transportation Research Board, 23–6 January, Washington, DC.
Boiteux-Orain, C. and Guillain, R. (2004) Changes in the Intrametropolitan location of producer services in Ile de France (1978–1997); do information technologies promote a more dispersed spatial pattern? *Urban Geography* 25, 550–78.
Bongaerts, J. (1999) Carbon dioxide emissions and cars: an environmental agreement at EU Level. *European Environmental Law Review* 8, 101–4.
Bonsall, P. and Dunkerley, C. (1997) Use of concessionary travel permits in London: results of a diary survey. Public Transport Planning and Operations, Proceedings of Seminar G, held at the PTRC European Transport Forum, London, 99–116.
Boucher, D. and Kelly, P. (eds) (1998) *Social Justice: from Hume to Walzer*. London: Routledge.
Bowen, J. (2004) The geography of freighter aircraft operations in the Pacific Rim. *Journal of Transport Geography* 12, 1–11.
Bowen, J. and Leinbach, T. (1995) The state and liberalization: the airline industry in the East Asian NICs. *Annals, Association of American Geographers* 85, 468–93.
Bowen, J. and Leinbach, T. (2006) Competitive advantage in global production networks: air freight services and the electronics industry in southeast Asia. *Economic Geography* 82, 147–66.
Brake, J., Nelson, J. and Wright, S. (2004) Demand responsive transport: towards the emergence of a new market segment. *Journal of Transport Geography* 12, 323–7.
Briggs, R. and McKelvey, D. (1975) Rural public transportation and the disadvantaged. *Antipode* 7, 31–6.
British Broadcasting Corporation (BBC) (2007) *Union Calls off BA Strike Action*. news.bbc.co.uk/1/hi/business/6309471.stm Accessed 29 January 2007.
British Waterways (2006) *Anderton Boat Lift*. Northwich: British Waterways. www.andertonboatlift.co.uk.
British Waterways Scotland (2006) *The Falkirk Wheel*. Falkirk: British Waterways Scotland. www.thefalkirkwheel.co.uk/index.asp.
Broughton, P. and Stradling, S. (2005) Why ride powered two-wheelers? In *Behavioural Research in Road Safety: Fifteenth Seminar*. London: Department for Transport, pp. 68–78.

Brown, B. and Laurier, E. (2005) Maps and car journeys: an ethnomethodological approach. *Cartographica* 4, 17–33.

Brown, B. and O'Hara, K. (2003) Place as a practical concern of mobile workers. *Environment and Planning A* 35, 1565–88.

Buchan, K. (1992) Enhancing the quality of life. In Roberts, J., Cleary, J., Hamilton, K. and Hanna, J. (eds), *Travel Sickness: the Need for Sustainable Transport Policy for Britain*. London: Lawrence and Wishart, pp. 7–17.

Buchanan, N.; Barnett, R.; Kingham, S. and Johnston, D. (2006) The effect of urban growth on commuting patterns in Christchurch, New Zealand. *Journal of Transport Geography* 14, 342–54.

Building Research Establishment (1993) *Effects of Environmental Noise on People at Home*. Information Paper 22/93. London: HMSO.

Bundesminsterium für Verkehr, Bau- und Wohnungswesen (2005) *Verkehr in Zahlen* 2004/2005. Berlin: BMVBW.

Burchardt, T., Le Grand, J. and Piachaud, D. (1999) Social exclusion in Britain 1991–1995. *Social Policy and Administration* 33, 227–44.

Bureau of Transportation Statistics (2006a) *National Transportation Statistics*. www.bts.gov/publications/national_transportation_statistics/. Accessed 25 January 2007.

Bureau of Transportation Statistics (2006b) *National Transportation Statistics: Average Length of Haul, Domestic Freight and Passenger Modes*. Washignton, DC: Bureau of Transportation. www.bts.gov/publications/national_transportation_statistics/html/table_01_35.html. Accessed 2 January 2007.

Bureau of Transportation Statistics (2006c) *National Transportation Statistics: Automobile Profile*. Washington, DC: Bureau of Transportation. www.bts.gov/publications/national_transportation_statistics/html/table_automobile_profile.html. Accessed 2 January 2007.

Bureau of Transportation Statistics (2006d) *National Household Travel Survey: Long Distance Travel Quick Facts*. Washington DC: Bureau of Transportation. www.bts.gov/programs/national_household_travel_survey/long_distance.html. Accessed 2 January 2007.

Bureau of Transportation Statistics (2006e) *Pocket Guide to Transportation Statistics 2006: Long-distance Trips by Gender, Area, and Age: 2001*. Washington, DC: Bureau of Transportation. www.bts.gov/publications/pocket_guide_to_transportation/2006/html/figure_06_table.html. Accessed 2 January 2007.

Burghouwt, G. (2005) *Airline Network Development in Europe and its Implications for Airport Planning*. Utrecht: Utrecht University.

Burghouwt, G., Hakfoort, J. and van Eck, J. (2003) The spatial configuration of airline networks in Europe. *Journal of Air Transport Management* 9, 309–23.

Burkett, N (2000) *Own Transport Preferred: Transport and Social Exclusion in the North East*. Newcastle: Low Pay Unit.

Burtenshaw, D., Bateman, M. and Ashworth, G. (1991) *The European City: a Western Perspective*. London: Fulton.

Bus Partnership Forum (2003) *Understanding Customer Needs*. London: Bus Partnership Forum.

Butcher, A. (ed.) (2006) *Railways Restored*, 27th edition. Hersham: Ian Allan Publishing.

Butler, R. (1997) Transportation innovations and island tourism. In: Lockhart, D. and Drakakis-Smith, D. (eds), *Island Tourism: Problems and Perspectives*. London: Mansell, pp. 36–56.

Butler, R. (1999) Sustainable tourism: a state-of-the-art review. *Tourism Geographies* 1, 7–25.

Button, K. (2001) Are current air transport policies consistent with a sustainable environment? In Feitelson E. and Verhoef, E. (eds), *Transport and Environment – in Search of Sustainable Solutions*. Cheltenham: Edward Elgar, pp. 54–72.

Button, K. and Gillingwater, D. (1986) *Future Transport Policy*. London: Croom Helm.

Button, K. and Stough, R. (1998) *The Benefits of Being a Hub Airport: Convenient Travel and High-tech Job Growth*. Fairfax: Institute of Public Policy, George Mason University.

Cairns, S., Sloman, L., Newson, C., Anable, J., Kirkbride, A. and Goodwin, P. (2004) *Smarter Choices: Changing the Way we Travel*. London: Department for Transport.

Callender, C. (1999) *The Hardship of Learning: Students' Income and Expenditure and Their Impact on Participation in Further Education*. Bristol: The Further Education Funding Council.

Cameron, A. (2006) Geographies of welfare and exclusion: social inclusion and exception. *Progress in Human Geography* 30, 396–404.

Cape Railway Enthusiasts' Association (2004) *Cape Railway Enthusiasts' Association*. Cape Town: CREA. www.capesteam.za.net/index.htm.

Capineri, C. and Leinbach, T. (2004) Globalization, E-economy and trade. *Transport Reviews* 24, 645–63.

CARLOS (2005) *Evaluation Pilotprojekt CARLOS*. Zürich: CARLOS.

Cass, N., Shove, E. and Urry, J. (2005) Social exclusion, mobility and access. *Sociological Review* 53, 539–55.

Castells, M. (1989) *The Informational City*. Oxford: Blackwell.

Castells, M. (1996) *The Information Age: Economy, Society and Culture: Vol. 1, The Rise of the Network Society*. Cambridge MA: Blackwell.

Centre for Mobilities Research (2006) www.lancs.ac.uk/fss/sociology/cemore. Accessed 14 December 2006.

Ceron, J.-P. and Dubois, G. (2003) Changes in leisure-tourism mobility facing the stake of global warming: the case of France. Paper presented at the International Geographical Union Conference *Human Mobility in a Globalising World*, 3–5 April, Palma de Mallorca, Spain.

Cervero, R. (2004) Job isolation in the US: narrowing the gap through job access and reverse-commute programs. In Lucas, K. (ed.), *Running on Empty*. Bristol: Policy Press, pp. 181–96.

Chadefaud, M. (1981) *Lourdes: un pèlerinage, une ville*. Edisud, Aix-en-Provence.

Charlton, C. and Gibb, R. (1998a) Transport deregulation and privatisation. *Journal of Transport Geography* 6, 85.

Charlton, C. and Gibb, R. (1998b) International surface passenger transport. In Hoyle, B. and Knowles, R. (eds), *Modern Transport Geography*, 2nd edition. Chichester: Wiley, pp. 291–310.

Chliaoutakis, J., Darviri, C. and Demakakos, P. (1999) The impact of young drivers' lifestyle on their road traffic accident risk in greater Athens area. *Accident Analysis & Prevention* 31, 771–80.

Chliaoutakis, J., Demakakos, P., Tzamalouka, G., Bakou, V., Koumaki, M. and Darviri, C. (2002) Aggressive behavior while driving as predictor of self-reported car crashes. *Journal of Safety Research* 33, 431–43.

Choi, H.-S. and Sirakaya, E. (2006) Sustainability indicators for managing community tourism. *Tourism Management* 27, 1274–89.

Christaller, W. (1933) *Die zentralen Orte in Süddeutschland*. Jena: Fischer.

Christaller, W. (1966) *Central Places in Southern Germany*. Translated by Baskin, C. Englewood Cliffs, NJ: Prentice Hall.

Church, A. and Frost, M. (1999) *Transport and Social Exclusion in London: Exploring Current and Potential Indicators*. London: London Transport Planning.

Church, A., Frost, M. and Sullivan, K. (2000) *Transport and Social Exclusion in London – Report Summary*. London: London Transport Planning.

Church, A., Frost, M. and Sullivan, K. (2001) Transport and social exclusion in London. *Transport Policy* 7, 195–205.

City of Edinburgh Council (2005) *A Vision for Capital Growth*. Edinburgh: City of Edinburgh Council. www.edinburgh.gov.uk/internet/Attachments/Internet/Environment/2040_Vision.pdf.

City of Portland (2004) *Portland Transit Mall: Urban Design Analysis & Vision*. Portland, OR: Bureau of Planning, City of Portland.

City of Portland (2006) *Central City District Plan*. Portland, OR: Bureau of Planning, City of Portland.

Claval, P. (1998) *An Introduction to Regional Geography*. Oxford: Blackwell.

Cloke, P., Cook, I., Crang, P., Goodwin, M., Painter, J. and Philo, C. (2004) *Practising Human Geography*. London and Thousand Oaks: Sage.

Cloke, P., Marsden, T. and Mooney, P. (eds) (2005) *Handbook of Rural Studies*. London: Sage.

Collier, A. (1994) *Principles of Tourism*, 3rd edition. Auckland: Longman Paul.

Colography Group (2004) *A World of Change: Global Transport in a Time Driven World*. www.colography.com/press/2004/whitepapermain.html. Accessed 2 February 2007.

Commission for Integrated Transport (CfIT) (2001a) *European Best Practice in Delivering Integrated Transport*. London: CfIT. www.cfit.gov.uk/docs/2001/ebp/ebp/key/pdf/key.pdf. Accessed 16 August 2006.

Commission for Integrated Transport (2001b) *Key Issues in Rural Transport*. Report for Commission for Integrated Transport Rural Working Group. www.cfit.gov.uk/docs/2001/rural/rural/key/index.htm. Accessed 14 December 2006.

Commission for Integrated Transport (CfIT) (2002) *Paying for Road Use*. London: CfIT.

Commission for Integrated Transport (CfIT) (2006) *CfIT's world review of road pricing phase 1: lessons for the UK*. London: CfIT. www.cfit.gov.uk/docs/2006/wrrp1/pdf/wrrp1.pdf. Accessed 16 August, 2006.

Commission for Rural Communities (CRC) (2005) *State of the Countryside 2005*. www.ruralcommunities.gov.uk//projects/stateofthecountrysidedata/overview. Accessed 14 December 2006.

Commission for Social Justice (1994) *Social Justice – Strategies for National Renewal*. Report of the Commission on Social Justice. London: Vintage-Random House.

Commission of the European Communities (CEC) (1985) Council Directive of 27 June 1985 on the assessment of the effects of certain public and private projects on the environment, 85/337/EEC. *Official Journal of the European Communities*, L 175, 05/07/1985 P. 0040–0048.

Commission of the European Communities (CEC) (1992) *Green Paper: the Impact of Transport on the Environment: a Community Strategy for 'Sustainable Mobility'*. Luxembourg: CEC Office for Official Publications.

Commission of the European Communities (CEC) (1995) *The Citizens' Network: Fulfilling the Potential of Public Passenger Transport in Europe*. COM(95)601, Brussels: CEC.

Commission of the European Communities (CEC) (1997) Council Directive 97/11/EC of 3 March 1997 on the assessment of the effects of certain public and private projects on the environment. *Official Journal of the European Communities*, L 073, 14/03/1997 P.5

Commission of the European Communities (CEC) (1999) Commission recommendation of 5 February 1999 on the reduction of CO_2 emissions from passenger cars. *Official Journal of the European Communities* L 40/50.

Commission of the European Communities (CEC) (2000) *The Auto-oil II Programme: a Report from the Services of the European Commission*. Brussels: CEC.

Commission of the European Communities (CEC) (2001a) Directive 2001/42/EC on the assessment of the effects of certain plans and programmes on the environment. *Official Journal of the European Communities* L 197, 0030–0037.

Commission of the European Communities (CEC) (2001b) *White Paper. European Transport Policy for 2010: Time to Decide*. Brussels: European Commission Directorate-General for Energy and Transport.

Commission of the European Communities (CEC) (2003) *Investing in Networks and Knowledge for Growth and Jobs: Final Report to the European Council*. COM (2003) 690 final/2, 21 November. Brussels: CEC.

Commission of the European Communities (CEC) (2005a) *Reducing the Climate Change Impact of Aviation*. COM (2005) 459 final, Brussels, 27.9.2005.

Commission of the European Communities (CEC) (2005b) *Trans-European Transport Network: TEN-T Priority Axes and Projects 2005*. Brussels: European Commission Directorate-General for Energy and Transport.

Commission of the European Communities (CEC) (2006a) *Transport and Environment: Legislation*. ec.europa.eu/environment/air/legis.htm#transport Accessed 9 June 2006.

Commission of the European Communities (CEC) (2006b) *Keep Europe Moving – Sustainable Mobility for Our Continent: Mid-term Review of the European Commission's 2001 Transport White Paper*. COM (2006) 314 final. Brussels: CEC.

Commission of the European Communities (CEC) (2006c) *Directive 2006/38/EC of the European Parliament and of the Council of 17 May 2006 amending Directive 1999/62/EC on the Charging of Heavy Goods Vehicles for the Use of Certain Infrastructure*. Brussels: CEC.

Committee on the Medical Effects of Air Pollutants (COMEAP) (1995) *Asthma and Outdoor Pollutants*. London: HMSO.

Community of European Railway and Infrastructure Companies (CER) (2006) *Annual Report 2005/2006*. Brussels: CER.

Cooke, P. and Morgan, K. (1998) *The Associational Economy: Firms, Regions and Innovation*. Oxford: Oxford University Press.

Couldry, N. (1998) The view from inside the simulacrum: visitors tales from the set of Coronation Street. *Leisure Studies* 17, 94–107.

Cox, W. (2001) *American Dream Boundaries: Urban Containment and its Consequences*. Georgia Public Policy Foundation. www.gppf.org/article.asp?RT=&p=pub/LandUse/Growth/american_dream_boundaries.htm. Accessed 28 November 2006.

Crafts, N. and Leunig, T. (2005) *The Historical Significance of Transport for Economic Growth and Productivity*. London: London School of Economics. www.dft.gov.uk/about/strategy/eddingtonstudy/researchannexes/researchannexesvolume1/historical significance.

Cramer, J. (2000) Early warning: integrating eco-efficiency aspects into the product development process. *Environmental Quality Management* Winter, 1–10.

Crang, M. (2002) Between places: producing hubs, flows and networks. *Environment and Planning A* 34, 569–74.

Cresswell, T. (2006) *On the Move: Mobility in the Western World*. London: Routledge.

Crime Concern (1999) *Personal Security Issues in Pedestrian Journeys*. London: DETR.

Cronon, W. (1991) *Nature's Metropolis: Chicago and the Great West*. New York: W.W. Norton.

Crozet, Y. (2005) Time and passenger transport. In: European Conference of Ministers of Transport Round Table 127. *Time and Transport*. Paris: ECMT, pp. 25–6.

Cullinane, K. and Khanna, M. (2000) Economies of scale in large containerships: optimal size and geographical implications. *Journal of Transport Geography* 8, 181–95.

Cullinane, S. (2003) Hong Kong's low car dependence: lessons and prospects. *Journal of Transport Geography* 11, 25–35.

Currie, G. (2006) *Perspectives on Transport and Access Issues of Young People in Australia*. Paper presented at Conference on Transport, Social Disadvantage and Wellbeing, Melbourne, Victoria, 5–6 April.

Curry, A., Hodgson, T. Kelnar, R. and Wilson, A. (2006). *Intelligent infrastructure futures. The scenarios – towards 2055*. London: Department of Trade and Industry.

Daigle, J. and Zimmerman, C. (2004) The convergence of transportation, information technology and tourist experience at Acadia National Park. *Journal of Tourism Research* 43, 151–60.

Daniels, P. and Warnes, A. (1980) *Movement in Cities*. London: Methuen.

Dann, G. (1994) Travel by train: keeping nostalgia on track. In: Seaton, A. (ed.), *Tourism: the State of the Art*. Chichester and New York: John Wiley & Sons, pp. 775–82.

Dargay, J. and Hanly, M. (2003). *Travel to Work: an Investigation Based on the British Household Panel Survey*. Paper presented at *NECTAR Conference No. 7*, Umeà, Sweden.

Davenport, J. and Davenport, J. (2006) The impact of tourism and personal leisure transport on coastal environments: a review. *Estuarine, Coastal and Shelf Science* 67, 280–92.

Davis, A. (1996) Medical thresholds or quality of life assessments? *Clean Air* 26, 61–4.

Davison, L. and Knowles, R. (2006) Bus quality partnerships, modal shift and traffic decongestion. *Journal of Transport Geography* 14, 177–94.

Debbage, K. (1991) Spatial behaviour in a Bahamian resort. *Annals of Tourism Research* 18, 251–68.

Debbage, K. and Daniels, P. (1998) The tourist industry and economic geography: missed opportunities? In: Ioannides, D. and Debbage, K. (eds), *The Economic Geography of the Tourist Industry*. London: Routledge, pp. 17–30.

Decker, M. (2004) *Structures et stratégies des compagnies aériennes à Bas Coûts*. Paris: L'Harmattan.

Deka, D. (2004) Social and environmental justice issues in urban transportation. In Hanson, S. and Giuliano, G. (eds), *The Geography of Urban Transportation*, 3rd edition. New York: Guilford Press.

Deloitte Research (2003) *Combating Gridlock: How Pricing Road Use Can Ease Congestion*. London: Deloitte Research.

Dempsey, P. (2000) *Airport Planning and Development Handbook: a Global Survey*. New York: McGraw-Hill.

Dempsey, P., Goetz, A. and Szyliowicz, J. (1997) *Denver International Airport: Lessons Learned*. New York: McGraw-Hill.

Dennis, N. (1994) Airline hub operations in Europe. *Journal of Transport Geography* 2, 219–33.

Department for Education and Employment (DfEE) (1998) *New Arrangements for Effective Student Support in Further Education*. Report of the Further Education Student Support Advisory Group. London: DfEE.

Department for Education and Employment (DfEE) (1999) *Jobs for All*. Report of the policy action team on jobs for the national strategy for neighbourhood renewal. London: DfEE.

Department for Education and Skills (DfES) (2003) *Travelling to School: an Action Plan.* London: DfES and DfT.

Department for Environment, Food and Rural Affairs (DEFRA) (2006) *Rural Services Review*. London: Defra.

Department for Transport (DfT) (2000) *Social Exclusion and the Provision and Availability of Public Transport.* London: Mobility and Exclusion Unit, DfT.

Department for Transport (DfT) (2003a) *The Future of Air Transport.* London: DfT.

Department for Transport (DfT) (2003b) *Evaluation of Rural Bus Subsidy Grant and Rural Bus Challenge.* London: DfT. www.dft.gov.uk/stellent/groups/dft_localtrans/documents/page/dft_localtrans_024814.hcsp Accessed 18 December 2006.

Department for Transport (DfT) (2004a) *Transport Statistics Great Britain.* London: DfT.

Department for Transport (DfT) (2004b): White Paper: *The Future of Transport: a network for 2030.* Cmd 6234. London: DfT.

Department for Transport (DfT) (2004c) *Strategic Environmental Assessment for Transport Plans and Programmes.* www.webtag.org.uk/webdocuments/2_Project_Manager/11_SEA/2.11.pdf.

Department for Transport (DfT) (2005) *Transport Statistics Great Britain.* London: DfT.

Department for Transport (DfT) (2006a) *Transport Statistics Great Britain 2005.* DfT, Scottish Executive and Welsh Assembly. London: TSO.

Department for Transport (DfT) (2006b) *Transport Trends Great Britain.* London: DfT.

Department for Transport (DfT) (2006c) *Transport Statistics Bulletin: National Travel Survey 2005.* London: DfT.

Department for Transport (DfT) (2006d) www.dft.gov.uk/stellent/groups/dft_localtrans/documents/page/dft_localtrans_0. Accessed 14 December 2006.

Department for Transport, Local Government and the Regions (DTLR) (2001) *Focus on Personal Travel: 2001 Edition.* London: The Stationery Office.

Department of the Environment, Transport and the Regions (DETR) (1998) *A New Deal for Transport: Better for Everyone.* London: The Stationery Office

Department of the Environment, Transport and the Regions (DETR) (1999) *Review of voluntary transport.* www.mobility-unit.detr.gov.uk/rvt/report/1.htm/

Department of the Environment, Transport and the Regions (DETR) (2000) *Social Exclusion and the Provision and Availability of Public Ttransport.* London: DETR.

Department of Trade and Industry (DTi) Teleworking in the UK, *Labour Market Trends June 2002.* London: DTi.

Derek Halden Consultancy (DHC) (2006) *Scoping the Impacts on Travel Behaviour in Scotland of E-working and Other ICTs.* Report for the Scottish Executive, Transport Research Series. Edinburgh: Scottish Executive Social Research Unit.

Dicken, P. (1998) *Global Shift*, 3rd edition. London: Sage.

Dicken, P. (2003) *Global Shift: Reshaping the Global Economic Map in the 21st Century*, 4th edition. New York: The Guilford Press.

Divall, C. (2002) Heritage railways as museums: occupations and landscapes. *Japan Railway and Transport Review* 30, 4–9.

Divall, C. and Scott, A. (2001) *Making Histories in Transport Museums.* Leicester: Leicester University Press.

Dobbs, L. (2005) Wedded to the car: women, employment and the importance of private transport. *Transport Policy* 12, 266–78.

Dobruszkes, F. (2005) Compagnies low cost et aéroports secondaires: quelles dépendences pour quel développement régional? *Les Cahiers Scientifiques du Transport* 47, 39–59.

Dobruszkes, F. (2006) An analysis of European low-cost airlines and their networks. *Journal of Transport Geography* 14, 249–64.

Docherty, I. (1999) *Making Tracks: the Politics of Local Rail Transport*. Aldershot: Ashgate.

Docherty, I. (2003) Policy, politics and sustainable transport: the nature of Labour's dilemma. In Docherty, I. and Shaw, J. (eds), *A New Deal for Transport? The UK's Struggle with the Sustainable Transport Agenda*. Oxford: Blackwell, pp. 3–29.

Docherty, I. and Shaw, J. (eds) (2003) *A New Deal for Transport? The UK's Struggle with the Sustainable Transport Agenda*. Oxford: Blackwell.

Docherty, I., Shaw, J. and Gather, M. (2004) State intervention in contemporary transport. *Journal of Transport Geography* 12, 257–64.

Docherty, I., Shaw, J. and Gray, D. (2007) Transport strategy in Scotland since devolution. *Public Money and Management* 27, 141–8.

Dodge, M. and Kitchin, R. (2004) Flying through code/space: the real virtuality of air travel. *Environment and Planning A* 36, 195–211.

Doganis, R. (2001) *The Airline Business in the 21st Century*. London: Routledge.

Donaghy, K., Poppelreuter, S. and Rudinger, G. (eds) (2005) *Social Dimensions of Sustainable Transport: Transatlantic Perspectives*. Aldershot: Ashgate.

Donald, R. and Pickup, L. (1991) The effects of local bus deregulation in Great Britain on low income families: the case of Merseyside. *Transportation Planning and Technology* 15, 331–47.

Douglas, Y. (2001) *Steam Railways: Britain's Preservation Railways and Museums*. Navigator Guides, Melton Constable.

Downward, P., Rhoden, S. and Lumsdon, L. (2006) Transport for tourism: can public transport encourage a modal shift in the day tourist market. *Journal of Sustainable Tourism* 14, 139–56.

Drewry Shipping Consultants (2006) *Annual Review of Global Terminal Operators*. London: Drewry Shipping Consultants.

Druva-Druvaskalne, I., Ãbols, I. and Šara, A. (2006) Latvia tourism: decisive factors and tourism development. In: Hall, D.; Smith, M. and Marciszewska, B. (eds), *Tourism in the New Europe: the Challenges and Opportunities of EU Enlargement*. Wallingford: CABI Publishing, pp. 170–82.

Dublin Transportation Office (2001) *A Platform for Change – Summary Report*. www.dto.ie/platform1.pdf. Accessed 18 December 2006.

Dubois, G. and Ceron, J.-P. (2006) Tourism/leisure greenhouse gas emission forecasts for 2050: factors for change in France. *Journal of Sustainable Tourism* 14, 172–91.

Dudleston, A., Hewitt, E., Stradling, S. and Anable, J. (2005) *Public Perceptions of Travel Awareness – Phase Three*. Edimburgh: Scottish Executive Central Research Unit.

Dudley, G. and Richardson, J. (2000) *Why does Policy Change? Lessons from British Transport Policy 1945–99*. London: Routledge.

Eckey, H.-F. and Kosfeld, R. (2004) *New economic geography. Critical reflections, regional policy implications and further developments*. Volkswirtschaftiche Diskussionsbeiträge Nr. 65/04 des Fachbereiches Wirtschaftswissenschaften der Universität Kassel. Kassel: Universität Kassel.

Eddington, R. (2006a). *Transport's Role in Sustaining the UK's Productivity and Competitiveness*. London: The Stationery Office.

Eddington, R. (2006b) *Understanding the Relationship: How Transport Can Contribute to Economic Success*. The Eddington Report, Volume I. London: The Stationery Office.

Eddington, R. (2006c) *The Eddington Transport Study: The Case for Action. Sir Rod Eddington's Advice to Government*. London: The Stationery Office.

Elby, D. and Molnar, L. (2001) Importance of scenic byways in route choice: a survey of driving tourists in the United States. *Transportation Research Part A* 36, 96–106.

Eliot-Hurst, M. (ed.) (1973) *Transportation Geography: Comments and Readings*. Toronto: McGraw Hill Ryerson, Ltd.

Ellaway, A., MacIntyre, S., Hiscock, R. and Hearns, A. (2003) In the driving seat: psychosocial benefits from private motor vehicle transport compared to public transport. *Transportation Research Part F: Traffic Psychology and Behaviour* 6, 217–31.

Endzina, I. and Luneva, L. (2004) Development of a national branding strategy: the case of Latvia. *Place Branding* 1, 94–105.

Engwicht, D. (1993) *Reclaiming our Cities and Towns (Towards an Eco-city)*. Philadelphia: New Society Publishers.

Environmental Change Institute (2006) *Predict and Decide: Aviation, Climate Change and UK Policy*. Oxford: Environmental Change Institute.

ESRC Transport Unit (1995) *Car Dependence*. London: RAC Foundation.

EU Business (2005) Luxury hotels cash in on tourism boom in new EU countries. *EU Business* 21 June. www.eubusiness.com/East_Europe/050622041934.nndypr28.

EurActiv (2007) *EU to Forge Ahead with Car-emissions Cap*. www.euractiv.com/en/climate-change/eu-forge-ahead-car-emissions-cap/article-161436 Accessed 6 February 2007.

Euractiv.com (2006) *Road charging (Eurovignette)*. www.euractiv.com/en/transport/road-charging-eurovignette/article-117451. Accessed 9 January 2007.

European Bank of Reconstruction and Development (2001) *Estonian Railway Privatisation – Project Summary Documentation* www.ebrd.com/projects/psd/psd2001/3810.htm Accessed 25 July 2006.

European Commission (2003a) *TAPESTRY Travel Awareness, Publicity and Education Supporting a Sustainable Transport Strategy in Europe*. Brussels: European Commission. www.euptapestry.org.

European Commission (2003b) *Tourism and the European Union*. Brussels: European Commission. europa.eu.int/comm/enterprise/services/tourism/tourismeu.htm#future.

European Commission (2003c) *Panorama of Transport: Statistical Overview of Transport in the European Union, Part 2*. epp.eurostat.ec.europa.eu/cache/ITY_OFFPUB/KS-DA-04-001-2/EN/KS-DA-04-001-2-EN.PDF Accessed 28 February 2007.

European Commission (2006) *Energy and Transport in Figures 2006 – Part 3: Transport*. Eurostat.

European Conference of Ministers of Transport (ECMT) (1990) *Transport Policy and the Environment*. Paris: ECMT/OECD.

European Conference of Ministers for Transport (ECMT) (2002) *Implementing Sustainable Urban Travel Policies*. Paris: OECD.

European Environment Agency (EEA) (2005) *Annual European Community Greenhouse Gas Iinventory 1990–2003 and Inventory Report 2005*. EEA Technical report No 4/2005. Copenhagen: EEA.

European Parliament (2006) *European Parliament Resolution on Reducing the Climate Change Impact of Aviation*, P6_TA-PROV(2006)0296. Strasbourg: European Parliament.

European Train Enthusiasts – Eastern New England (2004) *European Train Enthusiasts – Eastern New England Chapter*. ETE-ENE, Boston. www.ete-ene.org/.

Eurostat (2005) Total length of motorways. In *Europe in Figures: Eurostat Yearbook 2005*. epp.eurostat.cec.eu.int/cache/ITY_OFFPUB/KS-CD-05-001-6/EN/KS-CD-05-001-6-EN.PDF. Accessed 2 January 2007.

Eurostat (2006a) Eurostat Transport Homepage. epp.eurostat.ec.europa.eu/portal/page?_pageid=0,1136228,0_45572948&_dad=portal&_schema=PORTAL. Accessed 18 December 2006.

Eurostat (2006b) *Transport Portal.* epp.eurostat.ec.europa.eu. Accessed 14 December 2006.

Eurostat (2006c) *Statistics in focus: passenger transport in the European Union.* Eurostat. www.eustatistics.gov.uk/Download.asp?stats%20in%20focus%209-2006_tcm90-39383.pdf. Accessed 2 January 2007.

Evans, D. and Smyth, A. (1997) Fully scheduled or dial-a ride? The future direction of accessibility policy for local public transport. Public Transport Planning and Opera-tions. Proceedings of Seminar G, PTRC European Transport Forum, London, pp. 129–40.

Fan, T. (2006) Improvements in intra-European inter-city flight connectivity, 1996–2004. *Journal of Transport Geography* 14, 249–64.

Farrell, S. (1999) *Financing Europe's Transportation Future: Policies and Practice.* Basingstoke: Macmillan.

Farrington, J. (2007) The new narrative of accessibility: its potential contribution to discourses in (transport) geography. *Journal of Transport Geography* 15, 319–30.

Farrington, J. and Farrington, C. (2005) Rural accessibility, social inclusion and social justice: towards conceptualisation. *Journal of Transport Geography* 13, 1–12.

Farrington, J., Gray, D., Roberts, D. and Martin, S. (1998) *Car dependence in rural Scotland.* Edinburgh: The Scottish Office.

Farrington, J., Shaw, J., Leedal, M., Maclean, M., Halden, D., Richardson, T. and Bristow, G. (2004) *Settlements, Services and Access: the Development of Policies to Promote Accessibility in Rural Areas in Great Britain.* HM Treasury, Countryside Agency, Scottish Executive and Welsh Assembly Government. WAG, Cardiff.

Featherstone, M. (2004) Automobilities. An introduction. *Theory, Culture and Society* 21, 1–24.

Feitelson, E. (2002) Introducing environmental equity dimensions into the sustainable transport discourse: issues and pitfalls. *Transportation Research Part D: Transport and Environment* 7, 99–118.

Fleming, D. and Hayuth, Y. (1994) Spatial characteristics of transportation hubs: centrality and intermediacy. *Journal of Transport Geography* 2, 3–18.

Flexibility (2006) *Future Travel Virtual Call Centre.* www.flexibility.co.uk/cases/future-travel.htm. Accessed 14 December 2006.

Florida, R. (2005) *Cities and the Creative Class.* New York: Routledge.

Flyvbjerg, B., Bruzelius, N. and Rothengatter, W. (2003) *Megaprojects and Risk: An Anatomy of Ambition.* Cambridge: Cambridge University Press.

Foley, J (ed.) (2004) *Sustainability and Social Justice.* London: Institute for Public Policy Research.

Forsyth, P. (2003) Low-cost carriers in Australia: experiences and impacts. *Journal of Air Transport Management* 9, 277–84.

Foster, C. (1992) *Privatization, Public Ownership and the Regulation of Natural Monopoly.* Oxford: Blackwell.

Foster, M. (1981) *From Streetcar to Superhighway: American City Planners and Urban Transportation.* Philadelphia: Temple University Press.

Francis, G., Humphreys, I., Ison, S. and Aicken, M. (2006) Where next for low cost airlines? A spatial and temporal comparative study. *Journal of Transport Geography* 14, 83–94.

Frank, A. (1967) *Development and Underdevelopment in Latin America: Historical Studies of Chile and Brazil.* New York and London: Monthly Review Press.

Friedman, M. (1962) *Capitalism and Freedom.* London: University of Chicago Press.

Friends of the Earth (F0E) (2001) *Environmental justice: mapping transport and social exclusion*. London: FoE.

Froebel, F., Heinrichs, J. and Kreye, O. (1980) *The New International Division of Labour*. Cambridge: Cambridge University Press.

Fröidh, O. (2005) Market effects of regional high-speed trains on the Svealand line. *Journal of Transport Geography* 13, 352–61.

Fuellhart, K. (2003) Inter-metropolitan airport substitution by consumers in an asymmetrical airfare environment: Harrisburg, Philadelphia, and Baltimore. *Journal of Transport Geography* 11, 285–96.

Fujita, M., Krugman, P. and Venables, A. (1999) *The Spatial Economy: Cities, Regions and International Trade*. Cambridge, MA: MIT Press.

Fuller, R., Bates, H., Gormley, M., Hannigan, B., Stradling, S., Broughton, P., Kinnear, N. and O'Dolan, C. (2006) Inappropriate high speed: who does it and why? In *Behavioural Research in Road Safety: Sixteenth Seminar*. London: DfT, pp. 70–84.

Fullerton B. (1990) Deregulation in a European context – the case of Sweden. In Bell, P. and Cloke, P. (eds), *Deregulation and Transport: Market Forces in the Modern World*. London: David Fulton, pp. 125–40.

Funazaki, A., Taneda, K., Tahara, K. and Inaba, A. (2003) Automobile life cycle assessment issues at end-of-life and recycling. *Japanese Society of Automotive Engineers Review* 24, 381–6.

Furgala-Selezniow, G., Turkowski, K., Nowak, A., Skrzypczak, A. and Mamcarz, A. (2003) The Ostroda–Elblag canal – its past and future in aquatic tourism. In Härkönen, T. (ed.), *International Lake Tourism Conference 2–5 July*. Savonlinna, Finland: Savonlinna Institute for Regional Development and Research, University of Joensuu, pp. 55–72.

Gaffron, P., Hine, J. and Mitchell, F. (2001) *The Role of Transport in Social Exclusion in Urban Scotland: Literature Review*. Edinburgh: Scottish Executive Central Research Unit.

Gant, R. (2002) *Shopmobility* at the millennium: 'Enabling' access in town centres. *Journal of Transport Geography* 10, 123–33.

Gardiner, R. (ed.) (1992) *The Shipping Revolution*. London: Conway.

Garling, T. (2005) Changes of private car use in response to travel demand management. In Underwood, G. (ed.), *Traffic and Transport Psychology: Theory and Application. Proceedings of the ICTTP 2004*. Oxford: Elsevier.

Gatersleben, B. (2004) *Affective, Social and Instrumental Aspects of the Commute to Work: Comparing Perceptions of Drivers, Public Transport Users, Walkers and Cyclists*. Paper presented at the 3rd International Conference on Traffic & Transport Psychology, Nottingham.

Gather, M. (1998) *Liberalisierung der Verkehrsmärkete und nachhaltige Mobilität*. Universität St Gallen.

Gather, M., Kagermeier, A. and Lanzendorf, M. (eds) (2001) *Verkehrsentwicklung in den Neuen Bundeslandern*. Proceedings of the Annual Meeting of the Transport Research Group of the German Geographical Society. Erfurt: Erfurter Geographische Studien.

Gereffi, G. (1999) International trade and industrial upgrading in the apparel commodity chain. *Journal of International Economics* 48, 37–70.

Geurs, K. and van Wee, B. (2004) Accessibility evaluation of land use and transport strategies: review and research direction. *Journal of Transport Geography* 12, 127–40.

Gibb, K., Kearns, A., Keoghan, M., MacKay, D. and Turok, I. (1998) *Revising the Scottish Area Deprivation Index (Volume 1)*. Edinburgh: Scottish Office.

Gibb, R. (ed.) (1994) *The Channel Tunnel: a Geographical Perspective*. London: Belhaven Press.

Giddens, A. (2000) *The Third Way and its Critics*. London: Polity Press.
Giorgi, L. and Schmidt, M. (2005) Transalpine transport: a local problem in search of European solutions or a European problem in search of local solutions? *Transport Reviews* 25, 201–19.
Giuliano, G. (2004) The land use impacts of transportation investments: highway and transit. In Hanson, S. and Giuliano, G. (eds), *The Geography of Urban Transportation*, 3rd edition. New York: Guilford Press.
Giuliano, G. (2005) Low income, public transit and mobility. *Transportation Research Record* 1927, 63–72.
Giuliano, G. and Gillespie, A. (1997) Research issues regarding societal change and transport. *Journal of Transport Geography* 5, 165–76.
Giuliano, G. and Gillespie, A. (2002) Research issues regarding societal change and transport: an update. In Black, W. and Nijkamp, P. (eds) *Social Change and Sustainable Transport*. Bloomington: Indiana University Press, pp. 27–34.
Giuliano, G. and Small, K. (1999) The determinants of growth in employment subcentres. *Journal of Transport Geography* 7, 189–202.
Givoni, M. (2006) Development and impact of the modern high-speed train: a review. *Transport Reviews* 26, 593–611.
Givoni, M. (2006) Development and impact of the modern high-speed train: a review. *Transport Reviews* 26, 593–612.
Glaeser, E. (2004) *Four Challenges for Scotland's Cities*. Allender Series, University of Strathclyde.
Glaister, S. (2001) *UK Transport Policy 1997–2001*. Paper delivered to the Economics Section of the British Association for the Advancement of Science, Glasgow, 4 September.
Glaister, S. (2004) *London – on the Move or in Jam?* London: Development Securities plc.
Glaister, S. and Graham, D. (2000) *The Effect of Fuel Prices on Motorists*. Basingstoke: The AA Motoring Policy Unit.
Gleave, M. (1991) The Dar es Salaam transport corridor: an appraisal. *African Affairs* 91, 249–67.
Goetz, A. (2002) Deregulation, competition and antitrust implications in the US airline industry. *Journal of Transport Geography* 10, 1–18.
Goetz, A. and Graham, B. (2004) Air transport globalization, liberalization and sustainability: Post-2001 Policy Dynamics in the United States and Europe. *Journal of Transport Geography* 12, 265–76.
Goetz, A. and Sutton, C. (1997) The geography of deregulation in the U.S. airline industry. *Annals, Association of American Geographers* 87, 238–63.
Goetz, A. and Szyliowicz, J. (1997) Revisiting transportation planning and decision-making theory: the case of Denver International Airport. *Transportation Research A* 31, 263–80.
Goetz, A., Ralston, B., Stutz, F. and Leinbach, T. (2004) Transportation geography. In Gaile, G. and Willmott, C. (eds), *Geography in America at the Dawn of the 21st Century*. Oxford: Oxford University Press, pp. 221–36.
Goh, M. (2002) Congestion management and electronic road pricing in Singapore. *Journal of Transport Geography* 10, 29–38.
Goldman, T. and Gorham, R. (2006) Sustainable urban transport: four innovative directions. *Technology in Society* 28, 261–73.
Goodwin, P. (1997a) Asking questions. *Transport Policy* 4, 69–71.
Goodwin, P. (1997b) *Solving Congestion (When We Must Not Build Roads, Increase Spending, Lose Votes, Damage the Economy or Harm the Environment, and Will Never Find*

Equilibrium). Inaugural Lecture for the Professorship of Transport Policy, University College London, 23 October.

Goodwin, P. (1999) Transformation of transport policy in Great Britain. *Transportation Research Part A* 33, 655–69.

Goodwin, P. (2002) Are fuel prices important? In Lyons, G. and Chatterjee, K. (eds), *Transport Lessons from the Fuel Tax Protests of 2000*. Aldershot: Ashgate.

Goodwin, P. (ed.) (1995) *Car Dependence*. London: RAC Foundation for Motoring and the Environment.

Goodwin, P., Hallet, S., Stokes, P. and Kenny, G. (1991) *Transport: the New Realism*. Oxford: Transport Studies Unit, University of Oxford.

Gordon, P. and Richardson, H. (1996) Beyond polycentricity: the dispersed metropolis, Los Angeles, 1970–1980. *Journal of the American Planning Association* 62, 289–95.

Gordon, P., Richardson, H. and Yu, G. (1998) Metropolitan and non-metropolitan employment trends in the US: recent evidence and implications. *Urban Studies* 35, 1037–58.

Gorter, C., Nijkamp, P. and Rietveld, P. (1993) Barriers to employment – entry and reentry possibilities of unemployed job seekers in the Netherlands. *The Economist* 141, 70–95.

Goss, J. (2005) Consumption geographies. In Cloke, P., Crang, P. and Goodwin, M. (eds), *Introducing Human Geographies*, 2nd edition. London: Arnold, pp. 253–66.

Gössling, S. (2000) Sustainable tourism development in developing countries: some aspects of energy-use. *Journal of Sustainable Tourism* 8, 410–25.

Gössling, S. (2002a) Global environmental consequences of tourism. *Global Environmental Change* 12, 283–302.

Gössling, S. (2002b) Human-environmental relations within tourism. *Annals of Tourism Research* 29, 539–56.

Gössling, S., Borgström Hansson, C., Hörstmeier, O. and Saggel, S. (2002) Ecological footprint analysis as a tool to assess tourism sustainability. *Ecological Economics* 43, 199–211.

Gössling, S., Peeters, P., Ceron, J-P., Dubois, G., Patterson, T. and Richardson, R. (2005) The eco-efficiency of tourism. *Ecological Economics* 54, 417–34.

Graham, A. (2001) *Managing Airports: an International Perspective*. Oxford: Butterworth Heinemann.

Graham, B. (1995) *Geography and Air Transport*. Chichester: John Wiley.

Graham, B. (1998) Liberalization, regional economic development and the geography of demand for air transport in the European Union. *Journal of Transport Geography* 6, 87–104.

Graham, B. (2003) Air transport policy: reconciling growth and sustainability? In Docherty, I. and Shaw, J. (eds), *A New Deal for Transport? The UK's Struggle with the Sustainable Transport Agenda*. Oxford: Blackwell, pp. 198–225.

Graham, B. and Vowles, T. (2006) Carriers within carriers: a strategic response to low-cost airline competition. *Transport Reviews* 26, 105–26.

Graham, D. (2006) *Wider Economic Benefits of Transport Improvements: Link Between Agglomeration and Productivity*. London: Department for Transport.

Graham, S. and Marvin, S. (2001) *Splintering Urbanism*. London: Routledge.

Gray, D. (2004) Rural transport and social exclusion: developing a rural transport typology. *Built Environment* 3, 172–81.

Gray, D., Farrington, J., Shaw, J., Martin, S. and Roberts, D. (2001) Car dependence in rural Scotland: transport policy, devolution and the impact of the fuel duty escalator. *Journal of Rural Studies* 17, 113–25.

Gray, D., Shaw, J. and Farrington, J. (2006) Community transport, social capital and social exclusion in rural areas. *Area* 38, 89–98.

Gray, D., Shaw, J. and Illingworth, L. (2006) *Potential Impacts of National Road User Charging in Rural Areas: Scoping Study*. Report for Rural Evidence Research Centre. www.rerc.ac.uk/findings/documents_transport/T2RoadPricing_Final.pdf. Accessed 14 December 2006.

Grayling, T. (2001) *Transport and Social Exclusion*. Paper presented to the Transport Statistics User Group, January.

Greene, D. and Wegener, M. (1997) Sustainable Transport. *Journal of Transport Geography* 5, 177–90.

Greene, D., Hopson, J. and Li, J. (2003) *Running Out Of and Into Oil: Analyzing Global Oil Depletion and Transition Through 2050*. Oak Ridge: US Department of Energy.

Grieco, M. (2002) A comment on the limitations of transport policy. *Transport Reviews* 22, 509–10.

Grieco, M. and Turner, J. (1997) *Gender, Poverty and Transport: a Call for Policy Action*. Address delivered at UN International Forum on Urban Poverty, Florence.

Grieco, M., Pickup, L. and Whipp, G. (1989) *Gender, Transport and Employment*. Aldershot: Gower.

Grieco, M., Turner, J. and Hine, J. (2000) Transport, employment and social exclusion. *Local Work*, 26.

Griffiths, C. and Fitzpatrick, J. (eds) (2001) *Geographic Variations in Health*. London: National Statistics Office.

Grubler, A. and Nakicenovic, N. (1991) *Evolution of Transport Systems*. Vienna: IIASA Laxenburg.

Guest, A. (1999) *Some Evolving Thoughts on Leo F. Schnore as a Social Scientist*. Center for Demography and Ecology University of Wisconsin-Madison. CDE Working Paper No. 99-07.

Gurran, N., Squires, C. and Blakely, E. (2005) *Meeting the Sea Change Challenge: Best Practice Models of Local and Regional Planning for Sea Change Communities in Coastal Australia*. University of Sydney, Sydney.

Gutiérrez, J. and Garcia Palomares, J. (2007) New spatial patterns of mobility within the metrolpolitan area of Madrid: towards more complex and dispersed flow networks. *Journal of Transport Geography* 15, 18–30.

Gwilliam, K. (2003) Urban transport in developing countries. *Transport Reviews* 23, 197–216.

Haggett, P. (2001) *Geography: a Global Synthesis*. Harlow: Prentice Hall.

Halden, D. (2002) Using accessibility measures to integrate land use and transport policy in Edinburgh and the Lothians. *Transport Policy* 9, 313–24.

Hall, C. and Müller, D. (eds) (2004) *Tourism, Mobility and Second Homes: Between Elite Landscape and Common Ground*. Clevedon: Channel View.

Hall, D. (ed.) (1993) *Transport and Economic Development in the New Central and Eastern Europe*. London: Belhaven Press.

Hall, D. (1999) Conceptualising tourism transport: inequality and externality issues. *Journal of Transport Geography*, 7, 181–8.

Hall, D. (2004a) Towards a gendered transport geography. *Journal of Transport Geography* 12, 245–7.

Hall, D. (2004b) Transport and tourism: equity and sustainability issues. In Lumsdon, L. and Page, S. (eds), *Tourism and Transport: Issues and Agenda for the New Millennium*. Amsterdam: Elsevier, pp. 45–55.

Hall, D. (2004c) Branding and national identity: the case of Central and Eastern Europe. In Morgan, N., Pritchard A. and Pride, R. (eds), *Destination Branding: Creating the Unique Destination Proposition*, 2nd edition. Amsterdam: Elsevier, pp. 111–27.

Hall, D. (2005) Transport tourism. In Novelli, M (ed.) *Niche Tourism: Contemporary Issues, Trends and Cases*. Amsterdam: Elsevier, pp. 89–98.
Hall, D. and Brown, F. (2006) *Tourism and Welfare: Ethics, Responsibility and Sustained Well-being*. Wallingford: CABI Publishing.
Hall, P., Hesse, M. and Rodrigue, J.-P. (2006) Reexploring the interface between economic and transport geography: guest editorial. *Environment and Planning A* 38, 1401–8.
Halsall, D. (1992) Transport for tourism and recreation. In Hoyle, B. and Knowles, R. (eds), *Modern Transport Geography*. London: Belhaven Press, pp. 155–77.
Halsall, D. (2001) Railway heritage and the tourist gaze: Stoomtram Hoorn-Medemblik. *Journal of Transport Geography* 9, 151–60.
Hamilton, K. and Jenkins, L. (1992) Women and Transport. In Roberts, J., Clearly, J., Hamilton, K. and Hanna, J. (eds), *Travel Sickness: the Need for a Sustainable Transport Policy for Britain*. London: Lawrence and Wishart, pp. 57–74.
Hamilton, K., Hoyle, S. and Jenkins, L. (2000) *The Public Transport Gender Audit*. London: TSO.
Hamilton, K., Jenkins, L., Hodgson, F. and Turner, J. (2005) *Promoting Gender Equality in Transport*. Working Paper Series No. 34. Manchester: Equal Opportunities Commission.
Hanlon, S. (1996) *Where do Women Feature in Public Transport?* Washington, DC: US Department of Transportation, Federal Highway Administration.
Hansen, L. (2005) Impacts of infrastructure investment on logistics and transport – examples from the fixed links of the Great Belt and Oresund in Denmark. In Thomsen, T., Nielsen, L. and Gudmundsson, H. (eds), *Social Perspectives on Mobility*. Aldershot: Ashgate, pp. 67–88.
Hanson, S. (1998) Off the road? Reflections on transportation geography in the information age. *Journal of Transport Geography* 6, 241–50.
Hanson, S. (2004) The context of urban travel: concepts and recent trends. In Hanson, S. and Giuliano, G. (eds), *The Geography of Urban Transportation*, 3rd edition. New York: Guilford Press.
Hanson, S. (ed.) (1986) *The Geography of Urban Transportation*. New York: The Guilford Press.
Hanson, S. and Giuliano, G. (eds) (2004) *The Geography of Urban Transportation*, 3rd edition. New York: The Guilford Press.
Haq, G. (1997) *Towards Sustainable Transport Planning: a Comparison between Britain and the Netherlands*. Aldershot: Ashgate.
Harris, A. (1971) *Handicapped and Impaired in Great Britain, Part 1*. London: HMSO.
Harvey, D. (1973) *Social Justice and the City*. London: Edward Arnold.
Harvey, D. (1982) *The Limits to Capital*. Oxford: Blackwell.
Harvey, D. (1989) *The Condition of Postmodernity*. Oxford: Blackwell.
Harvey, D. (2005) *A Brief History of Neoliberalism*. Oxford: Oxford University Press.
Hashimoto, Y. (2005) The spread of cellular phones and their influence on young people in Japan. In: Kim, S. (ed.), *When Mobile Came: the Cultural and Social Impact of Mobile Communication*. Seoul: Communication Books, pp. 198–211.
Hass-Klau, C. (1993) Impact of pedestrianization and traffic calming on retailing. A review of the evidence from Germany and the UK. *Transport Policy* 1, 21–31.
Hay, A. (1973) *Transport for the Space Economy: a Geographical Study*. London: Macmillan.
Hay, A. (1995) Concepts of equity, fairness and justice in geographical studies. *Transactions of the Institute of British Geographers* 20, 500–8.

Hay, A. (2000) Transport geography. In Johnston, R., Gregory, D., Pratt, G. and Watts, M. (eds), *The Dictionary of Human Geography*. Oxford: Blackwell.
Haydock, D. (1995) *High Speed in Europe*. Sheffield: Platform 5 Publishing.
Haydock, D. (2007) New Brussels infrastructure speeds Eurostar and Thalys. *Today's Rilways Europe* 134, 19–21.
Hayek, F. (1960) *The Constitution of Liberty*. London: Routledge.
Hayek, F. (1973) *Economic Freedom and Representative Government*. London: Wincott Foundation/Institute of Economic Affairs.
Haynes, K., Gifford, J., Pelletiere, D., Lakshmanan, T. and Anderson, W. (2005) Sustainable transportation institutions and regional evolution: global and local perspectives. *Journal of Transport Geography* 13, 207–21.
Healey, P. (2001) Personal communication. Cited in Graham, S. and Marvin, S. (2001) *Splintering Urbanism*. London: Routledge, London, p. 205.
Heaver, T. (2002) The evolving roles of shipping lines in international logistics. *International Journal of Maritime Economics* 4, 210–30.
Helminen, V. and Ristimäki, M. (2007) Relationships between commuting distance, frequency and telework in Finland. *Journal of Transport Geography* 15, 331–42.
Hensher, D. and Button, K. (eds) (2003) *Handbook of Transport and the Environment*. Amsterdam: Elsevier.
Hensher, D., Button, K., Haynes, K. and Stopher, P. (2004) *Handbook of Transport Geography and Spatial Systems*. Oxford: Elsevier Science.
Herod, A. (1998) Discourse on the docks: containerization and inter-union work disputes in US ports, 1955–1985. *Transactions of the Institute of British Geographers* 23, 177–91.
Herod, A. (2003) Scale: the local and the global. In Holloway, S., Rice, S. and Valentine, G. (eds), *Key Concepts in Geography*. London: Sage, pp. 213–35.
Hesse, M. (2004) Logistics and freight transport policy in urban areas: a case study of Berlin Brandenburg/Germany. *European Planning Studies* 12, 1035–53.
Hesse, M. and Rodrigue, J.-P. (2004) The transport geography of logistics and freight distribution. *Journal of Transport Geography* 12, 171–84.
Hesse, M. and Rodrigue, J.-P. (2006) Transportation and global production networks. *Growth and Change* 37, 599–609.
Hickman, R. and Banister, D. (2006) *Visioning and Backcasting for UK Transport Policy*. London: DfT.
Hilling, D. (1996) *Transport in Developing Countries*. London: Routledge.
Hilling, D. and Browne, M. (1998) Ships, ports and bulk freight transport. In Hoyle, B. and Knowles, R. (eds) *Modern Transport Geography*, 2nd edition. Chichester: Wiley.
Hillman, M. (2004) *How Can We Save the Planet?* Harmondsworth: Penguin.
Hillman, M., Henderson, I. and Whalley, A. (1976) *Transport Realities and Planning Policy*. Political and Economic Planning, Broadsheet 567, London.
Himanen, V., Lee-Gosselin, M. and Perrels, A. (2005) Sustainability and the interactions between external effects of transport. *Journal of Transport Geography* 13, 23–8.
Hine, J. and Grieco, M. (2003) Scatters and clusters in time and space: implications for delivering integrated and inclusive transport. *Transport Policy* 10, 299–306.
Hine, J. and Mitchell, F. (2001) *The Role of Public Transport in Social Exclusion*. Edinburgh: Scottish Executive Central Research Unit.
Hine, J. and Mitchell, F. (2003) *Transport Disadvantage and Social Exclusion: Exclusionary Mechanisms in Transport*. Aldershot: Ashgate.
Hine, J. and Scott, J. (2001) Seamless, accessible travel: users' views of the public transport journey and interchange. *Transport Policy* 7, 217–26.

Hjorthol, R. (2000) Same city – different options. An analysis of the work trips of married couples in the metropolitan area of Oslo. *Journal of Transport Geography* 8, 213–20.

HM Treasury (2006) *Stern Review on the Economics of Climate Change*. London: HM Treasury. www.hm-treasury.gov.uk/independent_reviews/stern_review_economics_climate_change/stern_review_report.cfm. Accessed 1 February 2007.

HM Treasury and Department for Transport (2003) *Aviation and the Environment*. London: DfT.

Hodge, D. and Koski, H. (1997) Information and communication technologies and transportation: European–US collaborative and comparative research possibilities. *Journal of Transport Geography* 5, 191–7.

Höhmann, R. (2002) The future of heritage railways and rail conservation in Germany. *Japan Railway and Transport Review* 30, 20–2.

Höjer, M. and Mattsson, L.-G. (2000) Determinism and backcasting in future studies. *Futures* 2, 613–34.

Holden, E. and Høyer, K. (2005) The ecological footprints of fuels. *Transportation Research Part D: Transport and Environment* 10, 395–403.

Holding, D. and Kreutner, M. (1998) Achieving a balance between 'carrots' and 'sticks' for traffic in national parks: the Bayerischer Wald Project. *Transport Policy* 5, 175–83.

Holzer, H. (1991) The spatial mismatch hypothesis: what has the evidence shown? *Urban Studies* 28, 105–22.

Horner. M. and Casas, I. (2006) An introduction to assessments of research needs in transport geography. *Journal of Transport Geography* 14, 228–9. See also pages 230–242 and vol. 14, 384–400.

House of Lords Select Committee on the European Communities (1997) Seventh Report, Session 1997–98, Community Railway Strategy. *HL Paper 46*. London: HMSO.

Houston, D. (2001) Testing the spatial mismatch hypothesis in the United Kingdom using evidence from firm relocations. *European Research in Regional Science* 11, 134–51.

Høyer, K. (2001) Sustainable tourism or sustainable mobility? The Norwegian case. *Journal of Sustainable Tourism* 8, 147–60.

Høyer, K. (2006) *Sustainable Mobility in European Transport Politics: the Role of Technology*. Keynote address delivered at the 11th International Conference of the Hong Kong Society for Transportation Studies, 9–11 December.

Hoyle, B. (1994) A rediscovered resource: comparative Canadian perceptions of waterfront redevelopment. *Journal of Transport Geography* 2, 19–29.

Hoyle, B. (ed.) (1996) *Cityports, Coastal Zones and Regional Change: International Perspectives on Planning and Management*. Chichester: Wiley.

Hoyle, B. and Knowles, R. (eds) (1992) *Modern Transport Geography*. London: Belhaven Press.

Hoyle, B. and Knowles, R. (eds) (1998) *Modern Transport Geography*, 2nd edition. Chichester: Wiley.

Hoyle, B. and Smith, J. (1998) Transport and development: conceptual frameworks. In Hoyle, B. and Knowles, R. (eds), *Modern Transport Geography*, 2nd edition. Chichester: Wiley, pp. 13–40.

Hughes, M. (2006) High speed: TGV Est set to hit the headlines in June. Railway Gazette December 2006. www.railwaygazette.com/Articles/2006/12/01/3378/High+speed+TGV+Est+set+to+hit+the+headlines+in+June.html. Accessed 15 December 2006.

Hughes, P. (1993) *Personal Travel and the Greenhouse Effect*. London: Earthscan.

Hull, A. (2005) Integrated transport planning in the UK: from concept to reality. *Journal of Transport Geography* 13, 318–28.
Hunter, C. and Shaw, J. (2007) The ecological footprint as a key indicator of sustainable tourism. *Tourism Management* 28, 46–57.
Hunter, C., Farrington, J. and Walton, W. (1998) Transport and the environment. In Hoyle, B. and Knowles, R. (eds), *Modern Transport Geography*, 2nd edition. Chichester: Wiley, pp. 97–114.
Hurni, A. (2005) *Transport and social exclusion in western Sydney*. Paper presented at the Australasian Transport Research Forum, NSW, Sydney.
Huws, U., Jagger, N. and O'Regan, S. (1999) *Teleworking and Globalisation*. IES Report 358. Brighton: Institute for Employment Studies.
Ieromonachou, P., Potter, S. and Warren, J. (2006) Norway's urban toll rings: evolving towards congestion charging. *Transport Policy* 13, 29–40.
Ihlandfeldt, K. and Sjoquist, D. (1998) The spatial mismatch hypothesis: a review of recent studies and their implications for welfare reform. *Housing Policy Debate* 9, 849–92.
Imashiro M. (1997) Changes in Japan's transport market and JNR privatization. *Japan Railway and Transport Review* September, 50–3.
Indian Railways Fan Club (2005) *Welcome to the IRFCA Server!* IRFCA. www.irfca.org/
Interessengemeinschaft Eisenbahn (2006) *Eisenbahn-Souvenirartikel*. IGE Bahntouristik, Hersbruck. www.bahntouristik.de/cms/?cms_p=20l&cms_c=81&cms_a=81.
Intergovernmental Panel on Climate Change (IPCC) (1999) *Aviation and the Global Atmosphere*. Cambridge: Cambridge University Press.
International Air Transport Association (IATA) (2006) *New Financial Forecast, September 2006*. Montreal: IATA.
Jaakson, R. (2004) Beyond the tourist bubble? Cruiseship passengers in port. *Annals of Tourism Research* 31(1), 44–60.
Jackson, C. (2005) Three railway packages herald uncertainty across the continent. *Railway Gazette* January, 25–8.
Jacobs, J. (1961) *The Death and Life of Great American Cities*. New York: Random House.
Jacobs, J. (1968) *The Death and Life of Great American Cities*. London: Penguin.
Jacobsen, J. and Haukeland, J. (2002) A lunch with a view: motor tourists' choices and assessments of eating places. *Scandinavian Journal of Hospitality and Tourism* 2, 4–16.
Jain, J. (2004) *Networks of the Future: Time, Space and Rail Travel*. Unpublished PhD thesis, Lancaster University.
Jain, J. and Lyons, G. (2005) Travel time: gift or burden? *Proceedings of 37th Universities Transport Study Group Conference*. Bristol, 5–7 January 2005, 3B1.1–3B1.2.
Jain, J. and Lyons, G. (2008) The gift of travel time. *Journal of Transport Geography*. In press.
James, P. (2003) *Is Teleworking Sustainable? An Analysis of Its Economic, Environmental and Social Impacts*. www.sustel.org. Accessed 10 December 2006.
Janelle, D. (1969) Spatial reorganisation: a model and concept. *Annals, Association of American Geographers* 59, 348–64.
Janelle, D. (1997) Sustainable transportation and information technology: suggested research issues. *Journal of Transport Geography* 5, 39–40.
Janelle, D. and Beuthe, M. (1997) Globalization and research issues in transportation. *Journal of Transport Geography* 5, 199–206.
Janelle, D. and Gillespie, A. (2004) Space-time constructs for linking information and communication technologies with issues in sustainable transport. *Transport Reviews* 24, 665–78.

Jensen, A. (2005) The institutionalisation of European transport policy from a mobility perspective. In Thomsen, T., Nielsen, L. and Gudmundsson, H. (eds), *Social Perspectives on Mobility*. Aldershot: Ashgate, pp. 127–53.

Jensen, M. (1999) Passion and heart in transport: a sociological analysis on transport behaviour. *Transport Policy* 6, 19–33.

Jessop, B. (1990) Regulation theories in retrospect and prospect. *Economy and Society* 19, 153–216.

Jessop, B. (1997) Capitalism and its futures: remarks on regulation, government and governance. *Review of International Political Economy* 4, 561–81.

Jessop, B. (2004) Hollowing out the 'nation-state' and multilevel governance. In: *A Handbook of Comparative Social Policy*. Cheltenham: Edward Elgar Publishing, pp. 11–25.

Johnston, R. (2003) Order in space: geography as a discipline in distance. In Johnston, R. and Williams, M. (eds), *A Century of British Geography*. Oxford: Oxford University Press.

Johnston, R., Taylor, P. and Watts, M. (eds) (1995) *Geographies of Global Change: Remapping the World in the Late Twentieth Century*. Oxford: Blackwell.

Jones, D. (1990) Noise, stress and human behaviour. *Environmental Health* 98, 206–8.

Jopson, A. (2004) *Can psychology help to reduce car use revisited.* 36th UTSG annual conference. University of Newcastle upon Tyne, Volume 1.

Jun, M.-J. (2004) The effects of Portland's urban growth boundary on urban development patterns and commuting. *Urban Studies* 41, 1333–48.

Kagermeier, A. (ed.) (2004) *Verkehrssystem- und Mobilitätsmanagement im ländlichen Raum*. Mannheim: Verlag MetaGIS Infosysteme.

Kain, J. (1968) Housing segregation, negro unemployment and metropolitan segregation. *Quarterly Journal of Economics* 82, 175–97.

Kain, J. (1992) The spatial mismatch hypothesis: three decades later. *Housing Policy Debate* 3, 371–460.

Kapp, K. (1988) *Soziale Kosten der Marktwirtschaft: d. klass. Werk d. Umwelt-Ökonomie*. Frankfurt am Main: Fischer.

Keeling, D. (2007) Transportation geography: new directions on well-worn trails. *Progress in Human Geography* 31, 217–25.

Kenyon, S. (2003). Greater social inclusion: what are the consequences for transport? *Proceedings of the 35th Universities Transport Study Group Annual Conference*, Loughborough, 6–8 January.

Kenyon, S., Lyons, G. and Rafferty, J. (2002) Transport and social exclusion: investigating the possibility of promoting inclusion through virtual mobility. *Journal of Transport Geography* 10, 207–19.

Kenyon, S., Rafferty, J. and Lyons, G. (2003) Social exclusion and transport: a role for virtual accessibility in the alleviation of mobility-related social exclusion? *Journal of Social Policy* 32, 317–38.

Kerr, D. (2005) *Building a Health Service Fit for the Future*. Edinburgh: Scottish Executive.

Kessel, A. (2006) *Air, the Environment and Public Health*. Cambridge: Cambridge University Press.

Kesselring, S. (2004) *The Making of the Cosmobilities Network*. www.cosmobilities.net/downloads/Cosmobilities%20Workshops/2005/Kesselring_making_cosmobnet_oct_2005.pdf. Accessed 29 January 2007.

Kidder, A. (1989) Passenger transportation problems in rural areas. In Gillis, W. (ed.), *Profitability and Mobility in Rural America*. University Park: Pennsylvania State University Press, pp. 131–44.

Kilvington, R. and Cross, A. (1986) *Deregulation of Express Coach Services in Britain*. Aldershot: Gower.

Kingham, S., Zant, T. and Johnston, D. (2004) The impact of the minimum driver licensing age on mobility in New Zealand. *Journal of Transport Geography* 12, 301–14.

Kitagawa, T. (2005) Extending the Shinkansen network. *Japan Railway and Transport Review* 40, 14–17.

Klein, S. and Loebbecke, C. (2003) Emerging pricing strategies on the web: lessons from the airline industry. *Electronic Markets* 13, 46–58.

Knowles, R. (1993) Research agendas in transport geography for the 1990s. Editorial introduction. *Journal of Transport Geography* 1, 3–11.

Knowles, R. (2000) The Great Belt Fixed Link and Denmark's transition from inter-island sea to land transport, *Geography* 85, 345–54.

Knowles, R. (2004) Impacts of privatising Britain's rail passenger services – franchising, refranchising and Ten Year Transport Plan targets. *Environment and Planning A* 36, 2065–87.

Knowles, R. (2006a) Transport shaping space: differential collapse in time-space. *Journal of Transport Geography* 14, 407–25.

Knowles, R. (2006b) Transport impacts of the Øresund (Copenhagen to Malmö) fixed link. *Geography* 91, 227–40.

Knowles, R. and Hall, D. (1998) Transport deregulation and privatization. In Hoyle, B. and Knowles, R. (eds), *Modern Transport Geography*, 2nd edition. Chichester: John Wiley, pp. 75–96.

Knox, P. and Agnew, J. (1994) *The Geography of the World Economy*, 2nd edition. London: Edward Arnold.

Knox, P., Agnew, J. and McCarthy, L. (2003) *The Geography of the World Economy*, 4th edition. London: Arnold.

Ko, T. (2005) Development of a tourism sustainability assessment procedure: a conceptual approach. *Tourism Management* 26, 431–45.

Kohn, T. (1997) Island involvement and the evolving tourist. In Abram, S., Waldren, J. and Macleod, D. (eds), *Tourists and Tourism: Identifying with People and Places*. Oxford: Berg, pp. 13–28.

Konings, R. and Remmelt, T. (2001) Foldable containers: a new perspective on reducing container-repositioning costs. *European Journal of Transport and Infrastructure Research* 1, 333–52.

Kotkin, J. (2000) *The New Geography: How the Digital Revolution is Reshaping the American Landscape*. New York: Random House.

Kotler, P., Haider, D. and Rein, I. (1993) *Marketing Places: Attracting Industry, Investment and Tourism to Cities, States and Nations*. Glencoe, IL: The Free Press.

Kovacs, G. and Spens, K. (2006) Transport infrastructure in the Baltic States post-EU succession. *Journal of Transport Geography* 14, 426–36.

Kresl, P. and Singh, B. (1999) Competitiveness and the urban economy: twenty four large US metropolitan areas. *Urban Studies* 36, 1017–27.

Krugman, P. (1991) *Geography and Trade*. Cambridge, MA: Harvard University Press.

Kumar, S. and Hoffmann, J. (2002) Globalization: the maritime nexus. In Grammenos, C. (ed.), *The Handbook of Maritime Economics and Business*. London: Lloyd's, pp. 35–62.

Kwan, M.-P. (2006) *Mobile Communications and Transport Geography*. Paper presented at the Annual Meeting of the Association of American Geographers, Chicago, 7–11 March.

Lajunen, T., Parker, D. and Stradling, S. (1998) Dimensions of driver anger, aggressive and Highway Code violations and their mediation by safety orientation in UK drivers. *Transportation Research Part F: Traffic Psychology and Behaviour* 1, 107–21.

Lamb, B. and Davidson, S. (1996) Tourism and transportation in Ontario, Canada. In Harrison, L. and Husbands, W. (eds), *Practising Responsible Tourism*. Chichester and New York: John Wiley & Sons, pp. 261–76.

Lang, R. (2003) *Edgeless Cities: Exploring the Elusive Metropolis*. Washington, DC: Brookings Institution Press.

Lang, R. and Dhavale, D. (2005) America's megapolitan Areas. *Land Lines*. July, 1–4.

Larsen, J., Urry, J. and Axhausen, K. (2006a) *Social Networks and Future Mobilities*. Report to the Horizons Programme of Department for Transport. Department of Sociology, University of Lancaster and IVT. Lancaster and Zürich: ETH Zürich.

Larsen, J., Urry, J. and Axhausen, K. (2006b) *Mobilities, Networks, Geographies*. Aldershot: Ashgate.

Larsen, O. (1995) The toll cordons in Norway: an overview. *Journal of Transport Geography* 3, 187–97.

Lauber, V. (2002) The sustainability of freight transport across the Alps: European Union policy in controversies on transit traffic. In Lenschow, A. (ed.), *Environmental Policy Integration: Greening Sectoral Policies in Europe*. London: Earthscan, pp. 153–74.

Laurier, E. (2004) Doing office work on the motorway. *Theory, Culture & Society* 21, 261–77.

Law, R. (1999) Beyond 'women and transport': towards new geographies of gender and daily mobility. *Progress in Human Geography* 23, 567–88.

Lawless, P. (1995) Inner-city and suburban labour markets in a major English conurbation – processes and policy implications. *Urban Studies* 32, 1097–125.

Lawton, T. (2002) *Cleared for Take-off: Structure and Strategy in the Low Fare Airline Business*. Aldershot: Ashgate.

Lawton, T. (2003) Managing proactively in turbulent times: insights from the low-fare airline business. *Irish Journal of Management* 24, 173–93.

Lawton, T. and Solomko, S. (2005) When being the lowest cost is not enough: building a successful low-fare business model in Asia. *Journal of Air Transport Management* 11, 355–62.

Le Clercq, F. and Bertolini, L. (2003) Achieving sustainable accessibility: an evaluation of policy measures in the Amsterdam area. *Built Environment* 29, 36–47.

Le Grand, J. (1991) *Equity and Choice*. London: HarperCollins.

Lee, P. and Murie, A. (1999) *Literature Review of Social Exclusion*. Edinburgh: Scottish Office.

Lee, Y. (2006) On the wrong track: China's transportation revolution and urban air pollution. *Harvard International Review* hir.harvard.edu/articles/1435/. Accessed 25 January 2007.

Leinbach, T. (1995) Transportation and third world development: review, issues and prescription. *Transport Research* 20 A, 337–44.

Leinbach, T. (2000) Mobility in development context: changing perspectives, new interpretations and the real issues. *Journal of Transport Geography* 8, 1–9.

Lenschow, A. (2002) Greening the European Union. In Lenschow, A (Ed) *Environmental Policy Integration: Greening Sectoral Policies in Europe*. London: Earthscan, pp. 3–21.

Letherby, G. and Reynolds, G. (2005) *Train Tracks: Work, Play and Politics on the railways*. Oxford: Berg.

Leyshon, A. (1992) The transformation of regulatory order – regulating the global economy and environment. *Geoforum* 23, 249–67.

Leyshon, A. (1995) Annihilating space? The speed-up of communications. In Allen, J. and Hamnett, C. (eds), *A Shrinking World? Global Unevenness and Iinequality*. Oxford: Oxford University Press, Oxford, pp. 11–54.

Leyshon, A. and Thrift, N. (1995) Geographies of financial exclusion: financial abandonment in Britain and the United States. *Transactions of the Institute of British Geographers* 20, 312–41.

LGV Est européene (2006) LGV Est européene website: www.lgv-est.com. Accessed 22 December 2006.

LGV Rhin-Rhône website (2006) www.lgvrhinrhone.com/english.php. Accessed 22 December 2006.

Limtanakool, N., Dijst, M. and Schwanen, T. (2006a) On the participation in medium- and long-distance travel: a decomposition analysis for the UK and the Netherlands. *Tijdschrift voor Economische en Sociale Geografie* 97, 389–404.

Limtanakool, N., Dijst, M. and Schwanen, T. (2006b) The influence of socioeconomic characteristics, land use and travel time considerations on mode choice for medium- and longer-distance trips. *Journal of Transport Geography* 14, 327–41.

Ling, D. and Mannion, R. (1995) Improving older people's mobility and quality of life: an assessment of the economic and social benefits of dial-a-ride. Proceedings of 7th International Conference Mobility and Transport for Elderly and Disabled People, 16–19 July, Reading, UK, 331–9.

Link, H. (2005) Transport accounts – methodological concepts and empirical results. *Journal of Transport Geography* 13, 41–57.

Linneker, B. (1997) *Transport Infrastructure and Regional Economic Development in Europe: a Review of Theoretical and Methodological Approaches*. TRP 133. Dept of Town and Regional Planning, University of Sheffield.

Liverman, D. (2004) Who governs, at what scale and at what price? Geography, environmental governance, and the commodification of nature. *Annals, Association of American Geographers* 94, 734–8.

Loo, B. and Liu, K. (2005) A geographical analysis of potential railway load centers in China. *Professional Geographer* 57(4), 558–79.

Lubchenco, J. (1998) Entering the century of the environment: a new social contract for science. *Science* 279, 491–7.

Lucas, K. (ed.) (2004) *Running on Empty: Transport, Social Exclusion and Environmental Justice*. Bristol: The Policy Press.

Lucas, K., Grosvenor, T. and Simpson, R. (2001) *Transport, the Environment and Social Exclusion*. York: York Publishing Services.

Lumsdon, L. (2000) Transport and tourism: a sustainable tourism development model. *Journal of Sustainable Tourism* 8, 1–17.

Lumsdon, L. (2006) Factors affecting the design of tourism bus services. *Annals of Tourism Research* 33, 748–66.

Lumsdon, L. and Page, S. (2004) Progress in transport and tourism research: reformulating the transport-tourism interface and future research agendas. In: Lumsdon, L. and Page, S. (eds), *Tourism and Transport: Issues and Agenda for the New Millennium*. Amsterdam: Elsevier, 1–28.

Lutter, H. (1980) Raumwirksamkeit von Fernstraßen. Eine Einschätzung des Fenrstraßenbaus als Instrument zur Raumentwicklung unter heutigen Bedingungen. Forschungen zur Raumentwicklung Bd 8. Bonn: Bundesforschungsanstalt für Landeskunde und Raumordnung.

Lyons, G. (2004). Transport and society. *Transport Reviews* 24, 485–509.

Lyons, G. (2005) It's time we tried to understand more about what people do with their travel time. *Local Transport Today* 411, 18.

Lyons, G. and Chatterjee, K. (2007) A human perspective on the daily commute: costs, benefits and trade-offs. *Proceedings of the 39th Universities Transport Study Group Annual Conference*. Harrogate, 3–5 January.

Lyons, G. and Chatterjee, K. (eds) (2002) *Transport Lessons from the Fuel Tax Protests of 2000*. Aldershot: Ashgate.

Lyons, G. and Urry, J. (2005) Travel time use in the information age. *Transportation Research Part A* 39, 257–76.

Lyons, G., Chatterjee, K., Marsden, G. and Beecroft, M. (2000) *Society and Lifestyles*. The first of eight reports from the Transport Visions Network. London: Landor Publishing.

Lyons, G., Jain, J. and Holley, D. (2007) The use of travel time by rail travellers in Great Britain. *Transportation Research Part A* 41, 107–20.

MacKinnon, D. and Cumbers, A. (2007) *An Introduction to Economic Geography: Globalisation, Uneven Development and Place*. Harlow: Pearson.

Maddison, D., Johansson, O. and Pearce, D. (1996) *Blueprint 5: the True Costs of Road Transport*. London: Earthscan.

Mageean, J. and Nelson, J. (2003) The evaluation of demand responsive transport services in Europe. *Journal of Transport Geography* 11, 255–70.

Marchetti, C. (1994) Anthropological invariants in travel behavior. *Technological Forecasting and Social Change* 47, 75–88.

Marell, A. and Westin, K. (2002) The effects of taxicab deregulation in rural areas of Sweden. *Journal of Transport Geography* 10, 135–44.

Martin, J., Meltzer, H. and Eliot, D. (1988) *OPCS Surveys of Disability in Great Britain: Report 1, the Prevalence of Disability among Adults*. Office of Population Census and Surveys. London: HMSO.

Martin, R. (1989) The new regional economics and the politics of regional restructuring. In Albrechts, L., Moulaert, F., Roberts, P. and Swyngedouw, E. (eds), *Regional Policy at the Crossroads: European Perspectives*. London: Jessica Kingsley, pp. 27–51.

Massey, D. (1984) *Spatial Divisions of Labour: Social Structures and the Geography of Production*. London: Macmillan.

Massey, D. (2005) *For Space*. London: Sage.

Matthiessen, C. (2000) Bridging the Øresund: potential regional dynamics. Integration of Copenhagen (Denmark) and Malmö–Lund (Sweden). A cross-border project on the European metropolitan level. *Journal of Transport Geography* 8, 171–80.

MÁV Nostalgia (2000) *MÁV Nostalgia Tourist, Trading and Services Ltd*. Budapest: MÁV Nostalgia Tourist, Trading and Services Ltd. www.miwo.hu/partner/old_trains/eindex.htm.

MÁV Nostalgia (nd) *Trip to the Gypsy Village of Solt*. Budapest: MÁV Nostalgia Tourist, Trading and Services Ltd.

Mawdsley, E. and Rigg, J. (2003) The world development report II: continuity and change in development orthodoxies. *Progress in Development Studies* 3, 271–86.

Maxwell, S. (2001) Negotiating car use in everyday life. In Miller, D (ed.), *Car Cultures*. Oxford: Berg.

McCormick, J. (2001) *Environmental Policy in the European Union*. Basingstoke: Palgrave.

McGee, T. (1994) Labour force change and mobility in the extended metropolitan regions of Asia. In Fuchs, R., Brennan, E., Chamie, J., Lo, F.-C. and Uitto, J (eds), *Mega City Growth and the Future*. Tokyo: United Nations University Press, pp. 62–102.

McGregor, A., Fitzpatrick, I. and Glass, A. (1998) *Regeneration Areas and Barriers to Employment*. Edinburgh: Scottish Office.

McLellan, R. (1997) Bigger vessels: how big is too big? *Maritime Policy and Management* 24, 193–211.

Meinig, D. (1998) *The Shaping of America: a Geographical Perspective on 500 Years of History. Volume 3: Transcontinental America, 1850–1915*. New Haven, CT: Yale University Press.

Mesner, S. (2002) Teleworking promises fewer car trips, but what is it actually delivering? *Local Transport Today*, Issue 336.

Metz, D. (2004) Human mobility and transport policy. *Ingenia*, 18, 37–42.

Meyer, J. and Gomez-Ibanez, J. (1981) *Autos, Transit and Cities*. Cambridge, MA: Harvard University Press.

Ministry of Transport (1963) *Traffic in Towns*. London: HMSO.

Ministry of Transport (1964) *Road Pricing: the Economic and Technical Possibilities*. London: HMSO.

Minogue, K. (1998) Social justice in theory and practice. In Boucher, D. and Kelly, P. (eds), *Social Justice: from Hume to Walzer*. London: Routledge, pp. 253–66.

Mintel (2005a) *Airlines, Leisure Intelligence*. London: Mintel.

Mintel (2005b) *No-frills/Low-cost Airlines – UK, Leisure Intelligence*. London: Mintel.

Minten, B. and Kyle, S. (2002) The effect of distance and road quality on food collection, marketing margins, and traders' wages: evidence from the former Zaire. *Journal of Development Economics* 60, 467–95.

Mitchell, C. (1988) *Features on Buses to Assist Passengers with Mobility Handicaps*. Research Report RR137. Crowthorne: TRRL.

Mobile Herald (2006) *2006 Mobile Phone Statistics and Projections*. December 2006. www.mobileherald.com (accessed 16 December 2006).

Mokhtarian, P. (1990) A typology of relationships between telecommunications and transportation. *Transportation Research Part A: Policy and Practice* 24, 231–42.

Mokhtarian, P. (2000) Telecommunications and travel. *Transportation Research Board A1C08 Committee on Telecommunications and Travel Behavior*. January 2000. Publication No. UCD-ITS-RP-00-02.

Mokhtarian, P. and Salomon, I. (2001) How derived is the demand for travel? Some conceptual and measurement considerations. *Transportation Research Part A* 35, 695–719.

Monk, S., Dunn, J., Fitzgerald, M. and Hodge, I. (1999) *Finding Work in Rural Areas: Barriers and Bridges*. York: York Publishing Services.

Moorhouse, G. (1988) *Imperial City: the Rise and Rise of New York*. London: Hodder and Stoughton.

Morrell, P. (2005) Airlines within airlines: an analysis of US network airline responses to low cost carriers. *Journal of Air Transport Management* 11, 303–12.

Moseley, M. (1979) *Accessibility: the Rural Challenge*. London: Methuen.

Muller, P. (1997) The suburban transformation of the globalizing American city. *Annals, American Academy of Political and Social Science* 551, 44–58.

Murayama, Y. (1994) The impact of railways on accessibility in the Japanese urban system. *Journal of Transport Geography* 2, 87–100.

Murphy, R. (1974) *The American City: An Urban Geography*, 2nd edition. New York: McGraw-Hill.

Murray, M. and Graham, B. (1997) Exploring the dialectics of route-based tourism: the Camino de Santiago. *Tourism Management* 18, 513–24.

Naish, J. (2004) Baltics braced for new kids on the Bloc. *The Times Online* 1 May www.timesonline.co.uk/article/0,160-1092380,00.html.

Nash, C. (2005) Privatisation in transport. In Button, K. and Hensher, D. (eds), *Handbook of Transport Strategy, Policy and Institutions*. Amsterdam: Elsevier, pp. 97–113.

Nash, C. (2006) Europe: alternative models for restructuring. In Gómez Ibañez, J. and de Rus, G. (eds), *Competition in the Rail Industry: an International Comparative Analysis*. Cheltenham: Edward Elgar, pp. 25–48.

National Geospatial Intelligence Agency (2005) *World Port Index*. 18th edition.

Neff, J. (1998) *Government Investment in Transit before 1940*. Paper presented at the Annual Meeting of the Association of American Geographers, Boston.

Nelson A. and Moore, T. (1993) Assessing urban growth management: the case of Portland, Oregon, the USA's largest urban growth boundary. *Land Use Policy* 10, 293–302.

Nelson, J. (2006) Personal communication.

Newman, P. and Kenworthy, J. (1999) *Sustainability and Cities: Overcoming Automobile Dependence*. Washington, DC: Island Press.

Nicholls, S. (2004) Climate change and tourism. *Annals of Tourism Research* 31, 238–40.

Niskanen, W. (1973) *Bureaucracy: Servant or Master?* London: Institute of Economic Affairs.

Njenga, P. and Davis, A. (2003) Drawing the road map to rural poverty reduction. *Transport Reviews* 23, 217–41.

Noble, B. (2000) Strategic environmental assessment: What is it? What makes it strategic? *Journal of Environmental Assessment Policy and Management* 2, 203–24.

Notteboom, T. (2004) Container shipping and ports: an overview. *Review of Network Economics* 3, 86–106.

Notteboom, T. (2006) The time factor in liner shipping services. *Maritime Economics & Logistics* 8, 19–39.

Notteboom, T. and Merckx, F (2006) Freight integration in liner shipping: a strategy serving global production networks. *Growth and Change* 37, 550–69.

Notteboom, T. and Rodrigue, J.-P. (2005) Port regionalization: towards a new phase in port development. *Maritime Policy and Management* 32, 297–313.

Nuhn, H. and Hesse, M. (2006) *Verkehrsgeographie*. Paderborn: Schöningh.

Nutley, S. (1996) Rural transport problems and non-car populations in the USA: A UK perspective. *Journal of Transport Geography* 4, 93–106.

Nutley, S. (1998) Rural areas: the accessibility problem. In Hoyle, B. and Knowles R. (eds), *Modern Transport Geography*, 2nd edition. Chichester: Wiley, pp. 185–215.

Nutley, S. (2003) Indicators of transport and accessibility problems in rural Australia. *Journal of Transport Geography* 11, 55–71.

Nutley, S. (2005) Monitoring rural travel behaviour: a longitudinal study in Northern Ireland 1979–2001. *Journal of Transport Geography* 13, 247–63.

O'Connell, J. and Williams, G. (2005) Passengers' perceptions of low cost airlines and full service carriers: a case study involving Ryanair, Aer Lingus, AirAsia and Malaysia Airlines. *Journal of Air Transport Management* 11, 259–72.

O'Connor, K. (1995) Airport development in southeast Asia. *Journal of Transport Geography* 3, 269–79.

O'Connor, K. (2003) Global air travel: towards concentration or dispersal. *Journal of Transport Geography* 11, 83–92.

O'Connor, K. and Daniels, P. (2001) The geography of international trade in services: Australia and the APEC region. *Environment and Planning A* 33, 281–96.

O'Connor, K. and Healy, E. (2002) *The Links between Labour Markets and Housing Markets in Melbourne. Final Report*. Melbourne: Australian Housing and Urban Research Institute.

O'Reilly, D. (1989) *Concessionary Fares and Children's Travel Patterns: an Analysis Based on the 1978/1979 National Travel Survey*. Research Report 203. London: DoT.

O'Reilly, D. (1990) *An Analysis of Concessionary Bus Fare Schemes for OAPs Using the 1985/86 National Travel Survey*. Research Report 291. London: DoT.

Oberholzer-Gee, F. and Weck-Hannemann, H. (2002) Pricing road use: political-economic and fairness considerations. *Transportation Research Part D: Transport and Environment* 7, 357–71.

Office of National Statistics (ONS) (2002) www.statistics.gov.uk.cci/nugget.asp?id=8. Accessed 14 December 2006.

Office for National Statistics (ONS) (2005a) *Expenditure, Food and Family Spending 2004*. London: ONS.

Office for National Statistics (ONS) (2005b) *Home Based Working Using Communication Technologies*. London: ONS.

Office for National Statistics (ONS) (2006) *Internet access. Figures from National Statistics Omnibus Survey, Northern Ireland omnibus survey and survey of Internet service provider*. Published online August 2006. www.statistics.gov.uk (accessed 16 December 2006).

Office of Science and Technology (OST) (2006) *Intelligent Infrastructure Futures: Project Overview*. London: Department of Trade and Industry.

Olivier, D. and Slack, B. (2006) Rethinking the port. *Environment and Planning A* 38, 1409–27.

Ollivier-Trigalo, M. (2001) The implementation of major infrastructure projects: conflicts and co-ordination. In Giorgi, L. and Pohoryles, R. (eds), *Transport Policy and Research: What Future?* Aldershot: Ashgate, pp. 17–43.

Olsthoorn, X (2001) Carbon dioxide emissions from international aviation: 1950–2050. *Journal of Air Transport Management* 7, 87–93.

Olvera, L., Plat, D. and Pochet, P. (2003) Transportation conditions and access to services in a context of urban sprawl and deregulation. The case of Dar es Salaam. *Transport Policy* 10, 287–98.

Organisation for Economic Cooperation and Development (OECD) (1993) *Cars and Climate Change*. Paris: OECD.

Organisation for Economic Co-operation and Development (OECD) (2002a) *OECD Guidelines Towards Environmentally Sustainable Transport*. Paris: OECD.

Organisation for Economic Co-operation and Development (OECD) (2002b) *Policy Instruments for Achieving Environmentally Sustainable Transport*. Paris: OECD.

Organisation for Economic Cooperation and Development (OECD) (2006a) *Decoupling the Environmental Impacts of Transport from Economic Growth*. Paris: Organisation for Economic Cooperation and Development.

Organisation for Economic Cooperation and Development (OECD) (2006b) Broadband Statistics to June 2006. www.oecd.org/sti/ict/broadband (accessed 16 December 2006).

Orient Express (2006) *Venice Simplon-Orient-Express*. Venice Simplon Orient-Express London. www.orient-express.com/web/vsoe_a2a_home.jsp.

Ostróda-Elbląg Navigation (2003) *Ostróda-Elbląg Navigation 1860–2003*. Ostróda-Elbląg Navigation, Ostróda www.zegluga.com.pl/ang/index_ag.htm.

Owens, S. (1995) From 'predict and provide' to 'predict and prevent'? Pricing and planning in transport policy. *Transport Policy* 2, 43–50.

Oxley, P. (1977) Dial a ride in the UK, a general study. In Transport and Road Research Laboratory (eds), *Symposium on Unconventional Bus Services*. WP 23. Crowthorne: TRRL.

Oxley, P. and Benwell, M. (1985) *An Experimental Study of the Use of Buses by Elderly and Disabled People*. Research Report RR33. Crowthorne: TRRL.

Page, S. (1998) Transport for recreation and tourism. In Hoyle, B. and Knowles, R. (eds), *Modern Transport Geography*, 2nd edition. Chichester, UK and New York: John Wiley & Sons, pp. 217–40.

Page, S. (2005) *Transport and Tourism*, 2nd edition. Harlow: Pearson Educational.

Pain, R. (1997) Social geography of women's fear of crime. *Transactions of the Institute of British Geographers* 22, 231–44.

Painter, J. (1995) Regulation theory, post-fordism and urban politics. In Judge, D; Stoker, G and Wolman, H (1995) *Theories of Urban Politics*. London: Sage, pp. 276–96.

Palmer-Tous, T., Riera-Font, A. and Rosselló-Nadal, J. (2007) Taxing tourism: the case of rental cars in Mallorca. *Tourism Management* 28, in press.

Panayides, P. (2006) Maritime logistics and global supply chains: towards a research agenda. *Maritime Economics & Logistics* 8, 3–18.

Panigiua, M. (2004) *Integrating Economic Growth, Social Justice and Environmental Sustainability:a Review of Political Economy Theory*. Working paper presented ODPM/ESRC workshop on governance and institutions, University of York, 7 September.

Panou, M., Bekiaris, E. and Papakostopoulos, V. (2005) Modeling driver behaviour in EU and international projects. In Macchi, L., Re, C. and Cacciabue, P. (eds), *Proceedings of the International Workshop on Modelling Driver Behaviour in Automotive Environments*, Ispra, 25–7 May. Luxembourg: Office for Official Publication of the European Communities, 5–21.

Parkinson, M., Hutchins, M., Simmie, J. and Clark, G. (2004) *Competitive European Cities – How are the Core Cities Doing?* London: ODPM.

Patterson J. (1999) *Urban Growth Boundary Impacts on Sprawl and Redevelopment in Portland Oregon*. Working Paper, University of Wisconsin-Whitewater.

Patterson, T., Bastianoni, S. and Simpson, M. (2006) Tourism and climate change: two-way street, or vicious/virtuous circle? *Journal of Sustainable Tourism* 14, 339–48.

Pawasarat, J. and Stetzer, F. (1998) *Removing Transportation Barriers to Employment: Assessing Driver's License and Vehicle Ownership Patterns of Low Income Populations*. University of Wisconsin, Milwaukee. www.uwm.edu/Dept/ETI/dot.htm. Accessed 16 December 2006.

Pearce, D. (1992) *Tourism Organizations*. Harlow: Longman.

Pearce, D. (1993) Foreword. In Banister, D. and Button, K. (eds), *Transport, the Environment and Sustainable Development*. London: Chapman and Hall, pp. xiii–xiv.

Pearce, D., Markandya, A. and Barbier, E. (1989) *Blueprint for a Green Economy*. London: Earthscan.

Peck, J. (2001) Neoliberalising states: thin policies/hard outcomes. *Progress in Human Geography* 25, 445–55.

Peck, J. and Tickell, A. (2002) Neoliberalising space. *Antipode* 34, 380–404.

Pedersen, P. (2003) Development of freight transport and logistics in sub-Saharan Africa: Taaffe, Morrill and Gould revisited. *Transport Reviews* 23, 275–97.

Pedynowski, D. (2003) Science(s) – which, when and whose? Probing the metanarrative of scientific knowledge in the social construction of nature. *Progress in Human Geography* 27, 735–52.

Peeters, P. (2003) The tourist, the trip and the earth. In NHTV Marketing and Communication Departments (ed.), *Creating a Fascinating World*. Breda: NHTV, pp. 1–8.

Peeters, P. and Schouten, S. (2006) Reducing the ecological footprint of inbound tourism and transport to Amsterdam. *Journal of Sustainable Tourism* 14, 113–15.

Perren, B. (2006) Thalys' 10th birthday. *Modern Railways* August, 65.

Philip, L. (1998) Combining quantitative and qualitative approaches to social research in human geography – an impossible mixture? *Environment and Planning A* 30, 261–76.

Philip, L. and Shucksmith, M. (2003) Conceptualising social exclusion in rural Britain. *European Planning Studies* 11, 461–80.

Pigou, A. (1932) *The Economics of Welfare*, 4th edition. London: Macmillan.

Pinder, D. and Slack, B. (eds) (2004) *Shipping and Ports in the Twenty-first Century: Globalisation, Technological Change and the Environment*. London: Routledge.

Pine, B. and Gilmore J. (1998) Welcome to the experience economy. *Harvard Business Review*, July–August, 97–105.

Pine, B. and Gilmore J. (1999) *The Experience Economy: Work is Theatre and Every Business a Stage*. Boston: Harvard Business School Press.

Pinkard, J. (2005) Managing travel demand in Scotland. *Scottish Transport Review* 30, 8.

Pirie, G. (1982) The decivilising rails: railways and underdevelopment in Southern Africa. *Tijdschrift voor Economische en Sociale Geografie* 73, 221–8.

Pirie, G. (2006) 'Africanisation' of South Africa's international air links, 1994–2003. *Journal of Transport Geography* 14, 3–14.

Pirie, M. (1982) *The Logic of Economics*. London: Adam Smith Institute.

Polk, M. (1996) *Swedish men and women's mobility patterns: issues of social equity and ecological sustainability*. Washington, DC: US Department of Transportation, Federal Highway Administration.

Polk, M. (2004) The influence of gender on daily car use and on willingness to reduce car use in Sweden. *Journal of Transport Geography* 12, 185–95.

Pooley, C. and Turnbull, J. (2000) Modal choice and modal change: the journey to work in Britain since 1890. *Journal of Transport Geography* 8, 11–24.

Pooley, C., Turnbull, J. and Adams, M. (2005) *A Mobile Century? Changes in Everyday Mobility in Britain in the Twentieth Century*. Aldershot: Ashgate.

Porter, G. (2002) Living in a walking world: rural mobility and social equity in sub-Saharan Africa. *World Development* 30, 285–300.

Porter, H. (2007) Don't ignore a million angry voices, Mr. Blair. *The Observer*, 11 February. www.guardian.co.uk/commentisfree/story/0,,2010571,00.html. Accessed 11 February 2007.

Porter, M. and Ketels, C. (2004) *UK Competitiveness: Moving to the Next Stage*. Framework paper to DTi and ESRC Cities Programme. www.isc.hbs.edu/econ-natlcomp.htm. Accessed 18 December 2006.

Potter, R., Binns, T., Elliott, J. and Smith, D. (2004) *Geographies of Development*, 2nd edition. Harlow: Pearson.

Potter, S. and Parkhurst, G. (2005) Transport policy and transport tax reform. *Public Money and Management* 25, 171–8.

Preston, J. (2001) Integrating transport with socio-economic activity – a research agenda for the new millennium. *Journal of Transport Geography* 9, 13–24.

Preston, J. (2003a) A 'thoroughbred' in the making? The bus industry under Labour. In Docherty, I. and Shaw, J. (eds), *A New Deal for Transport? The UK's Struggle with the Sustainable Transport Agenda*. Oxford: Blackwell, pp. 158–77.

Preston, J. (2003b) Making the Connections: transport studies, geography and the social sciences. *Paper presented to the Royal Geographical Society (with the Institute of British Geographers) Annual Conference, London.*

Preston, J. (2004) *Moving Transport Geography Forward*. Keynote address to the International Geographical Congress, 17 August, Glasgow.

Preston, J. and Rajé, F. (2005) *Thinking Outside the Box? Accessibility, Mobility and Transport-related Social Exclusion*. Paper presented to the Royal Geographical Society (with the Institute of British Geographers) Annual Conference, London.

Preston, J. and Rajé, F. (2007) Accessibility, mobility and transport-related social exclusion. *Journal of Transport Geography* 15, 151–60.

Price, M. (1999) A new approach to the appraisal of road projects. *Journal of Transport Economics and Policy* 33, 221–6.

Prideaux, B. (1999) Tracks to tourism: Queensland Rail joins the tourism industry. *International Journal of Tourism Research* 1, 73–86.

Prideaux, B. (2000) The role of transport in destination development. *Tourism Management* 21, 53–64.

Pridmore, A., Bristow, A., May, T. and Tight, M. (2003) *Climate Change, Impacts, Future Scenarios and the Role of Transport*. Tyndall Centre for Climate Change Research Working Paper 33. www.tyndall.ac.uk/publications/working_papers/wp33.pdf.

Priemus, H. and Zonneveld, W. (2003) What are corridors and what are the issues? Introduction to special issue: the governance of corridors. *Journal of Transport Geography* 11, 167–77.

Pucher, J. and Kurth, S. (2002) Verkehrsverbund: the success of regional public transport in Germany, Austria and Switzerland. *Transport Policy* 13, 279–91.

Pucher, J. and Lefevre, C. (1996) *The Urban Transport Crisis in Europe and North America*. London: Macmillan.

Purvis, M. (2004) Geography and sustainable development. In Purvis, M. and Grainger, A. (eds), *Exploring Sustainable Development: Geographical Perspectives*. London: Earthscan, pp. 33–49.

Quinet, E. and Vickerman, R. (2004) *Principles of Transport Economics*. CheltemahM; Edward Elgar.

Railway Gazette (2004) Raffarin sets infrastructure priorities. *Railway Gazette* February, 81–2.

Railway Gazette (2005) Network expansion plan aims to reach 100,000 km by 2020. *Railway Gazette*, August, 479–83.

Railway Technology.com (2006a) *InterCity Express high-speed rail network, Germany*. www.railway-technology.com/projects/germany/. Accessed 8 August 2006.

Railway Technology.com (2006b) *Thalys PBKA high-speed trains, Europe*. www.railway-technology.com/projects/belgium/. Accessed 8 August 2006.

Rajé, F. (2004) *Transport, Demand Management and Social Inclusion: the Need for Ethnic Perspectives*. Aldershot: Ashgate.

Rajé, F., with Grieco, M., Hine, J. and Preston, J. (2004) *Transport, Demand Management and Social Inclusion*. Aldershot: Ashgate.

Reid, J., Armitage, C. and Spencer, C. (2004) *The Theory of Planned Behaviour Applied to Reducing Single Occupancy Driving: a Feasibility Study*. Paper presented at the 3rd International Conference on Traffic & Transport Psychology, Nottingham.

Réseau Ferré de France (2006) *Annual Report 2005*. Paris: RFF.

Richards, G. (2006) Tourism education in the new Europe. In Hall, D., Smith, M. and Marciszewska, B. (eds), *Tourism in the New Europe: the Challenges and Opportunities of EU Enlargement*. Wallingford: CABI Publishing, pp. 52–64.

Richardson, B. (1999) Towards a policy on a sustainable transportation system. *Transportation Research Record* 1670, 27–34.

Richardson, B. (2005) Sustainable transport: analysis frameworks. *Journal of Transport Geography* 13, 29–39.

Richardson, H. and Gordon, P. (2001) Portland and Los Angeles: beauty and the beast. Paper presented at the Pacific Regional Science Conference (PRSCO), Portland, OR.

Richardson, T., Livingstone, K., Banister, D., Goodwin, P., Urry, J. and Siemiatycki, M. (2004) Interface. *Planning Theory & Practice* 5, 487–514.

Robbins, D. (1997) The relationship between scheduled transport operations and the development of tourism markets to peripheral island destinations. Paper presented at the International Tourism Research Conference: Peripheral Area Tourism, Research Centre of Bornholm, Nexø, Denmark.

Robbins, D. (2003) Public transport as a tourist attraction. In Fyall, A, (ed.), *Managing Tourist Attractions: New Directions*. London: Butterworth-Heinemann, pp. 86–120.

Roberts, G. and Harrison, M. (2006) Budget airlines spread their wings to Africa. *The Independent* 2 March, 3.

Robinson, H. and Bamford, C. (1978) *Geography of Transport*. Plymouth: Macdonald and Evans.

Robinson, R. (1976) *Ways to Move: the Geography of Networks and Accessibility*. Cambridge: Cambridge University Press.

Robinson, R. (2002) Ports as elements in value-driven chain systems: the new paradigm. *Maritime Policy and Management* 29, 241–55.

Rodrigue, J.-P. (2004) Straits, passages and chokepoints: a maritime geostrategy of petroleum distribution. *Les Cahiers de Geographie du Quebec* 48(135), 357–74.

Rodrigue, J.-P., Comtois, C. and Slack, B. (2006) *The Geography of Transport Systems*. London: Routledge.

Romein, A., Trip, J. and de Vries, J. (2003) The multi-scalar complexity of infrastructure planning: evidence from the Dutch-Flemish megacorridor. *Journal of Transport Geography* 11, 205–13.

Romer, P. (1986) Increasing returns and long-run growth. *Journal of Political Economy* 95, 1002–38.

Ronedo (1998) *Adventures with Steam in Romania*. Ronedo, Piatra Neamţ.

Root, A. (ed.) (2003) *Delivering Sustainable Transport: a Social Science Perspective*. Oxford: Pergamon.

Rosenbloom, S. (1996) *Trends in Women's Travel Patterns*. Washington, DC: US Department of Transportation, Federal Highway Administration.

Rosiak, A. and Szarski, T. (1995) *Wenecja: Narrow Gauge Railway Museum*. Wrocław: ZET.

Rostow, W. (1960) *The Stages of Economic Growth*. Cambridge: Cambridge University Press.

Royal Commission on Environmental Pollution (RCEP) (1995) *Transport and the Environment*. Oxford: Oxford University Press.

Royal Commission on Environmental Pollution (RCEP) (2002a) *The Future Development of Air Transport in the United Kingdom: a National Consultation*. Response by the Royal Commission on Environmental Pollution. London: The Stationery Office.

Royal Commission on Environmental Pollution (RCEP) (2002b) *The Environmental Effects of Civil Aircraft in Flight*. London: Royal Commission on Environmental Pollution.

Rugg, J. and Jones, A. (1999) *Getting a job, Finding a home: rural youth transitions*. Bristol: Policy Press.

Rural Development Commission (1999) *Labour market detachment in rural England*. Rural Research Report Number 40. London: RDC.

Ryanair (2006a) *Ryanair Announces Biggest Ever Expansion at Dublin. 12 new European Routes Start from December*. Ryanair news release, 9 August. www.ryanair.com/site?EN/news.php?yr=06&month=aug&story=rte-en-090806.

Ryanair (2006b) *Ryanair Announces Long Term Agreement with the Government of Morocco*. Ryanair news release, 25 May www.ryanair.com/site?EN/news.php?yr=06&month=may&story=gen-en-250506.

Ryley, T. (2006) Use of non-motorised modes and life stage in Edinburgh. *Journal of Transport Geography* 14, 367–75.

Sachs, W. (1984) *For Love of the Automobile: Looking Back into the History of our Desires*. Berkeley: University of California Press.

Santos, G. and Bhakar, J. (2006) The impact of London congestion charging scheme on the generalised cost of car commuters to the City of London from a values of travel time savings perspective. *Transport Policy* 13, 22–33.

Savage, I. (1985) *The Deregulation of Bus Services*. Aldershot: Gower.

Saxenian, A. (1994) *Regional Advantage: Culture and Competition in Silicon Valley and Route 128*. Cambridge, MA: Harvard University Press.

Schaeffer, K. and Sclar, E. (1975) *Access for All: Transportation and Urban Growth*. London: Penguin.

Schafer, A. (1998) The global demand for motorized mobility. *Transportation Research Part A: Policy and Practice* 32, 455–77.

Schafer, A. and Victor, D. (2000) The future mobility of the world population. *Transportation Research A* 34, 171–205.

Schipper, Y., Rietveld, P. and Nijkamp, P. (2001) Environmental externalities in air transport markets. *Journal of Air Transport Management* 7, 169–79.

Schmocker, J., Fonzone, A., Quddus, M. and Bell, M. (2006) Changes in the frequency of shopping trips in response to a congestion charge. *Transport Policy* 13, 217–28.

Schreck, K., Meyer, H. and Strumpf, R. (1979) *S-Bahnen in Deutschland: Planung, Bau, Betrieb*. Cologne: Das Neue Köln.

Schumpeter, J. (1909) On the concept of social value. *Quarterly Journal of Economics* 23, 213–32.

Scott, A. (ed.) (2001) *Global City-regions: Trends, Theory, Policy*. Oxford: Oxford University Press.

Scottish Executive (2003) *Building Better Cities*. Edinburgh: Scottish Executive.

Scottish Executive (2004) *Scottish Household Survey Travel Diary Results for 2002*. Scottish Executive Statistical Bulletin Transport Series Trn / 2004 / 4, May 2004. Edinburgh: Scottish Executive National Statistics.

Scottish Executive (2006a) *Scotland's National Transport Strategy*. Edinburgh: Scottish Executive.

Scottish Executive (2006b) Household transport in 2005: some Scottish Household Survey results. *Scottish Executive Statistical Bulletin, October 2006*. Edimburgh: Scottish Executive.

Scottish Executive (2006c) *Review of Demand Responsive Transport in Scotland*. www.scotland.gov.uk/Publications/2006/05/18112606/3. Accessed 14 December 2006.

Searle, G. (1998) Changes in producer services location, Sydney: globalisation, technology and labour. *Asia Pacific Viewpoint* 39, 237–55.

Shaw, J. (2000) *Competition, Regulation and the Privatisation of British Rail*. Aldershot: Ashgate.

Shaw, J. and Walton, W. (2001) Labour's trunk road policy for England: an emerging pragmatic multimodalism? *Environment and Planning A* 31, 1131–56.

Shaw, J., Charlton, C. and Gibb, R. (1998) The competitive spirit reawakens the ghost of railway monopoly. *Transport Policy* 5, 39–47.

Shaw, J., Hunter, C. and Gray, D. (2006) Disintegrated transport policy: the multimodal studies process in England. *Environment and Planning C: Government and Policy* 24, 575–96.

Shaw, S. and Thomas, C. (2006) Social and cultural dimensions of air travel demand: hypermobility in the UK? *Journal of Sustainable Tourism* 14, 209–15.

Sheller, M. and Urry, J. (2006) The new mobilities paradigm. *Environment and Planning A* 38, 207–26.

Shepherd, E. and McMaster, R. (2004) *Scale and Geographic Inquiry: Nature, Society and Method*. Oxford: Blackwell.

Shoup, D. (2005) *The High Cost of Free Parking*. Chicago: Planners Press.

Shucksmith, M. and Philip, L. (2000) *Social Exclusion in Rural Areas: a Literature Review and Conceptual Framework*. Edinburgh: Scottish Executive.

Simmie, J. (2001) *Innovative Cities*. London. Spon Press.

Simon, D. (1989) Sustainable development: theoretical construct or attainable goal? *Environmental Conservation* 16, 41–8.
Simon, D. (1996) *Transport and Development in the Third World*. London: Routledge.
Sinclair, S. (2001) *Financial exclusion: an introductory survey*. Centre for Research into Socially Inclusive Services (CRSIS), Edimburgh: Edinburgh College of Art/Heriot-Watt University.
Skerratt, S. and Warren, M. (2004) Broadband in the countryside: the new digital divide. *Farm Management Journal* 11, 727–36.
Skuse, I. (2005) New airline ticket taxes – are airlines a soft touch? *Business Travel World* July, 45.
Slack, B. (1994) Pawns in the game: ports in a global transport system. *Growth and Change*, 24, 597–8.
Slack, B. (1998) Intermodal transportation. In Hoyle, B. and Knowles, R. (eds), *Modern Transport Geography*, 2nd edition. Chichester: Wiley, pp. 263–90.
Slack, B. and Fremont, A. (2005) Transformation of port terminal operations: from the local to the global. *Transport Reviews* 25, 117–30.
Slaven, A. (1986) Shipbuilding. In Langton, J. and Morris, R. (eds), *Atlas of Industrialising Britain 1780–1914*. London and New York: Methuen, pp. 136–9.
Smith, D. (1994) *Geography and Social Justice*. Oxford: Blackwell.
Smyth, A. (2003) Devolution and sustainable transport. In Docherty, I. and Shaw, J. (eds), *A New Deal for Transport? Labour's Struggle with the Sustainable Transport Agenda*. Oxford: Blackwell, pp. 30–50.
Social Exclusion Unit (SEU) (1998) *Bringing Britain Together: a National Strategy for Neighbourhood Renewal*. London: Cabinet Office.
Social Exclusion Unit (SEU) (2001) *National Strategy for Neighbourhood Renewal: Policy Action Team Audit*. London: Cabinet Office.
Social Exclusion Unit (SEU) (2003) *Making the Connections: Final Report on Transport and Social Exclusion*. London: Cabinet Office.
Sørensen, A. (1997) *Travel as Attraction*. Paper presented at the International Tourism Research Conference: Peripheral Area Tourism, Research Centre of Bornholm, Nexø, Denmark.
Stafford, B., Heaver, C., Ashworth, K., Bates, C., Walker, R., McKay, S. and Trickey, H. (1999) *Young Men's Experience of the Labour Market*. York: Joseph Rowntree Foundation.
Stahl, A. (1992) The provision of a community responsive public transportation in urban areas. Proceedings of 6th International Conference Mobility and Transport for Elderly and Disabled People, 31 May–3 June, Lyon, France, 160–7.
Stahl, A. and Brundell-Freij, K. (1995) The adaptation of the Swedish public transport system yesterday, today, tomorrow – an evaluation. Proceedings of 7th International Conference Mobility and Transport for Elderly and Disabled People, 16–19 July, Reading, UK, 23–34.
Staley, S., Edgens, J. and Mildner, G. (1999) *A Line in the Land: Urban-Growth Boundaries, Smart Growth, and Housing Affordability*. Reason Public Policy Institute, Policy Study No. 263.
Stanbridge, K. and Lyons, G. (2006) *Travel behaviour considerations during the process of residential relocation*. Paper presented at the 11th International Conference on Travel Behaviour Research, Kyoto, August.
Standing Advisory Committee on Trunk Road Assessment (SACTRA) (1994) *Trunk Roads and the Generation of Traffic*. London: HMSO.
Standing Advisory Committee on Trunk Road Assessment (SACTRA) (1999) *Transport and the Economy: Full Report*. London: The Stationery Office.

Starkie, D. (1972) *The Motorway Age*. Oxford: Pergamon.
Stead, D. and Banister, D. (2006) Decoupling transport growth and economic growth in Europe. In Rietveld, P., Jourquin, B. and Westin, K. (eds), *Towards Better Performing Transportation Networks*. London: Routledge.
Stefanova, A. (2006) *CEE Bankwatch Network comments on the HLG report on TEN-T Extension to the Neighbouring Countries: How to do it More Economically and Environmentally Acceptably*. www.bankwatch.org/documents/HLG_report_comments.pdf. Accessed 2 January 2007.
Steg, L. (2004) *Instrumental, Social and Affective Values of Car Use*. Paper presented at the 3rd International Conference on Traffic & Transport Psychology, Nottingham.
Steg, L. and Gifford, R. (2005) Sustainable transportation and the quality of life. *Journal of Transport Geography* 13, 59–69.
Steiner, T. and Bristow, A. (2000) Road pricing in national parks: a case study in the Yorkshire Dales national park. *Transport Policy* 7, 93–103.
Step Beyond (2006) *Drivers' Attitude Study*. Unpublished report to Midland Safety Camera Partnership Group, Stone.
Stopford, M. (1997) *Maritime Economics*, 2nd edition. London: Routledge.
Storper, M. (1997) *The Regional World: Territorial Development in a Global Economy*. New York: The Guilford Press.
Storper, M. (2000) Globalisation and knowledge flows: an industrial geographer's perspective. In Dunning, J. (ed.), *Regions, Globalisation and the Knowledge Based Economy*. Oxford: Oxford University Press, pp. 42–62.
Storper, M. and Venables, A. (2004) Buzz: face to face contact and the urban economy. *Journal of Economic Geography* 4, 351–71.
Stradling, S. (2002a) Transport user needs and marketing public transport. *Municipal Engineer* 151, 23–8.
Stradling, S. (2002b) Combating car dependence. In Grayson, G. (ed.) *Behavioural Research in Road Safety XII*. Transport Research Laboratory, Crowthorne, 174–87.
Stradling, S. (2003) Reducing car dependence. In Hine, J. and Preston, J. (eds) *Integrated Futures and Transport Choice: UK Transport Policy Beyond the 1998 White Paper and Transport Acts*. Aldershot: Ashgate, pp. 100–15.
Stradling, S. (2005) Readiness for modal shift in Scotland. *Scottish Geographical Journal* 120, 265–75.
Stradling, S. (2006) Cutting down and slowing down: changes in car use and speeding on Scotland's roads. In *Behavioural Research in Road Safety: Sixteenth Seminar*. DfT, London., 63–9.
Stradling, S. (2007) Determinants of car dependence. In Garling, T. and Steg, L. (eds), *Threats to the Quality of Urban Life from Car Traffic: Problems, Causes and Solutions*. Oxford: Elsevier, pp. 187–204.
Stradling, S., Campbell, M., Allan, I., Gorrell, R., Hill, J., Winter, M. and Hope, S. (2003) *The Speeding Driver: Who, How and Why?* Scottish Executive, Edinburgh.
Stradling, S., Carreno, M., Ferguson, N., Rye, T., Halden, D., Davidson, P., Anable, J., Hope, S., Alder, B., Ryley, T. and Wigan, M. (2005) *Scottish Household Survey Analytical Topic Report: Accessibility and Transport*. Edinburgh: Scottish Executive.
Stradling, S., Carreno, M., Rye, T. and Noble, A. (2007) *Acceptability of the Bus Travel Experience*. Napier University, Edinburgh. Unpublished.
Stradling, S., Meadows, M. and Beattly, S. (2000) Helping drivers out of their cars: integrating transport policy and social psychology. *Transport Policy* 7, 207–15.
Stradling, S., Meadows, M. and Beatty, S. (2001) Identity and independence: two dimensions of driver autonomy. In Grayson, G (ed.) *Behavioural Research in Road Safety*. Crowthrorne: Transport Research Laboratory.

Strangleman, T. (1999) The nostalgia of organisations and the organisation of nostalgia: past and present in the contemporary rail industry. *Sociology* 33, 725–46.
Stuber, N., Forster, P., Radel, G. and Shine, K. (2006) The importance of the diurnal and annual cycle of air traffic for contrail radiative forcing. *Nature* 441, 864–7.
Sudjic, D. (1992) *The 100 Mile City*. San Diego: Harcourt Brace & Company.
Sutton, J. (1988) *Transport Coordination and Social Policy*. Aldershot: Avebury.
Taaffe, E. and Gauthier, H. (1973) *The Geography of Transportation*. Englewood Cliffs, NJ: Prentice Hall.
Taaffe, E. and Gauthier, H. (1994) Transportation geography and geographic thought in the United States: an overview. *Journal of Transport Geography* 2, 155–68.
Taaffe, E., Gauthier, H. and O'Kelly, M. (1996) *Geography of Transportation*, 2nd edition. Upper Saddle River: Prentice Hall.
Taaffe, R., Morrill, R. and Gould, P. (1963) Transport expansion in underdeveloped countries: a comparative analysis. *Geographical Review* 53, 503–29.
Takel, R. (1981) The spatial demands of ports and related industry and their relationship with the community. In Hoyle, B and Pinder, D (eds), *Cityport Industrialisation and Regional Development: Spatial Analysis and Planning Strategies*. Oxford: Pergamon, pp. 47–68.
Taniguchi, E., Thompson, R., Yamada, T. and Van Duin, R. (2001) *City Logistics: Network Modeling and Intelligent Transport Systems*. Oxford: Elsevier.
Tapio, P. (2005) Towards a theory of decoupling in the EU and the case of road traffic in Finland between 1970 and 2001. *Transport Policy* 12, 137–51.
Taylor, C. (2005) France regains rail's blue riband. *Railway Gazette* November, 699–703.
Taylor, P. (2004) Is there a Europe of cities? World cities and the limitations of geographical scale analysis. In Sheppard, E. and McMaster, R. (eds), *Scale and Geographic Inquiry: Nature, Society and Method*. Oxford: Blcakwell, pp. 229–47.
Taylor, Z. (2006) Railways closures to passenger traffic in Poland and their social consequences. *Journal of Transport Geography* 14, 135–51.
Tenterden Railway Company (2006) *Kent & East Sussex Railway*. Tenterden: Tenterden Railway Company. www.kesr.org.uk.
Tertoolen, G., van Kreveld, D. and Verstraten, B. (1998) Psychological resistance against attempts to reduce private car use. *Transportation Research Part A* 32, 171–81.
Teufel, D., Bauer, P., Lippold, R., Brainfeld, S. and Schmidt, K. (1999) *Öko-Billanzen von Fahrzeugen*. Umwelt und Prognose Institut, Heidelberg. www.eu-transport.org/upi25.pdf Accessed 11 December 2006.
Therivel, R. and Partidario, M. (1996) *The Practice of Strategic Environmental Assessment*. London: Earthscan.
Thierstein, A. and Schnell, K.-D. (2002) Corporate strategies, freight transport and regional Development. *DISP* 148, 69–78.
Thomas, J. (1971) *A Regional History of the Railways of Great Britain, volume VI: Scotland, the Lowlands and the Borders*. Newton Abbot: David and Charles.
Thomsen, T., Nielsen, L. and Gudmundsson, H. (eds) (2003) *Social Perspectives on Mobility*. Aldershot: Ashgate.
Thomson, J. (1977a) *Modern Transport Economics*. Harmondsworth: Penguin.
Thomson, J. (1977b) *Great Cities and Their Traffic*. Harmondsworth: Penguin.
Tiebout, C. (1956) A pure theory of local expenditures. *Journal of Political Economy* 64, 416–24.
Tight, M., Bristow, A., Pridmore, A. and May, A. (2005) What is a sustainable level of CO_2 emissions from transport activity in the UK in 2050? *Transport Policy* 12, 235–44.

Tillman, J. (2002) Sustainability of heritage railways: an economic approach. *Japan Railway and Transport Review* 30, 38–45.

Timmermans, H., van der Waerden, P., Alves, M, Polak, J., Ellis, S., Harvey, A., Kurose, S. and Zandee, R. (2003) Spatial context and the complexity of daily travel patterns: an international comparison. *Journal of Transport Geography* 11, 37–46.

Tolley R. (ed.) (1990) *The Greening of Urban Transport: Planning for Walking and Cycling in Western Cities*. London: Belhaven Press.

Tolley, R. and Turton, B. (1995) *Transport Systems, Policy and Planning: a Geographical Approach*. Harlow: Longman.

Tolley, R. and Turton, B. (eds) (2001) *Global Transport Issues*. London: I.B. Tauris.

Tönnies, F. (1955) *Gemeinschaft Und Gesellschaft (Community and Association)*. London: Routledge.

Tonts, M. (2000), Restructuring Australia's rural communities. In Pritchard, W. and McManus, P. (eds), *Land of Discontent: the Dynamics of Change in Rural and Remote Australia*. Sydney: University of New South Wales Press, pp. 52–72.

Torrance, H. (1992) Transport for all: equal opportunities in transport policy. In Roberts, J., Clearly, J., Hamilton, K. and Hanna, J. (eds), *Travel Sickness: the Need for a Sustainable Transport Policy for Britain*. London: Lawrence and Wishart, London, pp. 48–56.

Transport for London (TfL) (2006) *Central London Congestion Charging. Impacts Monitoring, Fourth Annual Report Overview*. London: TfL. www.tfl.gov.uk/tfl/cclondon/pdfs/Fourth-Annual-Report-Overview.pdf. Accessed 16 August 2006.

Trench, S. and Lister, A. (1990) *Changes in Taxi Services – Can New Developments Help People with a Mobility Handicap?* Public Transport Planning and Operations, Proceedings of Seminar D, held at the PTRC Transport and Planning Annual Meeting, London, Unpaginated.

Tuan, Y.-F. (1977) *Space and Place: the Perspective of Experience*. Minneapolis: University of Minneapolis.

UNCTAD (2006) *Review of Maritime Transport*. Geneva: UNCTAD. www.unctad.org/. Accessed 27 February 2007.

United Nations Economic and Social Commission for Asia and the Pacific (2001) *Managing Sustainable Tourism Development*. New York: United Nations.

Upham, P., Maughan, J., Raper, D. and Thomas, C. (2003) *Towards Sustainable Aviation*. London: Earthscan.

Urban Task Force (1999) *Towards an Urban Renaissance*. London: HMSO.

Urry, J. (1995) *Consuming Places*. London and New York: Routledge.

Urry, J. (2002) Mobility and proximity. *Sociology* 36, 255–74.

Urry, J. (2003) Social networks, travel and talk. *British Journal of Sociology* 54, 155–75.

Urry, J. (2004a) Connections. *Environment and Planning D: Society and Space* 22, 27–37.

Urry, J. (2004b) *The New Mobilities Paradigm*. Paper presented at a workshop on mobility and the cosmopolitan perspective. Munich: Reflexive Modernisation Research Centre.

US Census Bureau (2005) *Computer and Internet Use in the United States: 2003*. Washington, DC: US Department of Commerce.

Van Reeven, P. (2005) In Button, K. and Hensher, D. (eds), *Handbook of Transport Strategy, Policy and Institutions*. Amsterdam: Elsevier, Amsterdam, pp. 707–24.

Vance, J. (1986) *Capturing the Horizon: Historical Geography of Transportation*. New York: Harper and Row.

Vance, J. (1991) Human mobility and the shaping of cities. In Hart, J. (ed.), *Our Changing Cities*. Baltimore: Johns Hopkins University Press.

Vickerman, R. (ed.) (1991) *Infrastructure and Regional Development*. London: Pion.
Vigar, G. (2002) *The Politics of Mobility: Transport, the Environment and Public Policy*. London: Spon.
Vowles, T. (2000) The effect of low-fare air carriers on airfares in the US. *Journal of Transport Geography* 8, 121–8.
Vowles, T. (2006) Airfare pricing determinants in hub-to-hub markets. *Journal of Transport Geography* 14, 15–22.
Vries, J. de and Priemus, H. (2003) Megacorridors in north-west Europe: issues for transnational spatial governance. *Journal of Transport Geography* 11, 223–33.
Walker, R. (1981) A theory of suburbanisation: capitalism and the construction of urban space in the United States. In Dear, M. and Scott, A. (eds), *Urbanisation and Planning in Capitalist Societies*. New York: Methuen, pp. 383–430.
Ward, D. (1964) A comparative historical geography of streetcar suburbs in Boston, Massachusetts and Leeds, England. *Annals of the Association of American Geographers* 54, 477–89.
Watson, S. (1999) City A/genders. In Watson, S. and Doyal, L. (eds), *Engendering Social Policy*. Buckingham: Open University Press.
Webb, S. and Webb, B. (1963) *The Story of the King's Highway*, New edition. London: Cass.
Weber, J. (2006) Reflections on the future of accessibility. *Journal of Transport Geography* 14, 399–400.
Webster, B. (2007) Runaway train fares could drive many passengers off the railways. *The Times*, 1 January, 8.
White, H. (1977) *The Geographical Approach to Transport Studies*. Discussion papers in geography No. 1. Salford: Salford University.
White, H. and Senior, M. (1983) *Transport Geography*. London: Longman.
Whitelegg, J. (ed.) (1981) *The Spirit and Purpose of Transport Geography*. Papers presented at the annual conference of the Transport Geography Study Group, Leicester.
Whitelegg, J. (ed.) (1992) *Traffic Congestion: Is There a Way Out?* Hawes: Leading Edge Press.
Whitelegg, J. and Cambridge, H. (2004) *Aviation and Sustainability*. Stockholm: Stockholm Environment Institute.
Whitelegg, J. and Haq, G. (2003) The global transport problem: same issues but a different place. In Whitelegg, J. and Haq, G. (eds), *The Earthscan Reader on World Transport Policy and Practice*. London: Earthscan, pp. 3–25.
Wicke, L. (1991) *Umweltoekonomie und Umweltpolitik*. Muncih: Deutscher Taschenbuch Verlag.
Williams, A. (1981) *Aims and Achievements of Transport Geography*. Paper presented at the annual conference of the Transport Geography Study Group, Leicester.
Wilson, A. (2000) *Complex Spatial Systems: the Modelling Foundations of Urban and Regional Analysis*. Harlow: Prentice Hall.
Wirth, L. (1938) Urbanism as a way of life. in Sennett, R. (ed.), *Classic Essays on the Culture of Cities*. London: Prentice-Hall.
Wit, R., Dings, J., Mendes de Leon, P., Thwaites, L., Peeters, P., Greenwood, D. and Doganis, R. (2002) *Economic Incentives to Mitigate Greenhouse Gas Emissions from Air Transport in Europe*. Delft: CE Delft.
Wolmar, C. (2005) *On the Wrong Line: How Ideology and Incompetence Wrecked Britain's Railways*. London: Aurum Press.
Wolmar, C. (2006) *Please Learn From Our Mistakes*. www.christianwolmar.co.uk/online_column/april06.shtml. Accessed 23 October 2006.

Wood, D. and Johnson, J. (1996) *Contemporary Transportation*, 5th edition. Upper Saddle River: Prentice Hall.
Woods, M. (2005) *Rural Geography: Processes, Responses and Experiences in Rural Restructuring*. London: Sage.
World Bank (1996) *Sustainable Transport: Priorities for Policy Reform*. www.worldbank.org/transport/pol_econ/tsr.htm Accessed 14 September 2004.
World Bank (2004) *Sub-Saharan Africa Transport Policy Programme*. Washington, DC: World Bank.
World Bank (2006) *World Development Indicators*. Washington, DC: World Bank.
World Coal Institute (2005) *Coal Facts 2005* (October). London: WCI.
World Commission on Environment and Development (1987) *Our Common Inheritance*. Oxford: Oxford University Press.
World Health Organisation (WHO) (1995) *Residential Noise: Concern for Europe's Tomorrow*. Stuttgart: Wissenschaftliche Verlagsgesellschaft mbH.
World Resources Institute (2005) *Transparency Issues with the ACEA Agreement: Are Investors Driving Blindly?* Sustainable Asset Management, Zurich. www.sam-group.com/downloads/studies/ACEA_Driving_Blindly.pdf. Accessed 16 November 2006.
World Tourism Organisation (1997) Tourism 2020 Vision. Madrid: WTO.
World Tourism Organisation (2005) *Tourism Facts and Figures: Highlights*. Madrid: WTO. www.world-tourism.org/facts/menu.html.
World Tourism Organisation (2006) *Tourism Facts and Figures: Highlights*. Madrid: WTO. www.world-tourism.org/facts/menu.html.
Wu, B. and Hine, J. (2002) *Analysis of Databases for Impacts of Road User Charging/Workplace Parking Levy on Social Inclusion/Exclusion: Gender, Ethnicity and Lifecycle Issues*. Unpublished report.
Wurzel, R. (2002) *Environmental Policy-making in Britain, Germany and the European Union: the Europeanisation of Air and Water Pollution Control*. Manchester: Manchester University Press.
Yago, G. (1984) *The Decline of Transit: Urban Transportation in German and US Cities, 1900–70*. Cambridge: Cambridge University Press.
York, I. and Balcombe, R. (1997) Accessible bus services: UK demonstrations. Public Transport Planning and Operations. Proceedings of Seminar G, European Transport Forum Annual Meeting, London, 117–28.
Young, R. (1999) Prioritising family health needs: a time-space analysis of women's health-related behaviours. *Social Science and Medicine* 48, 797–813.
Yunusa, M. and Shaibu-Imodagbe, A. (2002) Road rehabilitation: the impact on transport and accessibility in Nigeria. In Fernando, P. and Porter, G. (eds), *Balancing the Load: Women, Gender and Transport*. London: Zed, pp. 111–18.
Zelinsky, W. (1971) The hypothesis of the mobility transition. *Geographical Review* 61, 219–49.
Zito, A. (2000) *Creating Environmental Policy in the European Union*. Basingstoke: Macmillan.
Zonneveld, W. and Trip, J. (2003) *Megacorridors in North West Europe: Investigating a New Transnational Planning Concept*. Delft: DUP Science.
Zook, M. and Brunn, S. (2006) From Podes to Antipodes: positionalities and global airline geographies. *Annals, Association of American Geographers* 96, 471–90.

Index

Abahamse, W. 184
accessibility/mobility 7, 8, 50; and auto dependency 91; and car-centric thinking 7; crisis of 76, 80; current/future trends 8–9; definition 50; in developing world 25–6; elements/indicators of 50–1; first round effects 228–9; heightened 15; and individual transport patterns 183; and investment 13; and IURT sector 132–3; as key to urban transport 97; mixed methods research programmes on 7–8; opportunities/constraints 109–11; personal 7; and policies of 'predict and provide' 7; and role of social capital/networks 7; rural areas 102, 109–11, 118–19; and social justice 49–51; social/economic benefits of 97; spatial analysis 11; and technology 222–3; transitions 228–9; and transport provision 58, 61; urban 84–5, 232
Adam, B. 127
Adams, J. 234
AEA Technology 135
Aer Lingus 152
Agnew, J. 11
Air Deccan 38
Air France group 152
Air Transport Association 151
air transport 217; and air freight 38; Asia-Pacific 153–4; availability of 220; and CO2 emissions/atmospheric pollution 34–5, 37, 149; and congestion/shortage of airport-airspace capacity 149–51; costs 145, 148; demand for 90; deregulation of 134; development/growth rates 137, 155, 220; dispersal/fragmentation 147; Europe 152–3; free-riding aspects 46; global trends/patterns 138–40; and globalization 140–3; hub dominance 146–7; legacy (mainline) carriers 145, 153; and liberalization 143–6; and local/global overlap 121; and low cost carriers (LCCs) 120, 144–6, 151–4, 155, 203–5; network strategies 146–8; and noise 148–9; North America 151–2; and open-skies agreements 141–2; passenger expectations/profiles 145; and passenger numbers 37–8; rest of the world 154–5; sustainability 148–9, 155; and taxation 45–6; top airports/airlines 139, 140, 149; and transport freedoms 142
AirAsia 145
Airey, A. 26
AirTran 145
Åkerman, J. 148
Alamadari, F. 145
Albrechts, L. 134
Alderighi, M. 204
Allen, J. 121
American Airlines 152
Anable, J. 181, 186, 195

Aschauer, D. 234
Asia-Pacific 145, 153–4, 216–17, 218, 226
Atkins, S. 56
Australia 36; car ownership in 108, 112; coal trade 163; and rural public transport 112

Badland, H. 200
Bae, C. 98
Baggott, S. 5, 34
Bailey, I. 34
Baird, A. 171
Balcombe, R. 58
Bamberg, S. 184
Bamford, C. 227
Banister, D. 4, 5, 10, 11, 19, 21, 25, 28, 38, 84, 95, 99, 117, 132, 147, 195, 234, 235
Barker, K. 222
Barkham, P. 133
Barnes, E. 230
Barnes, I. 41
Barnes, P. 41
Barrett, S. 152
Baumol, W. 69, 77
Bavoux, J. 228
Baxter, J. 7
BBC 3
Becken, S. 201, 205, 206
Begg, I. 94
Behnen, T. 65
Bell, P. 74
Benwell, M. 54
Berechman, J. 5, 10, 11, 19, 21, 25, 28, 84, 95, 234
Bertolini, L. 235
Beuret, K. 60
Beuthe, M. 237
Bhakar, J. 93–4, 99
Bhalla, A. 51
Bieger, T. 204
Bird, J. 6
Bishop, S. 35
Bjerrgarrd, R. 32
Black, W. 4, 5, 62, 65, 68, 112, 131, 201, 227, 234
Blomgren, K. 199
Boardman, B. 108
Boile, M. 175
Boiteux-Oraim, C. 232
Bongaerts, J. 42
Bonsall, P. 58
Boucher, D. 49
Bowen, J. 141, 153
Brake, J. 113, 114
Briggs, R. 112
Bristow, A. 200
British Airways 3, 152
British Waterways 208
British Waterways Scotland 208
Brown, B. 6, 219, 236–7
Brown, F. 206
Brundell-Freij, K. 59
Brunn, S. 138, 143
Buchan, K. 68, 99
Buchanan Report (1963) *see Traffic in Towns*
Building Research Establishment (BRE) 32, 33
bulk cargo 157; break-bulk 157–8, 159; capacity standards 161, 162; characteristics 158–60; and choice of vessel size 160–1; coal trade 162–3; compatibility of 160; costs 160; destinations 159; dry goods 159; geographic stability of 158–9; grain trade 163; and load size 160; major/minor products 159; petroleum trade 161–2; and port activity 159, 161; and seasonal demand 161; and storage space 160–1; transport of 160–1
Burchardt, T. 51
Bureau of Transportation Statistics 83, 122
Burghouwt, G. 153
Burkett, N. 55
Burtenshaw, D. 84
Bus Partnership Forum 186
bus services 221; availability of 53–4; and car ownership 57; commercialization of 56–7; deregulation/privatization 69–71; and general/targeted subsidy 57–8; individual behaviour 186–8; Megabus inter-city services 133; as most used form of transport 187; network coverage 56; and physical accessibility 58; policies 56–8;

bus services (*cont'd*)
and Quality Contracts 75; quality of 188; in rural locations 103–4, 112, 113–14; and timetable schedules 57
Butcher, A. 209
Butler, R. 199, 202
Button, K. 29, 35, 84, 148, 230

Cairns, S. 190, 195
Callender, C. 55
Cambridge, H. 148
Cameron, A. 51
canals 13, 15, 36
Cape Railways Enthusiasts' Association (CREA) 208
Capineri, C. 141
CARLOS project (Switzerland) 115
Casas, J. 227
Cass, N. 132
Castells, M. 6, 121
CEE Bankwatch 130–1
Central Place Theory 11, 103
CER 134
Ceron, J.-P. 205
Cervero, R. 56
Chadefaud, M. 199
Charlton, C. 75, 123, 125
Chatterjee, K. 221
China, air transport in 150; economic development in 23; oil/coal trade 162, 163
Chliaoutakis, J. 185
Choi, H.-S. 202
Christaller, W. 11, 12, 103
Church, A. 50, 51
cities 230; asset offer 94–7; balancing demands of 97–100; and car ownership 88–90; Central Business District (CBD) 230, 232; and congestion 93–4; and culture of association 84; definition 83; foot and tracked 86, 88; impact of railways on 86–8; investment in infrastructure 95–7; and New Urbanist development 98–9; physical form of 86–8; and position of city centre 87; revitalization of 98; scale/density of economic activity in 83–4; social equity in 99; and social justice 84, 85; as social/cultural centres 84, 87; space/place perspective 84; transport problems 90–3; transport's role in 11, 84–6
City of Edinburgh Council 97
City of Portland 98
Claval, P. 121
Cloke, P. 74, 103, 121
Co-Op Travel Group *Future Travel* call centre 117
coal trade 162–3
Collier, A. 201
Colography Group 230
COMEAP 32
Commission of the European Communities (CEC) 41, 42, 46, 128, 130, 131
Commission for Integrated Transport (CfIT) 75, 76, 77, 118
Commission for Rural Communities 109
Commission on Social Justice 52
Committee on the Medical Effects of Air Pollutants (COMEAP) 32
congestion, airport 149–51; charges 22–3, 28, 36, 45, 47, 132, 202, 221–2; tackling 93–4
containerized transport 159–60; advantages of 163–4; capacity limits 175; construction landmarks 165; and 'empties' 175–6; flows/global imbalances 175–6; global port operators 167–70; growth in global trade 164; influence of technology on 164; and maritime freight 163–72; and offshore terminals 170–2; and pendulum services 173; and port accessibility 164, 166; and size of vessels 164; terminals 164, 166–7
Cooke, P. 220
Copenhagen Finger Plan (1947) 88, 97
Coppens, T. 134
Core Accessibility Indicators (UK) 116
Couldry, N. 209
Cox, W. 98
Crafts, N. 95
Cramer, J. 205
Crang, M. 3
Cresswell, T. 3
Crime Concern 56
critical theory, and annihilation of space by time 14; and cultural 'turn' 14–15; and fixed/immobile infrastructures 14;

and flows/places relationship 15; Marxist view 14; and mobility 14–15
Cronon, W. 6, 16
Cross, A. 69
Crozet, Y. 127
Cullinane, K. 164
Currie, G. 55
Curry, A. 224

Daigle, J. 200
Daniels, P. 86, 199, 230
Dann, G. 209
Dargay, J. 221
Davenport, J. 198
Davidson, S. 207
Davis, A. 24, 26, 27, 35
Davison, L. 56, 75
Debbage, K. 199, 200
Decker, M. 204
Deka, D. 91
Deloitte Research 23
demand–responsive transport (DRT) 59, 113–15, 118
Dempsey, P. 150
Dennis, N. 146
Department for Education and Employment (DfEE) 55
Department for Education and Skills (DfES) 35
Department of Environment, Food and Rural Affairs (DEFRA) 116–17
Department of the Environment, Transport and the Regions (DETR) 32, 49, 53, 54, 59, 60, 221
Department of Trade and Industry (DTI) 191
Department for Transport (DfT) 7, 33, 34, 41, 46, 69, 116, 122, 183, 187, 226
Department for Transport, Local Government and the Regions (DTLR) 219
Derek Halden Consultancy (DHC) 190–4, 195
developed countries, demand management measures in 22–3, 28; and enhancing capacity/efficiency of existing infrastructure 28; and new regionalism 17–19; production systems in 15–17; and spatial effects of investment 19–23, 28; transport/spatial development in 15–23
developing countries, and affordable/achievable transport improvements 27; and area infrastructure 25–6; and audibility of local voices 26–7; and colonial legacy 24–5; contrasts in 23; conventional thinking concerning 23–4; and infrastructure/local social needs 23–4; investment/political factors 25; and maximizing capacity of existing networks 28; new approach to transport provision 26; and paradox of urban transport 25; and small-scale local schemes 27, 28; transport/economic development in 23–7; and urban–rural/rich–poor divide 25
Dhavale, D. 230
Dicken, P. 143, 157, 229
Disability Discrimination Act (1995) 58
disabled 54, 58, 111, 112
Divall, C. 208, 209
Dobbs, L. 92
Dobruszkes, F. 153, 204
Docherty, I. 5, 6, 77, 86, 93, 134, 220, 221, 227, 234
Dodge, M. 6
Doganis, R. 145, 146
Donaghy, K. 203
Donald, R. 54, 58
Douglas, Y. 209
Downward, P. 200
Drewry Shipping Consultants 171
Druva-Druvaskalne, I. 204
Dublin Transportation Office (DTO) 95–6
Dubois, G. 205
Dudleston, A. 185, 186, 195
Dudley, G. 236
Dunkerley, C. 58

East/South-East Asia, air transport in 150, 153–4; and economic development 14; infrastructure improvements in 25; maritime freight 175; production systems in 17
Eastern Europe 5
easyJet 145, 152
Eckey, H.-F. 12
economic development, and critical theory 14–15; in developed countries 15–23; in developing countries 23–7;

economic development, and critical theory (*cont'd*)
and impacts of investment 10–11, 19–23; and modernization theory 12–14; and new regionalism 17–19; production systems/transport networks 15–17; and spatial analysis 11–12; theoretical frameworks 11–15; and transport linkage 10–11, 28
Eddington, R. 5, 11, 19, 21, 22, 23, 28, 121, 123, 131, 222, 234
Elby, D. 199
elderly people, and long-distance travel 133; rural vulnerability of 111, 112; and social justice 133; and travel time 183
Eliot-Hurst, M. 227
Ellaway, A. 184
Endzina, I. 204
environment, and activity patterns 31, 35–6; and air transport 37–8; assessing/managing transport impacts 38–46; and climate change 34–5; and CO2 emissions 5, 34–5, 37, 42–3, 46, 149; definition of 29–30; and emissions/waste production 30, 32–3, 34–5, 47; and energy use 31; and extraction, processing, manufacturing 31; first order effects 30, 32–5; global, regional, local aspects 30, 37–8; and global warming 5, 34–5, 194, 226; and governance 30, 38–46; and green belt/brownfield sites 40; and health effects 31, 32–3; and impact of transport on 30–2; and individual behaviour 194–5; and Life Cycle Analysis 31; and lifestyle health impacts 35; and maritime trade 162; natural/social dimensions 30; noise/atmospheric pollution 30, 33, 148–9; and planning agreements 40–1; relationship with transport 47; second order effects 30–1, 35–8; and settlement patterns/land use effects 31–6, 36–7; and state interventions 75–6; and sustainable systems 30; and technology 31; and urban transport 99
Environmental Change Institute 149
Environmental Impact Assessment (EIA) 41, 46, 47
ESRC Transport Unit 105
EU Business 204
EurActiv 43
European Bank of Reconstruction and Development (EBRD) 72, 73
European Commission 202; and Auto-Oil programme 42; Transport White Paper (2001) 127–2
European Conference of Ministers of Transport (ECMT) 5, 77, 234
European Environment Agency 29
European Investment Bank (EIB) 128
European Parliament 46
European Pollutant Release and Transfer Register (E-PRTR) 32
European Rail Traffic Management System (ERTMS) 130
European Train Enthusiasts-Eastern New England (ETE-ENE) 208
European Union (EU) 69; and air transport 142; air transport in 152, 152–3; Common Transport Policy 127; and EIA/SEA processes 41; enlargement of 152; governance 38–9; health issues 32–3; high-speed rail in 123–7; and infrastructural investment 46; and integrated environmental assessment 40–2; and IURT sector 122, 123–35; LCCs in 152–3; and legislation 39–40; market-based instruments 43–6; and new regionalism 18; oil trade 162; policies 38–9; public transport in 112; road systems in 89; and rural transport 103, 112; and TEN-T 127–31; transport systems in 18, 21, 28; and voluntary agreements 42–3
Eurostat 83, 117, 122
Evans, D. 58, 59
Extended Schools Programme 116
Eyles, J. 7

Fagan, S. 145
Fan, T. 153
Farrell, S. 132
Farrington, C. 7, 49–50, 52, 68, 109
Farrington, J. 7, 49–50, 52, 68, 105, 109
Featherstone, M. 184
Feitelson, E. 202
Fitzpatrick, J. 35
Fleming, D. 171
Flexibility 117

Florida, R. 84, 94, 230
flows/spaces 8; changing 4; and containerized transport 175–6; and critical theory 15; local/global 8; and mobility/accessibility 8; and transport choices/activities 8; urban focus 8
Flyvbjerg, B. 236
Foley, J. 49
Forsyth, P. 153
Foster, C. 66
Francis, G. 144
Frank, A. 24
freight 8, 63, 72–3, 121, 123, 135, 216, 217; air transport 31, 38, 46; and emissions 30, 34; expansion of 122; rail 127, 130, 134
Fremont, A. 167
Friedman, M. 68
Friends of the Earth 57
Froebel, F. 17
Fröidh, O. 126
Frost, M. 51
Fuellhart, K. 151
Fujita, M. 234
Fuller, R. 185
Fullerton, B. 72
Funazaki, A. 31
Furgala-Selezniow, G. 208

Gaffron, P. 51
Gant, R. 53, 59
Garcia Palomares, J. 126
Gardiner, R. 158
Garling, T. 180, 194
Gatersleben, B. 184
gateway 22, 123, 155, 159, 169, 177
Gather, M. 74, 228
Gauthier, H. 121, 227, 232
Ghana 13
Gibb, K. 75, 227
Gibbs, R. 123, 125
Giddens, A. 79
Gifford, R. 131, 202
Gillespie, A. 147, 237
Gillingwater, D. 84
Gilmore, J. 208
Giorgi, L. 135
Giuliano, G. 4, 91, 97, 147, 230, 237
Givoni, M. 126
Glaeser, E. 66
Glaister, S. 43, 76, 221
Gleave, M. 25
globalization, and air transport 140–3; and environment 37–8; and maritime transport 156; production systems 17
Goetz, A. 3, 4, 6, 140, 141, 143, 150, 151, 228
Goh, M. 99
Goldman, T. 202
Gomez-Ilbanez, J. 69
Goodwin, P. 5, 7, 43, 74, 75, 77, 93, 195, 220, 221, 222, 229
Gordon, P. 98, 230
Gorham, R. 202
Gorter, C. 55
Goss, J. 16
Gössling, S. 205, 206
governance, and car ownership 69, 74; centrality of 8; and congestion problems 74, 76, 79; and denationalization 79; and deregulation/privatization 69–74; and destatization 79; and devolved governments 77–8; economic/social imperatives 65; and environmental policy 76; European 38–9; and free market conditions 65; and geographical inequality 48; geographical nature of 38; hollowing out/filling in 75, 79; impact on transport geography 63; and infrastructural investment 46; and inter-state transfer of policy 79; and IURT 134–5; and legislation 39–40; and managing negative externalities 65; and market-based instruments 43–6; and military/national security 65; and monopoly in infrastructure provision 66; and need for demand management 76; neoliberal 68–9, 75, 79, 80; and planning/environment integration 39, 40–2; and public/private interdependence 75–9; reasons for state involvement 63–8; and redefining role of the state 75–9; and social equity 66, 68; and state interventions 62–3, 75, 76, 79; and state role as guarantor of last resort 76; and state-owned undertakings 69, 73, 74; and supra-national institutions 77; and technology/emissions 39; and theory of contestable markets 69; and third way policies 79;

governance, and car ownership (*cont'd*)
and traveller behaviour 39; types of measures 39; and variety of organizations 63; and voluntary agreements 42–3; and welfare economic model 74
Graham, A. 68, 92, 141, 146, 149
Graham, B. 140, 146, 154, 199
Graham, D. 43
Graham, S. 230
grain trade 163
Gray, D. 7, 105, 108, 110, 112, 118
Grayling, T. 35, 53, 58
Greece 45
Greene, D. 77, 226
Grieco, M. 50, 51, 54
Griffiths, C. 35
Grubler, A. 228
Guest, A. 189
Guillian, R. 232
Gurran, N. 36
Gutiérrez, J. 126
Gwilliam, K. 25

Haggett, P. 228
Halden, D. 50, 51
Hall, C. 204
Hall, D. 75, 92, 200, 201, 202, 204, 206, 207, 227
Hall, P. 3, 10, 15, 17, 28, 131
Halsall, D. 209
Hamilton, K. 54, 111
Hanlon, S. 54
Hanly, M. 221
Hansen, L. 128
Hanson, S. 4, 84, 91, 227, 233
Haq, G. 29, 131
Harris, A. 54
Harrison, M. 204
Harvey, D. 3, 14, 15, 17, 23, 69, 235
Hashimoto, Y. 223
Hass-Klau, C. 200
Haukeland, J. 199
Hay, A. 3, 6, 52, 227
Haydock, D. 127
Hayek, F. 68
Haynes, K. 202
Hayuth, Y. 171
Healy, E. 232
Heaver, T. 176
Helminen, V. 4, 190
Hensher, D. 4, 29, 232
Herod, A. 6, 120
Hesse, M. 4, 157, 228, 230
Hickman, R. 195
high-speed rail (HSR) 120; French (LGV/TGV services) 123–5; German 125; investment in 127; Italian 125; Japanese 123; planning dilemmas 126; Spanish (AVE service) 125; spatial discordance 126; Thalys/Eurostar network 125–6; and time-saving 127
Hilling, D. 5
Hillman, M. 50, 229
Himanen, V. 203
Hine, J. 49, 50, 51, 53, 54, 55, 56, 57, 58, 235
hinterland 103, 105, 113, 119, 149, 157, 159, 164, 166, 169, 171, 173, 174, 177
hitchhiking 115
Hjorthol, R. 93
HM Treasury 5, 46, 222
Hodge, D. 237
Hoffmann, J. 156
Höjer, M. 219
Holden, E. 205
Holding, D. 200
Holzer, H. 92
Horner, M. 227
House of Commons Select Committee on Education and Employment 55
House of Lords Select Committee on the European Communities 46
Houston, D. 68, 92
Høyer, K. 209, 210, 217
Hoyle, B. 3, 4, 5, 6, 7, 10, 13, 18, 19, 227
hub 16, 65, 138, 146–7, 149, 151–4, 164, 171, 172, 173
Hughes, M. 31, 124
Hunter, C. 36, 39, 205
Huws, U. 193

Iberia group 152
Ieromonachou, P. 45, 99
Ihlandfeldt, K. 92
Imashiro, M. 72
India 38, 127
Indian Railways Fan Club (IRFC) 208

individual behaviour, activity choice 180–1; average travel time 180; changing 194–5, 221; and destination accessibility 183; determinants of travel 194; environmental considerations 194–5; factors influencing 179, 180; gender, age, income 183; inclinations towards/away from buses 186–8; inclinations towards/away from car use 183–6; information, communication, travel substitution 188–94; and leisure trips 181; motivations 179–80; obligations 181; opportunities 181–3; personal costs 180; propulsion, compulsion, consumption 179–81; reasons for travel 217, 219–20; second order impacts 179; and transport adequacy/affordability 183; transport modes 180

information and communication technology (ICT) 117, 230, 232; affect on travel behaviour 189–90, 195; and air transport 147; benefits of 194; emergence of 6, 8; and fusion with travel
223; impact on travel habits 222; and importance of spatial location 188; major impacts 190; ownership of 188–9; and pace of change 226; and redefining accessibility 222–3; substitution versus complementarity 191; as supplement to travel 222–3; and teleworking/teleshopping 117, 190, 191–3, 194; and virtual worlds 117, 223–4; widespread effect of 4

Intelligent Infrastructure Systems (2006) 224–9

inter-urban and regional transport (IURT), advantages/disadvantages of 121; challenges for 135; definition 120, 121; and developing economies 135–6; development perspectives 131–5; diversity of 135; as domestic movements 121; enhanced systems as essential 127; EU policy 127–31; and governance/regulation 134–5; and high-speed rail 123–7; and increasing mobility 122–3; length of journey/trip distance 121, 122–3; as multi-stage 122; and new mobilities 132–3; overlap with other scales of transport 121; planning/management of 136; and social justice 133–4; and spatial discordance 126; and sustainability 131–2; and TEN-T projects 128–31; and time 'saved' 127

Interessengemeinschaft Eisenbahn (IGE) 208

intermediacy 153, 171

intermodal 153, 164, 172, 176

Intergovernmental Panel on Climate Change (IPCC) 140

International Air Transport Association (IATA) 137

International Monetary Fund (IMF) 69

Internet 117, 223

Jaakson, R. 200

Jackson, C. 134

Jacobs, J. 66, 84, 233

Jain, J. 6, 127, 219

James, P. 193

Janelle, D. 17, 147, 237

Jenkins, L. 54

Jensen, A. 127, 184

Jessop, B. 75, 79

JetBlue 145, 152

Johnson, J. 65

Johnston, R. 3, 232

Jones, A. 55

Jones, D. 33

Jopson, A. 237

Jun, M.-J. 98

Kagermeier, A. 111

Kain, J. 92

Kapp, K. 77

Keeling, D. 3, 4

Keep Europe Moving (2001) EC Transport White Paper 127–8

Kenworthy, J. 89, 216, 234

Kenyon, S. 220, 222, 233

Kessel, A. 29

Kesselring, S. 3

Ketels, C. 95

Khanna, M. 164

Kidder, A. 112

Kilvington, R. 69

Kingham, S. 111

Kitagawa, T. 123, 127

Kitchin, R. 6
Klein, S. 204
KLM 152
Knowles, R. 3, 4, 5, 6, 7, 11, 56, 69, 72, 74, 75, 93, 97, 108, 122, 123, 128, 131, 220, 227
Knox, P. 11, 17
Ko, T. 202
Kohn, T. 200
Konings, R. 176
Kosfeld, R. 12
Koski, H. 237
Kotkin, J. 92
Kotler, P. 209
Kovacs, G. 5
Kresl, P. 220
Kreutner, M. 200
Krugman, P. 12
Kumar, S. 156
Kurth, S. 79
Kwan, M.-P. 4
Kyle, S. 26
Kyoto Protocol 34, 35, 46

Lajunen, T. 184
Lamb, B. 207
landbridge 175, 177
Lang, R. 92, 230
Lapeyre, F. 51
Larsen, J. 3, 99, 132, 223
Latin America 17, 25
Lauber, V. 30
Laurier, E. 6, 191, 219, 236–7
Law, R. 92
Lawless, P. 55
Lawton, T. 144, 153
Le Clercq, F. 235
Le Grand, J. 99
Lee, P. 51
Lee, Y. 90
Lefevre, C. 5, 74, 91
Leinbach, T. 10, 25, 26, 141, 153
leisure 4, 6, 31, 37, 51, 60, 84, 85, 90, 132, 144, 147, 181, 184, 190; *see also* tourism
Lenschow, A. 39
Letherby, G. 6
Leunig, T. 95
Leyshon, A. 17, 66, 233
Life Cycle Analysis (LCA) 31–2
Limtanakool, N. 122
Ling, D. 59
Link, H. 203
Linneker, B. 21
Lister, A. 60
Liverman, D. 30
Local Transport Plans 116
location theory *see* spatial analysis
Loebbecke, C. 204
logistics 122, 157, 166, 169, 172–7, 201, 230, 232
Loo, B. and Liu, K. 217
Low Cost Carriers (LCCs) 120
Lubchenco, J. 29
Lucas, K. 49, 54, 235
Lufthansa 152
Lumsdon, L. 201, 207
Luneva, L. 204
Lyons, G. 6, 31, 127, 219, 220, 221, 223

Maastricht Treaty (1992) 127, 128
McCormick, J. 39
McGee, T. 230
McGregor, A. 55
McKelvey, D. 112
McLellan, R. 164
McMaster, R. 120
Maddison, D. 5, 77
Mannion, R. 59
Marchetti, C. 180
Marell, A. 112
maritime transport, break-bulk (general) cargo 157–8, 159; bulk cargo 157, 158–63; centrality of 156; containerized 163–72; and convergence with inland systems 176–7; fixed/flexible characteristics 156–7; and global production/consumption 230; globalization of 156; logistics of 157, 172–7; management strategies 157; and maritime/land interface 172–7; new services/networks 172–5; pendulum services 173–5; ports as crucial facilities 157; shipbuilding 15; types of cargo 158
market-based instruments, and aviation industry 45–6; and Energy Products Directive 44; and legislation 43; and road use charges 44–5, 47, 48; and taxes/fiscal incentives 43–6, 47

Martin, J. 54
Martin, R. 28
Marvin, S. 230
Marx, K. 14
Massey, D. 6, 11, 16
Matthiessen, C. 128
Mattsson, L.-G. 219
MÅV Nostalgia 209
Mawdsley, E. 26
Maxwell, S. 184
Meinig, D. 6
Metz, D. 180, 183
Meyer, J. 69
Middle East, air transport in 151, 154
Ministry of Transport 90, 93
Minogue, K. 49
Mintel 203
Minten, B. 26
Mitchell, C. 54
Mitchell, F. 49, 50, 51, 53, 54, 55, 56, 58, 235
mobility *see* accessibility/mobility
modernization theory, and catch-up investment 13; and developing/less developed countries 12–14; in-built assumptions 12–13, 14; and influence of Taaffe, Morrill, Gould model 13–14; and merging of transport/urban hierarchies 13; and social/spatial effects 28; and stages of economic growth 12–13; and transport as precondition for 'take off' 13
Mokhtarian, P. 9, 183, 190, 194, 222
Molnar, L. 199
Monk, S. 55
Moore, T. 98
Moorhouse, G. 86
Morgan, K. 220
Morrell, P. 146
Moseley, M. 7, 50, 55, 109
motor cars 221; and accessibility privileges 184; and accidents 185; attraction of 184; and CO2 emissions 42–3; dependence on 90–1, 94, 108; dominance of 216; as dominant mode of transport 187; driver segments 185–6, 194; and driving as skill-based activity 185; gendered accessibility to 111; (im)mobility of 183; independence/dependence link 219–20; individual behaviour 183–6; and journey times 184; and modern rural life 104–5, 108–9, 118; numbers of 183; ownership of 88–90, 91, 108, 116; policy measures 195; positive impact on countryside 108; and tourism 202–3
Müller, D. 204
Muller, P. 230
multinational corporations (MNCs) 17
Murayama, Y. 123
Murie, A. 51
Murphy, R. 89
Murray, M. 199

Naish, J. 204
Nakicenovic, N. 228
Nash, C. 134
National Geospatial Intelligence Agency (NGIA) 157
National Travel Survey 122
Neff, J. 68
Nelson, A. 98
New Economic Geography (NEG) 12, 15, 20
new mobilities paradigm 14–15; and demand for IURT 132–3; and emphasis on innovative/qualitative research methods 6–8; introduction of 3–4; and nature/production of space/place 6; and rejection of social science research 6; rural transport 117–18; and sustainability 235–6; and travel time 219
new regionalism, and developed countries 17–19; and devolution of political power 18; and high-technology clusters 18–19; and internal factors/conditions 18; rise of 17–18; and transport systems 18; *see also* regions
New Zealand 68, 72
Newman, P. 89, 216, 234
Nicholls, S. 206
Nigeria 13
Nijkamp, P. 201, 234
Niskanen, W. 68
Njenga, P. 24, 26, 27
Noble, B. 41
non-governmental organizations (NGOs) 48

North American Free Trade Area (NAFTA) 48, 142
Northwest airlines 152
Notteboom, T. 173, 176
Nuhn, H. 4, 228
Nutley, S. 108, 112, 118

Oates, W. 77
O'Connell, J. 145, 153
O'Connor, K. 147, 153, 230, 232
Office for National Statistics (ONS) 117, 191, 192
Office of Science and Technology (OST) 224; Foresight Programme 224
offshore terminals, advantages of 171; depth 171; hinterland accessibility 171; labour costs 171; land availability 171; location 171; logistics zones 172; major hubs 171; ownership 172; pendulum services 170–1
O'Hara, K. 6
Olivier, D. 167
Ollivier-Trigalo, M. 134
Olsthoorn, X. 203
Olvera, L. 27
O'Reilly, D. 58
Organisation for Economic Cooperation and Development (OECD) 31, 94, 132, 189, 202
Ostróda-Eblaąg Navigation 208
Owens, S. 236
Oxley, P. 54, 59

Page, S. 201, 207
Pain, R. 56
Painter, J. 79
Palmer-Tous, T. 201
Panama Canal 175
Panayides, P. 176
Panigiua, M. 66
Panou, M. 185
Parkhurst, G. 43
Partidario, M. 41
Patterson, J. 98
Patterson, T. 201
Pawasarat, J. 55
Pearce, D. 43, 77, 207
Peck, J. 69, 76
Pedersen, P. 14
Pedynowski, D. 30
Peeters, P. 205
Perren, B. 125
petroleum trade 161–2
Philip, L. 7, 102
Pickup, L. 54, 58
Pigou, A. 93
Pine, B. 208
Pinkard, J. 195
Pirie, G. 25, 154
Polk, M. 54, 93
Pooley, C. 6, 7, 228
Porter, G. 25, 27
Porter, H. 132
Porter, M. 95
ports, accessibility channel depth 164, 166; and competition 169–70; and concentration of ownership of 168; as elements of global logistical chains 177; emergence of mega-ports 164; financial assets 168; and gateway accessibility 169; global operators 167–70; and inland connections 166–7; investment in infrastructure 166; and leverage 169; and managerial expertise 168–9; private operators of 167–8; space consumption 166; and traffic capture 169; world's largest 167
Potter, R. 26
Potter, S. 43
Preston, J. 7, 69, 75, 237
Price, M. 234
Prideaux, B. 199
Pridmore, A. 195
Priemus, H. 134
production systems, agglomeration/concentration of 22; and canals/railways 15; and development of industrial centres 15–16; globalization of 17; and industrialization 15–16; and international division of labour 16–17; Japanese 18; and maritime transport 230; and new-Fordism 16; and regional sectoral specialization 16; and time-space convergence 17; and transport networks 15–17
public transport, exclusion from 91–3; feminist view 92–3; rural 105, 113–15, *see also* bus services; railways
Publicar service (Switzerland) 114

Pucher, J. 5, 74, 79, 91
Purvis, M. 30

Quinet, E. 43

railways, and CO2 emissions 34; development of 15; heritage attraction 209; impact of 86–8; investment in 13, 25; local/long-distance overlap 121; privatization of 72–4, 75–6; as recreational interest 208–9; restored 210; and rise of suburbia 86; and state interventions 75–6; UK companies 67
Rajé, F. 7, 54, 235, 237
regions, contested/ambiguous understanding of 120–1; development of 5, 21; and the environment 30, 37–8; and production systems 16; *see also* new regionalism
Reid, J. 184
Remmelt, T. 176
Reynolds, G. 6
Richards, G. 204
Richardson, B. 203, 234
Richardson, H. 98
Richardson, J. 236
Richardson, T. 45, 99
Rigg, J. 26
Rimmer, P. 227
Ristimäki, M. 4, 190
roads, congestion on 93–4, 97, 131; motorway building 88–9; negative impact of 94; *see also* congestion charges
Robbins, D. 199, 202
Roberts, G. 204
Robinson, H. 227
Robinson, R. 176, 227
Rodrigue, J.-P. 4, 63, 64, 65, 69, 112, 138, 157, 162, 176, 237
Romein, A. 134
Romer, P. 234
Ronedo 209
Root, A. 233
Rosenbloom, S. 54
Rosiak, A. 209
Rostow, W. 13
Royal Commission on Environmental Pollution (RCEP) 46, 149, 237
Rugg, J. 55
Rupp, S. 34
Rural Bus Challenge Fund 114
Rural Community Transport Initiative (Scotland) 115
Rural Development Commission 55
Rural LIFT service (Ireland) 114, 115
rural transport, accessibility/mobility aspects 102, 109–11, 118–19, 183; bus services for 103–4; capacity building 116–17; classification of 103; and commuting distances 103, 108; definitions of 'rural' 103–5; demand-responsive 118; and differences/similarities between areas 103; geographical/functional differences 104–5; heterogeneity of 105, 118; hinterlands 103, 119; ICT/virtual mobility developments 117; and informal mobility 117–18; national philosophies 111–12; and need to travel 117–18; planning/provision for 102, 116; policy 111–18; and population distribution 104; public 105, 113–15; and shopping/services accessibility 105, 108, 109–11; and social networks 108, 117–18; suite of possibilities for 119; typology of 105, 106–7; and use of cars 104–5, 108–9, 118
Ryanair 145, 147, 152, 204

Sachs, W. 184
Salomon, I. 183
Santos, G. 93–4, 99
Savage, I. 74
Saxenian, A. 230
Scandinavia 45, 59, 60
Schaeffer, K. 90
Schafer, A. 122, 219
Schipper, Y. 203
Schmidt, M. 135
Schmidt, P. 184
Schmocker, J. 36
Schnell, K.-D. 230
Schofield, G. 200
Schouten, S. 205
Schreck, K. 87
Schumpeter, J. 77, 93
Sclar, E. 90
Scott, A. 208, 230
Scott, J. 55, 56

Scottish Household Survey 103, 104, 181
Searle, G. 232
Senior, M. 4, 84, 227
services transport 116
Shaibu-Imodagbe, A. 26
Shaw, J. 6, 67, 69, 75, 77, 135, 205, 227
Shaw, S. 132, 148
Sheller, M. 3, 4, 6, 117, 132
Shepherd, E. 120
Shoup, D. 99
Shucksmith, M. 102
Simmie, J. 220
Simon, D. 10, 202, 234
Sinclair, S. 55
Singh, B. 220
Single European Act (1986) 39
Sirakaya, E. 202
Sjoquist, D. 92
Skerrat, S. 117
Slack, B. 167, 170
Slaven, A. 15
Small, K. 230
Smeed Report (1964) 93
Smith, D. 52
Smith, J. 10, 13, 18, 19
Smyth, A. 58, 59
social exclusion 61, 183, 220; definition of 51; dimensions/factors of 51; economic 52; facilities 51; fear-based 52; geographical 51; and household organization 52; individual 52; physical 51; space 52; time-based 52; and transport 51–2
Social Exclusion Unit (SEU) 49, 51, 55, 56, 91, 116, 235
social justice 61; accessibility 49–51; and consequences of transport disadvantage 55–6; as contested concept 49, 52–3; definition of 49; and IURT 133–4; and patterns of transport disadvantage 53–4; policies/practices 56–60, 61; and social exclusion 51–2; and transport 49
Solomko, S. 153
Sørensen, A. 199, 207
South Africa 36
South-East Asia *see* East/South-East Asia
Southern Africa Customs Union (SACU) 48
Southwest Airlines 145, 147, 151
Soviet Union 5, 17
space–time 6, 14, 17, 108, 122, 219
spatial analysis 11–12; and accessibility 11; and agglomeration advantages 12; and central place theory 11; and land use 11; and transport infrastructure 11; and travel costs 11; and value of location 11; von Thünen's theory 11
spatial mismatch hypothesis 92–3
specialist services, Demand Responsive Transport 59; Flexline 60; service route 59–60; shopmobility facilities 59; voluntary car schemes 59
Spens, K. 5
Stafford, B. 55
Stagecoach 133
Stahl, A. 59
Staley, S. 98
Standing Advisory Committee on Trunk Road Assessment (SACTRA) 5, 20, 21, 86, 95, 220
Starkie, D. 89
Stead, D. 4, 117, 132, 147
Steg, L. 131, 184, 202
Steiner, T. 200
Step Beyond 184
Stetzer, F. 55
Stopford, M. 161
Storper, M. 17–18, 230, 232
Stough, R. 230
Stradling, S. 179, 183, 184, 185, 186, 187, 195
Strangleman, T. 209
Strategic Environmental Assessment (SEA) 41–2, 46, 47
sub-Saharan Africa 14, 17, 23, 24–5, 27
suburbs 36–7, 230; gender aspects 92–3; rise of 86–7, 90, 92; and road-building 94; and urban transport 90
Sudjic, D. 230
sustainability 8, 28; as academic device 234; and air transport 148–9, 155; and climate change 206; and definition of sustainable transport 203; environment, society, economy dimensions 234–5; and geopolitical issues 206–7; identifying tourism transport 201–2; and IURT sector 131–2; as key to interdisciplinarity 234–6; and land-use

234; and less developed countries 202; and new mobilities paradigm 235–6; and tourism 198, 201–7; and transport policies 203, 206; and transport in tourism 202–3, 205–7; and travel distance 205–6
Sustainability Appraisal (SA) 42
Sutton, C. 151
Sutton, J. 69
Szarski, T. 209
Szyliowicz, J. 150

Taaffe, E. 121, 227, 232
Takel, R. 160
Taniguchi, E. 232
Tapio, P. 229
Tati, Jacques 37
taxis/minicabs, commercial firms 60; deregulation of 112; and non-car-owning households 53; shortcomings 60; and upgrading of termini 25; voluntary schemes 60
Taylor, C. 124, 126
Taylor, P. 121
Taylor, Z. 5
teleworking 117
Tenterden Railway Company 209
Tertoolen, G. 184
Teufel, D. 31
Therivel, R. 41
Thierstein, A. 230
Thomas, C. 132, 148
Thomas, J. 86
Thomsen, T. 4
Thomson, J. 76, 90
Tickell, A. 69
Tiebout, C. 68
Tight, M. 202
Tillman, J. 209
time–space *see* space–time
Tolley, R. 4, 62, 90, 227
Tönnies, F. 84
Tonts, M. 36
Torrance, H. 91
tourism, definition of 196; destination transport 200–1; domestic 196; environmental concerns 201; and facilitating travel along recreational route 199; and freedom to travel 197, 206–7; global context 197–9; host/guest relationship 199, 201; importance of 196; increase in 197, 198; and low cost carriers (LCCs) 203–5; mobility/accessibility 199–200; and place/activity relationship 200; policies for transport 197–8; sustainability issues 198, 201–7; transport roles 199–200, 211; and transport as tourism experience 207–9
Traffic in Towns (Buchanan Report) (1963) 93
Trans-European Transport Networks (TEN-T) 46, 120; accessibility to 128; community guidelines for development of 128; and cross-border operations 130; funding/costs of 128, 130; and interconnection/interoperability of 128; patterns of 123; progress of 128, 130; road/rail considerations 130–1
transport, academic interest in 3–4, 6; as added-value experience 208; analysis of 6–8; challenges/threats to 3; and changing flows/spaces 4; current trends 215–17; effect of ICTs on 4; enhancing capacity/efficiency of 22–3; enabling role of 21; freight 8, 34; future possibilities 223–5; importance/significance of 3, 4; and improving quality/range of visitor experience 209; and journey patterns 4; location/infrastructure 209; mobile paradigm 3–4; peculiarities of 65; and 'place promotion' 209; quality of life objectives 84, 100; reasons for travel 217, 219–20; and recreational sub-groups 208; responsibility for current trends 220–2; as support for society 220–1; taken for granted 3; and technology 222–3; as unique experience 208
Transport Act (1985) 57
transport costs 11; air transport 145, 148; bulk cargo 160; core–periphery aspects 12; iceberg model 12; and offshore terminals 171; personal 180; reduced 28; TEN-T 128, 130
transport disadvantage, and barriers to employment 55; and bus services 53–4; and car ownership 53; consequences of 55–6; and disabled 54; and exclusion from services 55; and fear/perceptions of fear 56;

transport disadvantage, and barriers to employment (*cont'd*)
and gender 54; and income 53; patterns of 53–4; and race/ethnicity 54; social groups 53; and taxi/minicab usage 53
transport geography, concept 4; critical/humanistic approaches 233; and development of cities, regions, countries 5; and distribution of social/economic activity 5, 6; East/West differences 5; fundamentals of 8; and global warming 5; and impact of transport 5; interdisciplinary future of 236–7; logical positivist approach 232–3; and place embeddedness 230; pluralistic methodologies 232–6; (post-) structuralist views 233; second round effects 229–30, 232; spatial/socio-economic transformations 229–30, 232; and supply chain 230; and sustainability 234–6; systems/infrastructures 4–5; Taaffe, Morrill, Gould model 13–14; textbooks on 227–8, 232; and travel space 6; and use of quantitative/qualitative research methods 6–8
transport investment, and accessibility 13; and agglomeration/concentration of production 22; and benefits of improved systems 19–20, 28; and 'catch up' in developing countries 13; and cause/effect ambiguity 21; diminishing impact of 20–1; and importance of canals/railways 13; locational pattern of 18–19; and regional economic development 21; and reliability of networks/services 21–2; secondary effects of 20–1; spatial effects of 19–23, 28
Transport for London (TfL) 45, 99, 195, 222
Transport White Papers (1998, 2004) 34, 221
travel time 6, 180, 183, 184, 219
Trench, S. 60
Trip, J. 134
Tuan, Y.-F. 84
Turner, J. 54
Turton, B. 4, 62, 90, 227
UK National Atmospheric Emissions Inventory 34
United Kingdom, bus services in 57–8; car ownership in 108; and CO2 emissions 46; congestion charging in 45; effects of EIA/SEA in 41–2; and impact of M25/Channel Tunnel on 21; neoliberalism in 68–9; and new regionalism 17–18; production systems in 15, 17; road building in 21, 88–9; rural transport in 116–17
United Nations Economic and Social Commissions for Asia and the Pacific (UNESCAP) 202
United Nations PRTR Protocol 32
United States 68; air transport in 139, 150, 151–2; car ownership in 108, 112; and environment 5; land use effects 36; oil trade 162; Portland Urban Growth Boundary (UGB) 98; production systems/transport networks 15–16, 17, 18; road building in 5, 89; and rural public transport 112; social exclusion in 91; trip distance in 122; and urban sprawl 89; voluntary agreements in 42
Upham, P. 140, 141, 148
urban transport, categorization of needs 84–5; challenges 97–9; competitiveness paradigm 94–7; and connectivity concept 86–8, 97; in contemporary cities 94–7; in developing countries 25; economic/social distinction 84–6; and the environment 84; equity in 99; focus on 8; impact of technology on 86; infrastructure 84–5; investment in 96–7; mobility/accessibility 84–5; and modernization theory 13; paradox of 25; pricing instruments 99; role of 84–6; and sustainability/environmental issues 99
Urbanism as a Way of Life (Wirth) 84
Urry, J. 3, 4, 6, 108, 117, 127, 132, 133, 209, 219, 223, 233
US Census Bureau 117

Van Reeven, P. 127
Vance, J. 10, 86
Venables, A. 232
Vickerman, R. 21, 43

Victor, D. 122
Vigar, G. 236
Virgin Blue 145
Vowles, T. 146, 151, 154

Walker, R. 16
Walton, W. 69
Ward, D. 86
Warnes, A. 86
Warren, M. 117
Watson, S. 93
Webb, B. 66
Webb, S. 66
Weber, J. 7
Webster, B. 133
Wegener, M. 77
Westin, K. 112
WestJet 145
White, H. 4, 5, 84, 227
Whitelegg, J. 131, 148, 227, 233
Wicke, L. 77
Williams, A. 236
Williams, G. 145, 153
Wilson, A. 234
Wirth, L. 84
Wittmer, A. 204
Wolmar, C. 72, 75, 76, 122
women, and accessibility to motor cars 111; and long-distance travel 133; and social justice 133
Wood, D. 65
Woods, M. 102
World Bank 28, 69, 234
World Coal Institute 162
World Commission on Environment and Development 77
World Conservation Strategy (1980) 234
World Health Organisation (WHO) 33
World Resources Institute 42
World Tourism Organisation 197, 202, 206
World Trade Organisation 48, 197, 202
Wu, B. 54, 57
Wurzel, R. 42

Yago, G. 66
York, I. 58
young people, rural vulnerability of 111, 112; and travel time 183
Young, R. 55
Yunusa, M. 26

Zelinsky, W. 229
Zimmerman, C. 200
Zito, A. 44
Zonneveld, W. 134
Zook, M. 138, 143